T0329566

PROBABILISTIC TRANSMISSION SYSTEM PLANNING

IEEE Press
445 Hoes Lane
Piscataway, NJ 08854

A complete list of titles in the IEEE Press Series on Power Engineering appears at the end of this book.

PROBABILISTIC TRANSMISSION SYSTEM PLANNING

Wenyuan Li, Fellow, IEEE, EIC
BC Hydro, Canada

Mohamed E. El-Hawary, *Series Editor*

IEEE-PRESS

A JOHN WILEY & SONS, INC., PUBLICATION

Published by John Wiley & Sons, Inc., Hoboken, New Jersey.
Published simultaneously in Canada.

Library of Congress Cataloging-in-Publication Data is available.
ISBN 978-0-470-63001-3

10 9 8 7 6 5 4 3 2 1

Dedicated to Jun and my family

CONTENTS

PREFACE AND ACKNOWLEDGMENTS

Transmission system planning is one of the most essential activities in the electric power industry. Billions of dollars are invested in electric utility systems through planning activities every year. In the past and at present, transmission system planning is basically dominated by deterministic criteria and methods. However, there are a considerable number of uncertain factors in transmission systems, and therefore probabilistic methods will provide planning solutions closer to reality. Fragmentary papers for probabilistic transmission planning have been published so far, but there has not been a book to systematically discuss the subject. The intent of this book is to fill the gap. It is important to appreciate that the purpose of introducing probabilistic models and techniques into transmission planning is not to replace but to enhance the existing deterministic criteria.

The book originated from my deep interest and involvement in this area. My technical reports and papers formed core portions of the book, although general knowledge had to be included to ensure its systematization. All basic aspects in transmission planning are covered, including load forecast and load modeling, conventional and special system analysis techniques, reliability evaluation, economic assessment, and data preparation and uncertainties, as well as various actual planning issues. The probabilistic concept is a main thread throughout the book and touches each chapter. It should be emphasized that probabilistic transmission planning is far beyond reliability evaluation, although the latter is one of most important procedures toward this direction. I have followed such a principle for book structure: any new contents associated with the subject are illustrated in detail, whereas for a topic for which readers can find more information in other sources, an outline that is necessary for the book to stand alone is provided.

Materials in both theory and actual applications are offered. The examples in the applications are all based on real projects that have been implemented. I believe that the book will meet the needs of practicing engineers, researchers, professors, and graduates in the power system field.

I am indebted to many friends and colleagues. My special thankfulness goes to Roy Billinton, Paul Choudhury, Ebrahim Vaahedi, and Wijarn Wangdee for their continuous support and encouragement in my daily work. The papers that I coauthored with them are parts of the materials used in the book. Some data and results in a few examples are based on Wijarn Wangdee's reports.

Drs. Roy Billinton, Lalit Goel, Murty Bhavaraju, and Wenpeng Luan reviewed the book proposal/manuscript and provided many helpful suggestions. I would also thank all the individuals whose publications are listed in the References at the end of the book.

I am grateful for the cooperation and assistance received from the IEEE Press and John Wiley & Sons, especially Mary Mann and Melissa Yanuzzi.

Finally, I would like to thank my wife, Jun Sun, for her sacrifices and patience in the quite long time period during which I worked on the book.

WENYUAN LI

Vancouver, Canada
February 2011

1

INTRODUCTION

1.1 OVERVIEW OF TRANSMISSION PLANNING

1.1.1 Basic Tasks in Transmission Planning

The fundamental objective of transmission planning is to develop the system as economically as possible and maintain an acceptable reliability level. The system development is generally associated with determination of a reinforcement alternative and its implementation time. A decision on retirement or replacement of aging system equipment is also an important task in planning.

There are many drivers for the development of a transmission system, and these mainly include

- Load growth
- New-generation sources
- Equipment aging
- Commercial opportunities
- Changes in export to and import from neighbor systems
- Variations in supply reliability requirements of customers

Probabilistic Transmission System Planning, by Wenyuan Li

- Access of new loads or independent power producers (IPPs)
- New wheeling requirements

Most transmission development projects are driven by the first three factors: load growth, new-generation sources, and equipment aging. Traditionally, a utility company was vertically organized. Generation, transmission, and distribution were owned and therefore planned by a single company. Generation and transmission have been unbundled in most countries since the deregulation of the power industry in the 1990s. In the deregulated environment, generation and transmission assets generally belong to different owners and thus are operated, planned, and managed separately by different companies. This book focuses on probabilistic planning of transmission systems, and the information on generation sources is treated as a known input.

Transmission planning can be divided into three stages in terms of timespan: long-term planning, medium-term planning, and short-term planning. *Long-term planning* is associated with a long period, such as 20–30 years. It often focuses on a high-level view of system development. The problems discussed in long-term planning are preliminary and may need significant changes or even redefinitions in subsequent planning stages because of very uncertain input data and information used. *Medium-term planning* can refer to a timeframe between 10 and 20 years. In this stage, preliminary considerations in long-term planning are modified according to actual information obtained in previous years, and study results are utilized to guide short-term projects. *Short-term planning* deals with the issues that have to be resolved within 10 years. Concrete alternatives must be investigated in depth and compared. Planning studies at this stage should lead to a capital plan for planning projects.

Transmission planning includes different tasks, such as

- Determination of voltage level
- Network enhancement
- Substation configuration
- Reactive resource planning
- Load or independent power producer (IPP) connection planning
- Equipment planning (spare, retirement, or replacement)
- Selection of new technologies [light high-voltage direct current (HVDC), flexible AC transmission system (FACTS), superconductive technology, wide-area measurement system (WAMS)-based technology]
- Special protection system scheme versus network reinforcement

A transmission planning project may be associated with one or more of the tasks listed above, and each task requires technical, economic, environmental, social, and political assessments. The technical assessment alone covers multiple considerations in space and time dimensions and requires numerous studies, which include load forecast, power flow calculation, contingency analysis, optimal power flow calculation, voltage and transient stability analysis, short-circuit analysis, and reliability evaluation. Essentially, the studies are operation simulations of future situations in many years. The purpose of the studies is to select and compare planning alternatives. It is necessary to identify which system situations can occur in the future and in which manner the transmission system can operate for each situation. The combinations of system states and operation manners will reach an astronomical figure. It is impractical to simulate all the cases. Obviously, some simplification is necessary in system modeling and selection of system states.

Transmission planning is an extremely complicated problem and is always broken into subproblems in system modeling. Coordination among subproblems is needed. The judgment of planning engineers and preselected feasible alternatives play an important role in coordination. Many optimization modeling approaches for transmission planning have been developed in the past. These approaches are merely techniques for solving one or more special subproblems. It is important to recognize that it is impossible to make a decision for a system reinforcement scheme based only on the result from a single optimization model. In fact, many constraints and considerations in environmental, social, and political aspects cannot be quantitatively modeled.

The load levels, network topologies, generation patterns, availability of system components, equipment ratings in different seasons, possible switching actions, and protection and control measures must be considered in selection of system states. There are two methods for the selection: deterministic and probabilistic. The traditional *deterministic* method has been used for many years. In this method, selection of system states relies on the judgment of planning engineers, and a planning decision depends only on consequences of selected system states. The *probabilistic* method is relatively new and has not been widely used yet in the planning practice of most utilities, although some efforts have been devoted to this area. A fundamental idea in the latter method is to stochastically select system states in terms of their probabilities of occurrence. Both probabilities and consequences of simulated system states are combined to create the results for a planning decision.

1.1.2 Traditional Planning Criteria

In order to ensure reliability and economy in system development, conventional transmission planning criteria have been established at a country, regional organization, or company level. The famous NERC (North American Electric Reliability Corporation) reliability standard [1] is a good example. It includes the following sections:

- BAL—resource and demand balancing
- CIP—critical infrastructure protection
- COM—communications
- EOP—emergency preparedness and operations
- FAC—facilities design, connections, and maintenance
- INT—interchange scheduling and coordination
- IRO—interconnection reliability operations (and coordination)
- MOD—modeling, data, and analysis
- NUC—nuclear

- PER—personnel performance, training, and qualifications
- PRC—protection and control
- TOP—transmission operation
- TPL—transmission planning
- VAR—voltage and reactive power

It can be seen that this standard covers a wide range of areas and is beyond transmission planning. However, it is important to appreciate that the criteria for transmission planning are not only limited within the TPL section but should also be associated with the sections of BAL, FAC, MOD, PRC, TOP, TPL, and VAR.

The contents of conventional transmission planning criteria should at least include, but not be limited to, the following aspects [1–3]:

1. *Deterministic Security Principle.* This basically refers to the $N - 1$ principle. The $N - 1$ criterion means that a transmission system must have a sufficient number of elements to ensure that the outage or fault disturbance of a single element in any system condition does not result in any system problem, including overloading, under- or overvoltage, disconnection of other components, unplanned load curtailment, transient instability, and voltage instability. The NERC criteria also include performance requirements for planning conditions associated with two or more element outages. However, only very few important multielement outage conditions can be assessed in actual practice since the number of such conditions is too large.
2. *Voltage Levels.* The voltage level is often chosen as the one in the existing system as long as it can provide required power transfer capability without incurring unreasonable losses. Where existing voltages do not provide the required capability at a reasonable cost, a new voltage level may be established on the basis of technical and economic analyses.
3. *Equipment Ratings.* Equipment ratings (including normal and emergency ratings) are essential inputs in planning and are generally specified in manufacturer designs or industry standards. The equipment in a transmission system includes transmission lines, underground or submarine cables, transformers, instrument transformers, shunt and series capacitors, shunt and series reactors, circuit breakers, switches, static VAR compensators (SVCs), static synchronous compensators (STATCOMs), HVDC devices, bus conductors, and protection relays.
4. *System Operating Limits.* In addition to equipment ratings, system operating limits are also essential inputs in planning and include limits on voltage, frequency, thermal capacity, transient stability, voltage stability, and small-signal stability. The limits are divided into two categories for pre- and postcontingencies, and are expressed in different measure units, such as megawatts (MW), megavolt-amperes reactive (MVAR), amperes, hertz, volts, or a permissible percentage. The violation of a system operating limit may cause a consequence of instability, system split, or/and cascading outages.

5. *Transfer Capability.* Transfer capability is the amount of electric power that can be moved on a cut-plane between two areas under a specified system condition. A cutplane is often a group of transmission lines. This is associated with two terms: total transfer capability (TTC) and available transfer capability (ATC) [4]. The latter can be determined by

$$\text{ATC} = \text{TTC} - \text{TRM} - \text{CBM} - \text{ETC}$$

   where TRM (transmission reliability margin) is the transfer capability margin required to ensure system security under a reasonable range of uncertainties and possible system conditions, CBM (capacity benefit margin) is the transfer capability reserved to ensure the access to generation from interconnected areas to meet generation reliability requirements, and ETC (existing transmission commitment) is the transfer capability scheduled for transmission services. Obviously, determination of TTC and ATC is associated with not only thermal but also stability limits.
6. *Connection Requirements.* As transmission system open access becomes a reality with deregulation in the power industry, various interconnections to a transmission system have greatly increased. These include the connections of generation, transmission, and end-user facilities. The term *generation facilities* refers to not only generators of generation companies but also regular IPPs and renewable sources. The connection projects require considerable feasibility, system impact, and facility studies, which have naturally become parts of transmission planning activities. The connection requirements are not only related to technical studies but also heavily associated with regulatory policies and business models.
7. *Protection and Control.* In some cases of transmission planning, a protection and control scheme can ensure system security while avoiding addition of primary equipment, resulting in a considerable saving of capital investment. In addition to traditional equipment protection and control schemes, the special protection system (SPS) and wide-area measurement system (WAMS) play a greater role in system security. The SPS, which is sometimes called a *remedial action scheme* (RAS), includes different schemes such as undervoltage load shedding, underfrequency load shedding, auto-VAR control, generation rejection, line tripping, and transient overvoltage control [5]. All the SPS schemes must meet reliability requirements and design principles. The WAMS, which can be also called a *wide-area control system* since its function is not limited to measurement, has been rapidly developed and applied in recent years. However, the reliability criteria for WAMS have not been well established so far.
8. *Data and Models.* All planning studies require adequate and accurate data and models, including those for static and dynamic simulations in both internal and external representations. The data and models for loads, generation sources, and customer connections need the coordination between transmission companies and their customers. The requirements of data management, including

validation, database, and historical data analysis, are essential to transmission planning.

9. *Economic Analysis Criteria.* Generally, more than one system alternative meet the technical planning criteria, and therefore economic analyses are conducted to select the least total cost alternative. The total cost includes capital and operating costs. In traditional planning, unreliability cost is not a component of economic analysis. The cost analysis must be carried out in a planning period of many years. Because of the unpredictability of economic parameters for the future, it is often necessary to perform sensitivity studies to examine the effects of parameter variations.

1.2 NECESSITY OF PROBABILISTIC TRANSMISSION PLANNING

The purpose of probabilistic planning is to add one more dimension enhancing the transmission planning process rather than to replace the traditional criteria summarized in Section 1.1.2. The majority of the traditional criteria will continue to be used in probabilistic planning with the exception of the following new ideas:

- The $N-1$ principle is no longer a unique security criterion. In addition to single contingencies, multicomponent outages (as many as possible) have to be considered.
- Not only the consequences but also the occurrence probabilities of outage events must be simulated.
- Uncertainties in network configurations, load forecast, generation patterns, and other parameters should be represented as possible using probabilistic or/and fuzzy modeling methods.
- On top of the traditional studies (power flow, optimal power flow, contingency analysis, and stability assessment), the probabilistic techniques (probabilistic power flow, probabilistic contingency analysis, and probabilistic stability assessments) should be conducted. In articular, probabilistic system reliability evaluation is performed and becomes a key step.
- Unreliability cost assessment is a crucial part of overall economic analysis and plays an important role in planning decisions. Introduction of the unreliability cost, which depends on various probabilistic factors, establishes a probabilistic feature in economic analysis. The uncertain factors in economic parameters can also be considered.

There are numerous reasons for doing probabilistic transmission planning [6–8]:

1. One major weakness of deterministic criteria is the fact that the probabilistic nature of outage data and system parameters is overlooked. For instance, an outage event, even if extremely undesirable, is of little consequence if it is so

unlikely that it can be ignored. A planning alternative based on such an event will lead to overinvestment. Conversely, if selected outage events are not very severe but have relatively high probabilities of occurrence, an option based only on the effects of such events will still result in a high-risk outcome. Probabilistic planning can recognize not only the severity but also the likelihood of occurrence of events.

2. The deterministic criterion is based on the worst-case study. The "worst case" may be missed. For example, system peak load is generally used as one of the worst conditions. However, some serious system problems may not necessarily occur at peak load. Also, even if a system withstands the "worst case," the system is still not risk-free. It is worthy to identify the risk level associated with the $N-1$ criterion. This is one of the tasks in probabilistic transmission planning.
3. Major outages are usually associated with multiple component failures or cascading events in real life. This suggests that the $N-1$ criterion is insufficient to preserve a reasonable level of system reliability. However, on the other hand, it is almost impossible for any utility to justify the $N-2$ or $N-3$ principle for all outage events in transmission planning. A better alternative is to bring risk management into planning practice and keep system risk within an acceptable level.

There is no conflict between deterministic and probabilistic planning criteria. A complete system planning process includes societal, environmental, technical, and economic assessments. Probabilistic economic assessment and reliability evaluation are suggested to add as a part of the whole process. Figure 1.1 gives a conceptual example in which seven candidates for planning alternatives are assumed at the beginning. Two

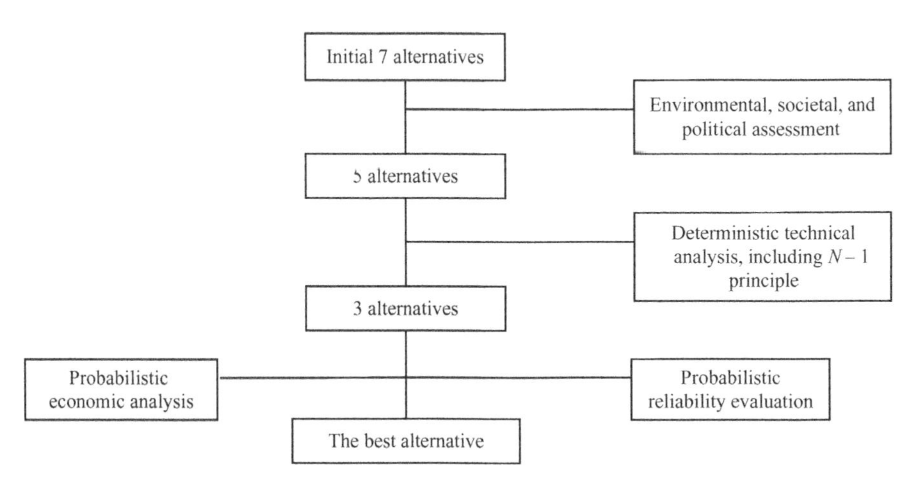

Figure 1.1. System planning process.

of them are excluded on the basis of environmental, societal, or political considerations. Deterministic technical criteria including the $N-1$ principle are applied to the remaining five alternatives. Two more alternatives are eliminated from the candidate list because of their inability to meet the deterministic technical criteria. Then probabilistic reliability evaluation and probabilistic economic analysis are performed to select the best scenario. Both the $N-1$ principle and probabilistic reliability criteria are satisfied. Other probabilistic techniques can also be applied to system analyses even in the domain of deterministic criteria.

Although the majority of the traditional criteria are still effective in probabilistic planning, introduction of the probability-related ideas (particularly the concept of unreliability cost) will significantly change the planning process and the philosophy in decisionmaking. Probabilistic transmission planning brings the missed (overlooked) factors in the traditional planning into studies and will definitely lead to a more reasonable decision in the sense of a tradeoff between reliability and economy.

1.3 OUTLINE OF THE BOOK

The book can be divided into four parts. The first part includes Chapters 2–7, which discuss the concepts, models, methods, and data that are used in probabilistic transmission planning. Chapter 8, as the second part, focuses on a special issue—how to deal with the uncertainty of data in probabilistic planning using fuzzy techniques. The third part, Chapters 9–12, addresses four essential issues in probabilistic transmission planning using actual utility systems as examples. The fourth part consists of three appendixes, which provide the basic knowledge in mathematics for probabilistic planning.

Chapter 2 presents the basic concepts of probabilistic transmission planning emphasizing the criteria and general procedure.

Chapter 3 addresses load modeling issues in both time and space perspectives. Load growth is a major driver in transmission planning. Various practical load forecast methods are discussed. Other aspects of load modeling include load clustering, uncertainty, and correlation of bus loads, as well as voltage and frequency characteristics of loads.

Chapter 4 focuses on system analysis techniques. The traditional analysis methods, which are still required in probabilistic transmission planning, are briefly summarized. These include power flow, optimal power flow, contingency analysis, and voltage and transient stability assessments. Probabilistic power flow, probabilistic optimization techniques, and risk-index-based contingency ranking are presented as new analysis methods.

Chapter 5 illustrates transmission reliability evaluation. This is a key step toward probabilistic transmission planning. Reliability indices and reliability worth assessment are explained. In the adequacy perspective, reliability evaluation methods for composite generation–transmission systems and substation configurations are discussed. In the security perspective, probabilistic voltage and transient stability assessments are proposed as new techniques. This chapter only provides a summary of the topic; more details can be found in the author's previous book, *Risk Assessment of Power Systems*, in the same IEEE Press Series on Power Engineering.

Chapter 6 discusses the methods for economic analysis, which is another key in probabilistic transmission planning. Three cost components for planning projects are outlined. Following the basic concepts on time value of money and depreciation methods, the economic assessment techniques for project investment and equipment replacement are presented in detail. A probabilistic method for the uncertainty of economic parameters is also proposed. Compared to books on engineering economics, a major feature is that the unreliability cost is incorporated in the proposed techniques.

Chapter 7 addresses data issues in probabilistic transmission planning. The data for system analysis include equipment parameters, ratings, system operation limits, and load coincidence factors. Preparing reliability data is an important step toward probabilistic planning. Both equipment outage indices and delivery point indices, which are based on outage statistics, are discussed with typical examples. Other data required in planning are also outlined.

Chapter 8 proposes a solution to data uncertainty in probabilistic transmission planning using combined fuzzy and probabilistic techniques. The uncertainty includes both randomness and fuzziness of loads and outage parameters, particularly for the data associated with weather conditions. Two examples for reliability assessment are provided to demonstrate applications of the techniques presented. Similar ideas can be extended to the uncertainty of other data and other system analyses.

Chapters 9–12 are devoted to practical applications that are day-to-day topics in transmission planning. The special concept, method, and procedure for each topic are developed, and actual planning examples from real utility transmission systems are provided. In particular, the results in the applications have been implemented in the utility's decisionmaking.

Chapter 9 discusses transmission network reinforcement planning. This is a task that transmission planners have to deal with every day. Two applications are developed using probabilistic planning methods. One is the reinforcement of a bulk power supply system; the other is the comparison among planning alternatives for a regional transmission network.

Chapter 10 copes with retirement planning of system components. The retirement timing of a system component is a challenge facing planners as a transmission system ages. New probabilistic planning methods are presented for two applications: (1) the retirement of an AC cable and (2) the replacement strategy of a DC cable.

Chapter 11 discusses substation planning. There are two types of problems in substation planning. One is selection of substation configuration; the other is determination of the number and timing of spare equipment shared by substations. Probabilistic approaches to both problems are established and demonstrated using actual utility application examples.

Chapter 12 is committed to the probabilistic planning method for radial single-circuit supply systems. This is a challenging issue that cannot be resolved using the traditional $N-1$ planning criterion. The method presented is based on historical outage indices and probabilistic economic analysis with predictive reliability evaluation. A utility example is provided.

There are three appendixes. Appendix A provides the basic concepts in probability theory and statistics, whereas Appendix B presents the elements in fuzzy mathematics. Appendix C gives the fundamentals in reliability evaluation, including both crisp and fuzzy reliability assessment methods.

Transmission planning covers a very wide range of topics. The book does not pretend to include all known and available materials in this area. The intent is to focus on the basic aspects and most important applications in probabilistic transmission planning.

2

BASIC CONCEPTS OF PROBABILISTIC PLANNING

2.1 INTRODUCTION

The most distinguishing feature of probabilistic planning is a combination of probabilistic reliability evaluation and economic analysis into the planning process. In traditional deterministic planning, system reliability is considered through some simple rules such as the $N - 1$ principle, whereas in probabilistic planning, system reliability is quantitatively assessed and expressed using one or more indices that represent system risks. Two essential tasks of probabilistic planning are (1) establishment of probabilistic planning criteria through reliability indices and (2) combination of quantitative reliability evaluation with probabilistic economic analysis to form a basic procedure. It is important to appreciate that the objective of introducing probabilistic planning is to enhance but not to replace traditional planning. The new procedure must be designed in such a way that both deterministic and probabilistic criteria are coordinated in a unified process.

Probabilistic planning also requires other power system analysis and assessment techniques in addition to reliability evaluation and economic analysis. These include probabilistic methods in load forecast and load modeling, power flow and probabilistic power flow, traditional and probabilistic contingency analyses, and optimal power flow

Probabilistic Transmission System Planning, by Wenyuan Li

and probabilistic search-based optimization methods, as well as conventional and probabilistic voltage and transient stability assessments. Another important task is preparation and management of various data required by probabilistic planning activities as well as treatment of data uncertainties. All of these will be discussed in Chapters 3–8.

This chapter focuses on the basic concepts of probabilistic transmission planning. The probabilistic criteria are presented in Section 2.2. The general procedure of probabilistic planning is illustrated in Section 2.3, and other relevant aspects are briefly outlined in Section 2.4.

2.2 PROBABILISTIC PLANNING CRITERIA

Although probabilistic planning criteria are not as straightforward as the deterministic $N - 1$ criterion, different approaches can be developed to establish probabilistic criteria [7–9]. Four possible approaches are presented in this section. Which one is used depends on the utility business model and the project to which the criterion is applied. For instance, different approaches can be used for bulk networks and regional systems, or for transmission-line addition and substation enhancement.

2.2.1 Probabilistic Cost Criteria

Reliability is one of multiple factors considered in probabilistic transmission planning. System unreliability can be expressed using unreliability costs so that system reliability and economic effects can be assessed on a unified monetary basis. There are two methods for incorporating the unreliability cost: the total cost method and the benefit/cost ratio method.

1. *Total Cost Method.* The basic idea is that the best alternative in system planning should achieve the minimum total cost:

   $$\text{Total cost} = \text{investment cost} + \text{operation cost} + \text{unreliability cost}$$

   The calculation of investment cost is a routine economic analysis activity in transmission planning. The operation cost includes OMA (operation, maintenance, and administration) expenditures, network losses, financial charges, and other ongoing costs. The unreliability cost is obtained using the EENS index (expected energy not supplied, in MWh/year) times the unit interruption cost (UIC, \$/kWh), which will be discussed in Chapter 5.
2. *Benefit/Cost Ratio Method.* The capital investment of a planning alternative is the cost, whereas the reduction in operation and unreliability costs is the benefit due to the alternative. The benefit/cost ratios for all preselected alternatives are calculated and compared. In other words, the alternatives can be ranked using their benefit/cost ratios. A project may be associated with multistage investments and a planning timeframe (such as 5–20 years) is always considered. All three cost components should be estimated on an annual basis to create their

cash flows on the timeframe first, and then a present value method is applied to calculate the benefit/cost ratios. The BCR method will be discussed in further detail in Chapter 6.

2.2.2 Specified Reliability Index Target

Many utilities have used reliability indices to measure the system performance and made an investment decision based on the metrics. One or more reliability indices can be specified as a target reliability level. For example, the target of an outage duration index such as the T-SAIDI (system average interruption duration index) or an outage frequency index such as the T-SAIFI (system average interruption frequency index) (where the T prefix denotes transmission) can be specified with a tolerant variance range. If the evaluated result exceeds the specified range, an enhancement is required. The definitions of transmission system reliability evaluation indices and historical performance-based indices will be discussed in Chapters 5 and 7, respectively.

The essence of this approach is to use a reliability index as a target. It is well known, for instance, that the LOLE (loss of load expectation) index of one day per 10 years has been used as a target index in generation planning for many years. Unfortunately, it is not easy to set an appropriate index target for transmission system reliability. Historical statistics can help in determination of an index target. On the other hand, caution should be taken regarding the coherent uncertainty and inaccuracy in historical records when this approach is used.

2.2.3 Relative Comparison

In most cases, the purpose of transmission planning is to compare different alternatives (including the doing nothing option). One major index or multiple indices (such as the EENS, probability, frequency, and duration indices) can be used in the comparison.

Performing a relative comparison is often better than using an absolute index target because

- Not only reliability indices but also economic and other aspects can be compared.
- Historical statistics and input data used in probabilistic reliability evaluation are always fraught with uncertainties.
- The historical system performance may not represent the future performance that a planning projects targets.
- There are computational errors in modeling and calculation methods, and errors can be offset in a relative comparison.

2.2.4 Incremental Reliability Index

If it is difficult to use unreliability cost in some cases, an incremental reliability index (IRI) can be applied. The IRI is defined as the reliability improvement due to per M$ (million dollars) of investment, which can be expressed as follows:

$$\mathrm{IRI} = \frac{\mathrm{RI}_B - \mathrm{RI}_A}{\mathrm{cost}}$$

The *cost* is the total cost for investment and operation (in M$) required for a reinforcement option. The RI_B and RI_A respectively are the reliability indices before and after the reinforcement. Conceptually, any appropriate reliability index (such as the EENS, probability, frequency, or duration index) can be used. In most cases, the EENS is suggested if it can be quantified since this index is a combination of outage frequency, duration, and severity, and carries more information than does any other single index.

The IRI can be used to rank projects or compare alternatives for a project. The disadvantage of the IRI approach is the fact that the "doing nothing" option cannot be included.

2.3 PROCEDURE OF PROBABILISTIC PLANNING

There are different ways to perform probabilistic transmission planning [7–9]. Figure 2.1 shows a general process that includes the criteria mentioned above. It also indicates how to combine the deterministic $N - 1$ principle with the probabilistic criteria.

The basic procedure of probabilistic transmission planning includes the following four major steps:

1. If the single-contingency criterion is a mandate, select the planning alternatives that meet the $N - 1$ principle. Otherwise, if the $N - 1$ principle is not considered as the strict criterion, select all feasible alternatives. In either case, system analysis using the traditional assessment techniques (power flow, optimal power flow, contingency analysis, and stability studies) is needed. These techniques will be summarized in Chapter 4.
2. Conduct probabilistic reliability evaluation and unreliability cost evaluation for the selected alternatives over a planning timeframe (such as 5–20 years) using a reliability assessment tool for transmission systems.
3. Calculate the cash flows and present values of investment, operation, and unreliability costs for the selected alternatives in the planning time period.
4. Select an appropriate criterion described in Section 2.2 and conduct an overall probabilistic economic analysis.

It can be seen that probabilistic reliability evaluation and economic analysis are two key steps, which are briefly discussed in Sections 2.3.1 and 2.3.2 and will be detailed in Chapters 5 and 6.

2.3.1 Probabilistic Reliability Evaluation

There are two fundamental methods for probabilistic reliability evaluation [6,10,11] of transmission systems: Monte Carlo simulation and state enumeration. The difference

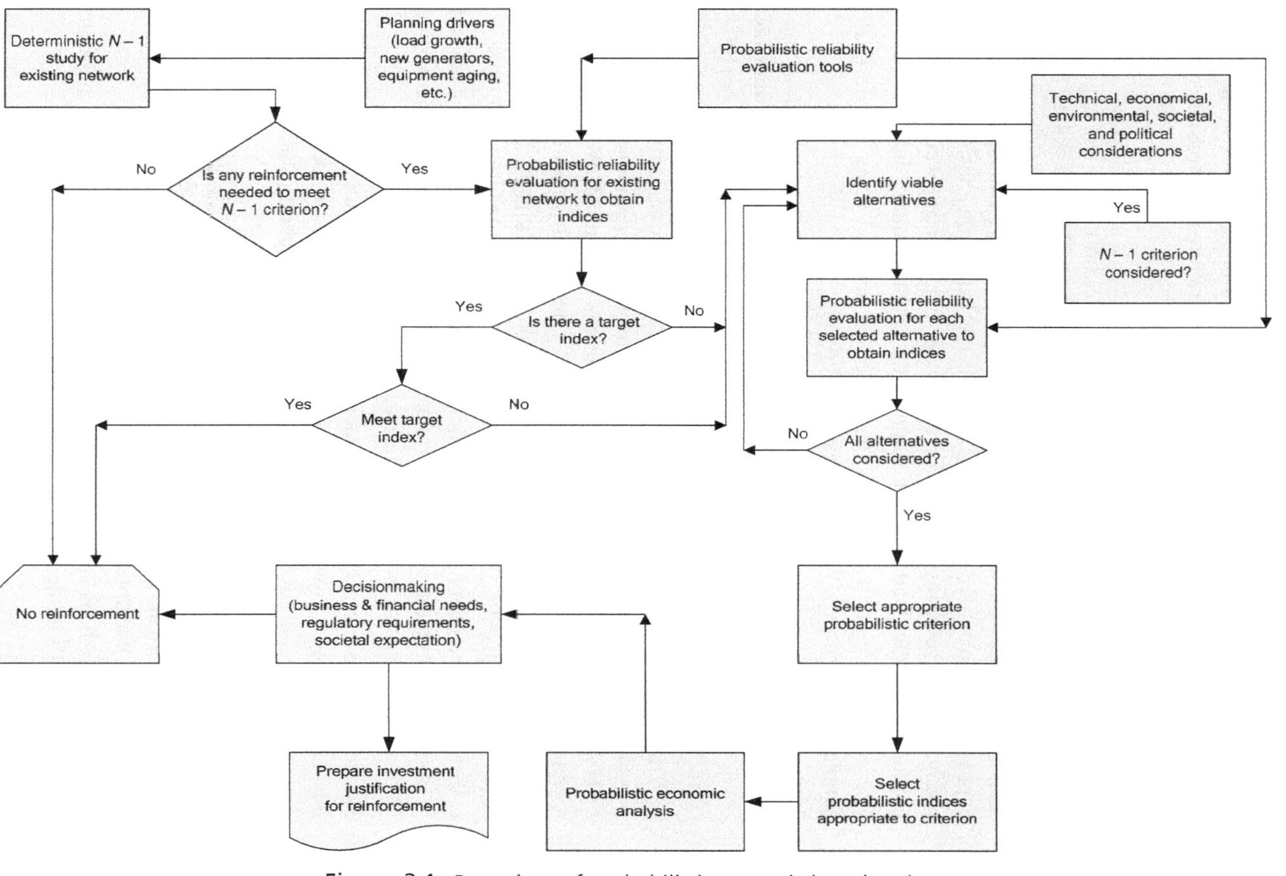

Figure 2.1. Procedure of probabilistic transmission planning.

between the two methods is associated with how to select system states, whereas the system analysis in assessing the consequences of selected outage states is the same. The probabilistic reliability assessment of composite generation–transmission systems using the Monte Carlo method is summarized as follows:

1. A multistep load model is created that eliminates the chronology and aggregates load states using hourly load records during one year. The uncertainty of load at each step can be modeled using a probability distribution if necessary. Annualized indices are calculated first by using only a single load level and are expressed on the one-year basis. All the load-level steps are considered successively, and the resulting indices for each load level that are weighted by the probability for that load level are summed up to obtain annual indices.
2. The system states at a particular load level are selected using Monte Carlo simulation techniques. This includes the following:
    a. Generally, generating unit states are modeled using multistate random variables. If generating units do not create different impacts on selected transmission planning alternatives, the generating units can be assumed to be 100% reliable.
    b. Transmission component states are modeled using two-state (up and down) random variables. For some special transmission components such as HVDC lines, a multistate random variable can be applied. Weather-related transmission-line forced outage frequencies and repair times can be determined using the method of recognizing regional weather effects. Transmission-line common-cause outages are simulated by separate random numbers.
    c. Bus load uncertainty and correlation are modeled using a correlative normal distribution random vector. A correlation sampling technique for the normal distribution vector is used to select bus load states.
3. System analyses are performed for each selected system state. In many cases, this requires power flow and contingency analysis studies to identify possible system problems. In some cases, transient and voltage stability studies may also be required.
4. An optimal power flow (OPF) model is used to reschedule generations and reactive sources, eliminate limit violations (line overloading and/or bus voltage violations), and avoid any load curtailment if possible or minimize the total load curtailment or interruption cost if unavoidable.
5. The reliability indices are calculated on the basis of the probabilities and consequences of all sampled system states.

If the state enumeration method is used, step 2 is carried out differently and all other steps basically remain the same. The reliability evaluation for substation configurations follows a simpler procedure since it is essentially a problem of connectivity between sources and load points, and no power-flow-based model is required. The probabilistic reliability evaluation will be discussed in more detail in Chapter 5.

2.3.2 Probabilistic Economic Analysis

There are three cost components in probabilistic economic analysis [6,10]: investment, operation, and unreliability costs.

Investment analysis is a fundamental part of the economic assessment in a planning process. The cash flow of annual investment cost can be created using the capital return factor (CRF) method and actual capital estimates. The parameters associated with economic analysis of capital investment (such as the useful life of a project, discount rate, and capital estimates) are usually given by deterministic numbers. However, a probabilistic method can be applied to capture the uncertainty of the parameters. For instance, a discrete probability distribution of discount rate can be obtained from historical statistics and considered in the calculation model. The concepts and methods used in the economic analysis will be illustrated in Chapter 6.

The cash flows of operation and unreliability costs are calculated through year-by-year evaluations. In addition to fixed cost components, the operation cost in a transmission system is also related to evaluation of network losses, simulation of system production costs, and estimation of energy prices on the power market. This is associated with considerable uncertainty factors, including load forecasts, generation patterns, maintenance schedules, and power market behaviors. In some cases, on the other hand, the planning alternatives selected may involve only limited modifications of network configuration and have basically the same or close operation cost. In such a situation, the operation cost may not have to be included in the total cost for comparison. This is case-dependent.

As mentioned earlier, the unreliability cost equals to the product of EENS and a unit interruption cost. Obviously, this cost component is a random number that depends on various probabilistic factors in a transmission system, particularly on random outage events. The EENS can be calculated by a probabilistic reliability evaluation method, whereas the unit interruption cost (UIC) can be estimated using one of the following four techniques. The first one is based on customer damage functions (CDFs) that can be obtained from customer surveys. A CDF provides the relationship curve between the average unit interruption cost and duration of power outage. The second one is based on a gross domestic product divided by a total electric energy consumption, which gives a dollar value per kilowatt-hour (kWh) reflecting the average economic damage cost due 1 kWh of energy loss. The third one is based on the relationship between capital investment projects and system EENS indices. The fourth one is based on the lost revenue to the utility due to power outages. This last technique would typically represent the lowest level of UIC. Utilities can select an appropriate technique that best aligns with their business objectives. The details of unreliability cost assessment will be given in Chapter 5.

2.4 OTHER ASPECTS IN PROBABILISTIC PLANNING

Probabilistic planning involves a wide range of study activities, including traditional deterministic analyses and new probabilistic assessments. The traditional analysis techniques used in transmission planning, including power flow, contingency analysis, optimal power flow, transient stability, and voltage stability, will continue to be important while new probabilistic assessment methods are introduced.

Load forecast and generation conditions are two crucial prerequisites of transmission planning. Load forecast has been performed using probabilistic methods even in the conventional planning practice because of the inherent uncertainty in predicting future loads. However, applications of new methods (such as neural network and fuzzy clustering) provide vehicles to improve accuracy in long-term load forecasting. Load forecast and other load modeling issues will be discussed in Chapter 3. Generation conditions, including types of generators, locations, capacities and availability in the future, are the outcome of generation planning, which is a complex task by itself. Detailed discussion of generation planning is beyond the scope of the book. An integrated planning of generation and transmission may be necessary in some cases.

Obviously, probabilistic methods are not limited to reliability evaluation and economic analysis. Probabilistic power flow and probabilistic contingency analysis are often useful tools in probabilistic transmission planning studies. In the optimization analysis for system planning, probabilistic search optimization techniques can provide solutions to some special problems. Probabilistic transient stability and voltage stability assessments should also be considered when necessary. Essentially, these are extensions of conventional transient and voltage stability analyses in order to incorporate probabilistic modeling of various uncertain factors. The basic procedure includes three main aspects: (1) a probabilistic method is used to select random factors for dynamic system states, (2) a stability analysis technique is used to conduct stability simulations of stochastically selected system states, and (3) a probabilistic index or its distribution is created to represent system instability risk. Generally, probabilistic stability assessments provide a deeper and broader insight into system dynamic behavior and instability risk. The additional probabilistic techniques will also be presented in Chapters 4 and 5.

Another crucial aspect is the data preparation for probabilistic transmission planning. This includes not only the regular data for conventional system analyses but also the data for probabilistic assessments. The data required in probabilistic planning will be discussed in detail in Chapter 7. Reliability data are obtained from historical statistical records, and a computerized database is required to collect, store, and manage outage data. Maintaining a high quality of data is a key for successful probabilistic planning.

There are two types of uncertainty in input data and modeling: randomness and fuzziness. Randomness is characterized by probability, whereas fuzziness is characterized by fuzzy variables. Dealing with the two uncertainties is an important issue in probabilistic transmission planning. This will be addressed in Chapter 8.

2.5 CONCLUSIONS

This chapter described the basic concepts of probabilistic transmission planning. The first step toward probabilistic planning is to establish and understand its criteria and procedure. The four probabilistic planning criteria have been discussed. Utilities may select one or more criteria in actual applications to suit their business requirements. A probabilistic planning flowchart is presented to illustrate the details of planning procedure and the coordination between probabilistic and deterministic $N-1$ criteria.

The most important tasks in probabilistic planning are quantified probabilistic reliability evaluation and probabilistic economic analysis. The basic ideas of the two tasks have been discussed. Besides, probabilistic studies in other aspects are also required depending on the particular cases and problems under study. These include probabilistic load forecast and load modeling, probabilistic power flow and contingency analysis, and probabilistic transient and voltage stability assessments. It is essential to recognize that the deterministic $N-1$ criteria must be used to select initial alternatives for probabilistic planning and the conventional system analysis techniques including power flow, contingency analysis, optimal power flow, and stability simulations are still important tools in the integrated planning process. Probabilistic planning methods are designed to complement and enhance traditional transmission planning.

This chapter provided only a high-level description of the subject. Detailed discussions of each topic will be developed in subsequent chapters.

3

LOAD MODELING

3.1 INTRODUCTION

Power system loads have two basic attributes: spatiotemporal characteristics and voltage/frequency dependence. Modeling the two attributes is the first fundamental task in probabilistic transmission planning. *Spatiotemporal characteristics* refer not only to the feature of variation with time and space but also to the uncertainty and correlation in variation. Spatiotemporal characteristics of power loads can be modeled using a probabilistic or fuzzy variable. The term *voltage/frequency dependence* refers to the fact that a bus load changes with voltage and frequency. This dependence is deterministic at a specific timepoint and a specific bus location. On the other hand, these two attributes are related to each other because voltage and frequency also change with time and space.

The purpose of load forecast is to predict a load at a given timepoint or load curve with multiple timepoints in the future. Essentially, the objective of load forecasting is to model the characteristics of loads in the time dimension. Load growth is one of main drivers for system reinforcement, and an accurate load forecast is a crucial prerequisite in transmission planning. Several applied load forecast methods are discussed in Section 3.2.

Probabilistic Transmission System Planning, by Wenyuan Li

A transmission system contains many buses (substations), and a load curve at a bus includes a considerable number of timepoints. It is often necessary to cluster bus loads or/and load curves into groups in the time or space dimension in order to simplify the load representation in power system analysis. Two load clustering techniques are described in Section 3.3.

In conventional power system analysis such as power flow, voltage stability, and transient stability calculations, bus loads are considered as deterministic variables. However, bus load uncertainty and correlation should be modeled in probabilistic power system analyses, which will be discussed in Chapters 4 and 5. Modeling the correlation of loads among multiple buses is still a challenge, although it has been recognized that the uncertainty of loads can be addressed using sensitivity studies or a probabilistic distribution. A modeling technique for both bus load uncertainty and correlation is presented in Section 3.4.

Dependence of loads on voltage and frequency is also an important aspect in system analysis for transmission planning. This topic is briefly summarized in Section 3.5.

3.2 LOAD FORECAST

Many mathematical methods have been available for probabilistic load forecast [12–19]. This section focuses on regression and time-series techniques, which are the most popular methods in load forecasting. A neural network is also introduced. It is necessary to emphasize the following points before we discuss the methods:

- A load can be predicted as a single peak value, or the energy consumption in a given period, or a load curve with multiple timepoints. It can be a forecast for an entire system, a region, a substation (bus), or a single customer. Regression techniques are often used to predict one or more values of MW load or energy consumption, whereas time-series methods are used to predict a load curve in a time specific period.
- As in any other prediction for the future, errors cannot be avoided in load forecast. For this reason, a confidence range for load forecast is always of importance. The range is not only an indicator of accuracy but also extremely useful for sensitivity or probabilistic studies in transmission planning.
- Some uncertain factors impacting loads may not follow any known statistical rule. Therefore, customer interviews or surveys are necessary in addition to the mathematical modeling methods. The information from customer interviews can be used in building or improving a probabilistic or fuzzy load forecasting model.

3.2.1 Multivariate Linear Regression

3.2.1.1 Regression Equation. The power load is affected by multiple factors in economical, political, and environmental (including weather-related) areas. The factors can be expressed using different quantified variables including gross domestic product, financial budget, population, number of customers in each sector, quantity of major products, temperature, and windspeed. The load forecast for a region or entire system often requires more variables, whereas the load forecast for a substation needs fewer variables. In some cases (such as at a distribution feeder level), only one single variable may be acceptable if no further information is available.

The relationship between the load and the multiple variables affecting the load can be represented using the following linear expression [12]:

$$y_i = a_0 + a_1 x_{i1} + a_2 x_{i2} + \cdots + a_m x_{im} + e_i \qquad (i = 1,2,\ldots,n) \tag{3.1}$$

Here y_i is the observed load; $x_{i1}, x_{i2}, \ldots, x_{im}$ are the ith set of observed values of m affecting variables in the history (normally in the ith year in the past); $a_0, a_1, a_2, \ldots, a_m$ are called the regression coefficients; e_i is the remainder error; and n is the number of sets of observed values.

By using a matrix form, we can rewrite Equation (3.1) as

$$\mathbf{Y} = \mathbf{XA} + \mathbf{E} \tag{3.2}$$

where

$$\mathbf{Y} = \begin{bmatrix} y_1 \\ y_2 \\ \vdots \\ y_n \end{bmatrix}, \quad \mathbf{X} = \begin{bmatrix} 1 & x_{11} & \cdots & x_{1m} \\ 1 & x_{21} & \cdots & x_{2m} \\ \vdots & \vdots & \vdots & \vdots \\ 1 & x_{n1} & \cdots & x_{nm} \end{bmatrix}, \quad \mathbf{A} = \begin{bmatrix} a_0 \\ a_1 \\ \vdots \\ a_m \end{bmatrix}, \quad \mathbf{E} = \begin{bmatrix} e_1 \\ e_2 \\ \vdots \\ e_n \end{bmatrix}$$

The least-squares technique can be applied to minimize the sum of squares of the remainder errors in Equation (3.2) in order to obtain the estimates of the regression coefficients:

$$\mathbf{E}^T\mathbf{E} = (\mathbf{Y} - \mathbf{XA})^T(\mathbf{Y} - \mathbf{XA}) \tag{3.3}$$

Letting the derivatives of Equation (3.3) be equal to zero, we obtain

$$\frac{\partial \mathbf{E}^T\mathbf{E}}{\partial \mathbf{A}} = -2\mathbf{X}^T\mathbf{Y} + 2\mathbf{X}^T\mathbf{XA} = \mathbf{0} \tag{3.4}$$

The estimate of **A** is obtained from Equation (3.4):

$$\mathbf{A} = (\mathbf{X}^T\mathbf{X})^{-1}\mathbf{X}^T\mathbf{Y} \tag{3.5}$$

3.2.1.2 Statistical Test of Regression Model. Mathematically, this model must meet the following assumptions:

- The load y_i and related variables x_{ik} ($k = 1,\ldots,m$) satisfy the linear relationship given in Equation (3.1).

- No linear relationship should exist between any two variables of x_{ik} and x_{il} (k, $l = 1,\ldots,m$ and $k \neq l$).
- The remainder error e_i follows the normal distribution with the following conditions: (1) its mean is equal to zero and (2) the covariance between any two e_i and e_j corresponding to two sets of different observed values is not equal to zero.

Before determining the variables impacting the power load, an engineer should make a judgment on the correlation between the load and selected variables. However, the human being's judgment cannot guarantee validation of the assumptions presented above. The following two statistical tests are necessary to verify the created regression model before it can be used for load forecast [13]:

1. t-*Test of Regression Coefficients.* The purpose of this test is to ensure a significant linear relationship between the load y_i and related variables x_{ik} ($k = 1,\ldots,m$). Construct the following t statistic of a regression coefficient a_k ($k = 1,\ldots,m$):

$$t_{a_k} = \frac{a_k}{S_{a_k}} \tag{3.6}$$

Here, S_{a_k} is the sample standard deviation of a_k and can be estimated by

$$S_{a_k} = \sqrt{S_e^2 C_{kk}} \tag{3.7}$$

where C_{kk} is the kth diagonal element of matrix $(\mathbf{X}^T\mathbf{X})^{-1}$ and S_e^2 is the sample variance of the remainder error e_i in Equation (3.1), which can be estimated by

$$S_e^2 = \frac{1}{n-m-1}\sum_{i=1}^{n}(y_i - \hat{y}_i)^2 \tag{3.8}$$

where y_i is the actually observed load in the ith year in the past and $\hat{y}_i$ is the load estimated using the regression method and historical data before the ith year.

If $|t_{a_k}| > t_{\alpha/2}(n-m-1)$, where $t_{\alpha/2}(n-m-1)$ is such a value that for the given significance level α, the integral of the t distribution with $(n - m - 1)$ degrees of freedom from $t_{\alpha/2}(n-m-1)$ to ∞ equals $\alpha/2$, then it indicates that the characteristic of $a_k \neq 0$ is valid. In other words, the linear relationship between the load y_i and the variable x_{ik} is acceptable, and this variable should be kept in the regression equation. Otherwise, $|t_{a_k}| \leq t_{\alpha/2}(n-m-1)$ indicates that the regression coefficient a_k is significantly close to zero and suggests that the variable x_{ik} should be excluded from the regression equation. There are two possibilities for the failure of the coefficient test: (1) it is true in real life that this variable may not have any significant impact on the power load to be predicted or (2) there may be collinearity between selected variables. In the latter case, a quantified correlation analysis between variables is necessary. Only one of multiple variables with collinearity should be kept in the regression equation. The regression coefficients need to be reestimated after one or more variables are removed from the model. Although a successive regression technique, in which invalid variables can be automatically filtered, is mathematically available [12], it may or may not provide a better solution.

2. F-*Test of Regression Model.* In addition to the (Student's) t-test on individual regression variables, the regression model as a whole still requires an F-test. Failure of the F-test suggests that either some important variables impacting the load have been missed from the equation, or the relationship between the load and related variables is nonlinear. Construct the following F statistic:

$$F = \frac{\sum_{i=1}^{n}(\hat{y}_i - \bar{y})^2 / m}{\sum_{i=1}^{n}(y_i - \hat{y}_i)^2 / (n - m - 1)} \tag{3.9}$$

Here, y_i and $\hat{y}_i$ are the same as defined in Equation (3.8), and $\bar{y}$ is the sample mean of $y_i (i = 1,\ldots,n)$, that is, $\bar{y} = \sum_{i=1}^{n} y_i / n$.

If $F > F_\alpha(m, n-m-1)$, where $F_\alpha(m, n-m-1)$ is the value of the F distribution with m degrees of freedom for the first parameter and $(n - m - 1)$ degrees of freedom for the second parameter at the given significance level α, this indicates that the regression is effective and the model test has been passed successfully. Otherwise, the created regression model cannot be used. In this case, it is necessary to investigate the possibility of using a nonlinear regression or other forecast method.

3.2.1.3 *Regression Forecast.* Once the regression coefficients are estimated and the two statistical tests are passed, the future value of the load can be predicted using the future values of selected variables by

$$\hat{y}_j = a_0 + a_1 x_{j1} + a_2 x_{j2} + \cdots + a_m x_{jm} \tag{3.10}$$

where the subscript j indicates the jth year in the future.

Equation (3.10) can be expressed in its matrix form:

$$\hat{y}_j = \mathbf{X}_j \mathbf{A} \tag{3.11}$$

Here, $\mathbf{X}_j = [1, x_{j1}, x_{j2}, \ldots, x_{jm}]$ represents the values of the selected variables affecting the load in the jth future year.

It can be proved that the sample variance of the forecasted load in the jth year can be estimated by [12]

$$S_{\hat{y}_j}^2 = S_e^2 \left[1 + \frac{1}{n} + \sum_{k=1}^{m} \sum_{l=1}^{m} (x_{jk} - \bar{x}_k)(x_{jl} - \bar{x}_l) C_{kl} \right] \tag{3.12}$$

where S_e^2 is the sample variance of remainder error given in Equation (3.8); $\bar{x}_k$ is the sample mean of observed values of the kth variable (i.e., $\bar{x}_k = \sum_{i=1}^{n} x_{ik} / n$), x_{jk} is the value of the kth variable in the jth future year, and C_{kl} is the element of matrix $(\mathbf{X}^T\mathbf{X})^{-1}$. Obviously, when the number of samples is relatively large and the value of each related variable is not far away from its mean, the sample variance of the forecasted load is close to the sample variance of the remainder error.

The upper and lower limits of the forecasted load for the jth year in the future can be estimated as

$$[\hat{y}_{j\min}, \hat{y}_{j\max}] = \begin{cases} \hat{y}_j \pm t_{\alpha/2}(n-m-1) \cdot S_{yj} & \text{if } n < 30 \\ \hat{y}_j \pm z_{\alpha/2} \cdot S_{yj} & \text{if } n \geq 30 \end{cases} \tag{3.13}$$

where α is the given significant level (such as 0.05), $t_{\alpha/2}(n-m-1)$ is a value such that the integral of the t distribution with $(n - m - 1)$ degree of freedom from $t_{\alpha/2}(n-m-1)$ to ∞ equals $\alpha/2$, and $z_{\alpha/2}$ is a value such that the integral of the standard normal density function from $z_{\alpha/2}$ to ∞ equals $\alpha/2$.

3.2.2 Nonlinear Regression

In many cases, the relationship between load and related variables is not linear and a nonlinear regression equation can be used. The basic procedure in the nonlinear regression includes three steps:

- Determining an appropriate nonlinear equation that represents the relationship between the load and related variables
- Converting the nonlinear relationship into a linear expression to estimate the parameters in the nonlinear regression model using the linear regression technique
- Conducting load forecast using the nonlinear regression model

3.2.2.1 Nonlinear Regression Models. The following nonlinear functions can be used to model the relationship between the electric power load and related variables:

- Polynomial model: $$y = \sum_{i=1}^{m} a_i x^i + e \tag{3.14}$$
- Exponential model: $$y = b \cdot \exp(ax) \cdot \exp(e) \tag{3.15}$$
- Power model: $$y = b \cdot x^a \cdot \exp(e) \tag{3.16}$$
- Modified exponential model: $$y = K - b \cdot \exp(-ax) \tag{3.17}$$
- Gompertz model: $$y = K \cdot \exp[-b \cdot \exp(-ax)] \tag{3.18}$$
- Logistic model: $$y = \frac{K}{1 + b \cdot \exp(-ax)} \tag{3.19}$$

When the nonlinear functions are used as regression forecast models, a_i, a, b, and K are the parameters that need to be estimated. Generally, all the parameters are positive since the load y normally increases with the variable x in load forecast. In Equation (3.14), the e is an error and denotes the deviation between the actual and forecasted load, whereas in Equations (3.15) and (3.16), the $\exp(e)$ denotes the difference between the actual and forecasted load in their ratio. No error term is explicitly given in Equations (3.17)–(3.19) since the error can be expressed in a different form depending on how the parameters are estimated. It is noted that although Equations (3.15)–(3.19) contain only one variable x, it is not difficult to introduce more variables with the same function form.

3.2.2.2 Parameter Estimation of Models. There are two methods for parameter estimation of a nonlinear regression model.

1. *Transformation into a Linear Expression.* By letting $x_1 = x, x_2 = x^2, \ldots, x_m = x^m$, we can directly transform the polynomial regression model (3.14) into the following multiple linear regression model:

$$y = a_1 x_1 + a_2 x_2 + \cdots + a_m x_m + e \tag{3.20}$$

The parameters a_i can be estimated using the method in Section 3.2.1.1. By taking logarithms at the both sides of Equation (3.15) and letting $y' = \ln y$ and $b' = \ln b$, we note that the exponential model becomes the following linear regression model:

$$y' = b' + ax + e \tag{3.21}$$

The parameters a and b' are estimated using the method in Section 3.2.1.1, and b is obtained by $b = \exp(b')$. By taking logarithms at the both sides of Equation (3.16) and letting $y' = \ln y$, $x' = \ln x$, and $b' = \ln b$, we obtain the following linear regression model for the power model:

$$y' = b' + ax' + e \tag{3.22}$$

Similarly, the parameters a and b' are estimated using the method in Section 3.2.1.1, and b is obtained by $b = \exp(b')$. The K in the modified exponential, Gompertz, or logistic model represents the limiting value on the load growth curve and is often known. In this case, the parameters a and b in the three models can be estimated through a similar transformation technique. Take the modified exponential model as an example. Adding an error term in Equation (3.17) yields

$$y = K - b \cdot \exp(-ax) \cdot \exp(e) \tag{3.23}$$

By moving K in Equation (3.23) to the left side, multiplying by -1 at the two sides, taking the logarithm and letting $y' = \ln(K - y)$ and $b' = \ln b$, the modified exponential model becomes the following linear regression model:

$$y' = b' - ax + e \tag{3.24}$$

A similar transformation technique can be applied to the parameter estimation for the Gompertz or logistic model.

2. *Taylor Expansion into a Successive Linear Estimation Process.* If the K in the three models of Equations (3.17)–(3.19) is unknown and also needs to be estimated, the Taylor expansion provides a general method. Assume the following nonlinear regression model

$$y = f(x_1,\ldots,x_m,a_1,\ldots,a_k) + e \tag{3.25}$$

where, x_i ($i = 1,\ldots,m$) are the variables affecting the load y for prediction; a_i ($i = 1,\ldots,k$) are the parameters to be estimated. A set of initial values of the parameters is selected. Equation (3.25) is expanded into the first-order Taylor series at the initial parameters and can be expressed as

$$y = \sum_{i=1}^{k} \frac{\partial f}{\partial a_i} \cdot (a_i - a_i^{(0)}) + e \tag{3.26}$$

where $a_i^{(0)}$ are their initial estimates and $\partial f/\partial a_i$ are the first-order derivatives, which can be calculated from the observed values of the affecting variables x_i and initial estimates of the parameters. The parameters are estimated using Equation (3.26), which is in the form of a linear regression model. Once the new estimates of the parameters are obtained, Equation (3.26) is updated with the new estimates until the following convergence rule is satisfied:

$$\min_i \left| \frac{a_i^{(j)} - a_i^{(j-1)}}{a_i^{(j-1)}} \right| < \varepsilon \tag{3.27}$$

Here, the superscript (j) indicates the iteration count.

3.2.3 Probabilistic Time Series

The historical load records form a sequence in the order of time, and therefore the time series techniques [12,14] can be used to predict a future load curve. Unlike the regression method that relies on other variables impacting the electrical load, the time series method uses only historical load data. The time series can be hourly, daily, weekly, or even monthly peak points depending on the purpose of load forecast.

In a deterministic time series technique, the value of a load at the current point is estimated using some weighted average of loads at past points. The exponential smoothing approach, in which the load value at a point near to the current time has a much higher weight than that at an earlier point, is a commonly used deterministic load forecast technique. The probabilistic time series treats a load time series as a stochastic process and has better prediction accuracy. The probabilistic time series method for load forecast includes the following aspects:

- The original load time series is converted into a stationary series by filtering trend and periodic components using differentiating operations.
- An appropriate prediction model is identified and then used to perform the prediction of the stationary time series.
- Load points in the future load curve are calculated using inverse differentiating operations.

3.2.3.1 Conversion to a Stationary Time Series.

An original load time series is assumed as follows:

$$\{y_t\} = \{y_1, y_2, \dots, y_n\} \tag{3.28}$$

In general, the load time series is not a stationary one since power loads contain a linear trend and one or more periodic cycles (daily, weekly, and seasonal). A stationary time series $\{w_t\}$ can be obtained by filtering the linear and periodic components in the load time series using the following differentiating operations:

$$w_t = \nabla_1^d \nabla_s^D y_t = (1-B)^d (1-B^s)^D y_t \tag{3.29}$$

Here, the subscript 1 or s denotes the step length in differentiation and the superscript d or D denotes the order of differentiation. B and B^s are backward differentiation operators:

$$By_t = y_{t-1} \tag{3.30}$$

$$B^s y_t = y_{t-s} \tag{3.31}$$

It should be noted that only one periodic differentiation operator ∇_s^D is contained in Equation (3.29) for simplicity. A load time series may simultaneously contain daily, weekly, and seasonal periodic cycles. It is not difficult to add several similar periodic differentiation operators in Equation (3.29) in programming.

The differentiated time series $\{w_t\}$ is generally stationary. If necessary, its stationariness can be tested using the following method. The time series is divided into several subseries with each having M points. Calculate thc following statistics of each subseries:

- Subseries mean:
$$\overline{w} = \frac{1}{M}\sum_{i=1}^{M} w_i \tag{3.32}$$
- Subseries variance:
$$C_0 = \frac{1}{M}\sum_{i=1}^{M} (w_i - \overline{w})^2 \tag{3.33}$$
- Autocorrelation function:
$$r_k = \frac{1}{C_0 M}\sum_{i=1}^{M-k} (w_i - \overline{w})(w_{i+k} - \overline{w}) \tag{3.34}$$
$$(k = 1, \dots, m; m \approx (0.1-0.2)M)$$

If the statistics (particularly the correlation function) of all the subseries do not have a significant difference, the stationariness of the time series $\{w_t\}$ is verified.

3.2.3.2 Model Identification.

There are three probabilistic time series models: autoregression (AR), moving average (MA), and autoregression moving average (ARMA).

The AR(p) model is expressed as follows:

$$w_t = \phi_1 w_{t-1} + \phi_2 w_{t-2} + \cdots + \phi_p w_{t-p} + e_t \tag{3.35}$$

Here, ϕ_k ($k = 1,\ldots,p$) are called the autoregression coefficients. This model suggests that the value of the tth point in the time series can be expressed using a linear combination of p prior points with a random remainder noise e_t.

The MA(q) model is expressed as follows:

$$w_t = e_t - \theta_1 e_{t-1} - \theta_2 e_{t-2} - \cdots - \theta_q e_{t-q} \tag{3.36}$$

Here, θ_k ($k = 1,\ldots,q$) are the moving average coefficients. This model suggests that the value of the tth point in the time series can be expressed using a linear combination of q prior white noises and the remainder noise e_t.

The ARMA(p,q) model is the combination of AR(p) and MA(q):

$$w_t = \phi_1 w_{t-1} + \phi_2 w_{t-2} + \cdots + \phi_p w_{t-p} + e_t - \theta_1 e_{t-1} - \theta_2 e_{t-2} - \cdots - \theta_q e_{t-q} \tag{3.37}$$

The purpose of model identification is to select an appropriate prediction model that fits historical data in the time series and the parameter p or q, or both. The selection principle is as follows. If the time series $\{w_t\}$ satisfies the condition of AR(p), this model should be selected first. Otherwise, the MA(q) is tested. If both AR(p) and MA(q) are not appropriate, the ARMA(p,q) is considered as the last option.

The autocorrelation function coefficients r_k of the whole time series $\{w_t\}$ are calculated using an approach similar to Equation (3.34) and substituted into the following recursive formulas to calculate the partial autocorrelation coefficients ϕ_{mm} and ϕ_{mi}:

$$\phi_{mm} = \begin{cases} r_1 & (m=1) \\ \dfrac{r_m - \sum_{i=1}^{m-1} \phi_{m-1,i} r_{m-i}}{1 - \sum_{i=1}^{m-1} \phi_{m-1,i} r_i} & (m = 2,3,\ldots) \end{cases} \tag{3.38}$$

$$\phi_{mi} = \phi_{m-1,i} - \phi_{mm} \cdot \phi_{m-1,m-i} \qquad (i = 1,\ldots,m-1) \tag{3.39}$$

The appropriate model is identified according to the following criteria:

- For a given p, if $\phi_{mm} \approx 0$ when $m > p$, the AR(p) model is appropriate. The condition of $\phi_{mm} \approx 0$ can be approximately judged as follows. Let N be an integer that is obtained from rounding of $\sqrt{n}$, where n is the number of load points in the time series $\{w_t\}$. Calculate N values of ϕ_{mm} after $m > p$. If about 68% in the N values satisfy $|\phi_{mm}| \le 1/\sqrt{n}$, then $\phi_{mm} \approx 0$ holds.

- For a given q, if $r_m \approx 0$ when $m > q$, the MA(p) model is appropriate. The condition of $r_m \approx 0$ can be approximately judged as follows. Calculate N values of r_m after $m > q$. If about 68% in the N values satisfy $|r_m| \le \sqrt{(1+2\sum_{i=1}^{q} r_i^2)/n}$, then $r_m \approx 0$ holds.
- Only if $\phi_{mm} \approx 0$ and $r_m \approx 0$ do not hold, the ARMA(p, q) model is selected. Several sets of p and q are considered to conduct preprediction. A posterior test of remainder errors (see Section 3.2.3.5) is used to determine the most appropriate values of p and q.

In an actual load forecast, p or q is normally smaller than 3.

3.2.3.3 Estimating Coefficients of Models. For the AR(p) model, the partial autocorrelation function coefficients $[\phi_{p1},\ldots,\phi_{pp}]$ calculated using Equations (3.38) and (3.39) are the autoregression coefficients $[\phi_1,\ldots,\phi_p]$.

For the MA(q) model, the following relationship holds between the autocorrelation function coefficients r_k and the moving average coefficients θ_k $(k = 1,\ldots,q)$:

$$r_k = \frac{-\theta_k + \theta_1\theta_{k+1} + \cdots + \theta_{q-k}\theta_q}{1+\theta_1^2+\theta_2^2+\cdots+\theta_q^2} \qquad (k = 1,2,\ldots,q) \tag{3.40}$$

This is a set of nonlinear equations and can be solved using an iterative method to obtain θ_k $(k = 1,\ldots,q)$.

For the ARMA(p,q) model, the sum of squares of the errors in the model can be expressed as a nonlinear function of the observed values w_{t-i} $(i = 1,\ldots,p)$ in the time series and the coefficients ϕ_i $(i = 1,\ldots,p)$ and θ_k $(k = 1,\ldots,q)$:

$$Q(\mathbf{w},\boldsymbol{\varphi},\boldsymbol{\theta}) = \sum_t e_t(\mathbf{w},\boldsymbol{\varphi},\boldsymbol{\theta})^2 \tag{3.41}$$

To achieve the minimum Q, the coefficients ϕ_i $(i = 1,\ldots,p)$ and θ_k $(k = 1,\ldots,q)$ should satisfy

$$\begin{aligned} \frac{\partial Q}{\partial \boldsymbol{\varphi}} &= \mathbf{0} \\ \frac{\partial Q}{\partial \boldsymbol{\theta}} &= \mathbf{0} \end{aligned} \tag{3.42}$$

The coefficients ϕ_i $(i = 1,\ldots,p)$ and θ_k $(k = 1,\ldots,q)$ can be estimated by solving the set of nonlinear Equations (3.42) or using the nonlinear least-squares technique (i.e., the successive linear estimation process), which is explained in Section 3.2.2.2. The estimation is an iterative process, and the initial estimates of the coefficients ϕ_i $(i = 1,\ldots,p)$ and θ_k $(k = 1,\ldots,q)$ are required. The error terms e_k $(k = t - 1, \ldots, t - q)$ can be estimated using the similar method presented in Equation (3.44), which will be described in the next section.

3.2.3.4 Load Forecast Equation. The forecast equation to predict load values at future timepoints l ($l = 1,\cdots,L$) of the differentiated load time series $\{w_t\}$ is as follows:

$$\hat{w}_{n+l} = \phi_1 \tilde{w}_{n+l-1} + \cdots + \phi_p \tilde{w}_{n+l-p} - \theta_1 e_{n+l-1} - \cdots - \theta_q e_{n+l-q} \tag{3.43}$$

1. If the AR(p) was selected in the model identification, the coefficients $\theta_k = 0$ ($k = 1,\ldots,q$); if the MA(q) was selected, the coefficients $\phi_i = 0$ ($i = 1,\ldots,p$).
2. The subscript n denotes the current timepoint, and l represents future timepoints. Equation (3.43) is used in a recursive manner. The load $\hat{w}_{n+1}$ is predicted first and then its predicted value is used as a known value in predicting $\hat{w}_{n+2}$, and so on. This means that any $\tilde{w}_{n+j}$ at the right side of Equation (3.43) is an observed value if $j \le 0$ and a predicted value obtained in the prior step of the recursive process if $j > 0$. Note that the w under the caret ^ denotes a value to be predicted and the w under the tilde ~ indicates that it is either an observed value or a value that has been predicted in the prior step.
3. All of the error terms e_{n+j} are the estimated values using the following recursive estimation equation:

   $$e_i = \tilde{w}_i - \phi_1 \tilde{w}_{i-1} - \cdots - \phi_p \tilde{w}_{i-p} + \theta_1 e_{i-1} + \cdots + \theta_q e_{i-q} \tag{3.44}$$

   It can be seen from Equation (3.43) that in order to predict future values at the current timepoint n, the values of q error terms ($e_{n-1},\ldots,e_{n-q}$) before the current point n must be available. These values can be estimated using Equation (3.44) and historical data before the timepoint ($n - q$) in the time series. In the first steps of the recursive estimation process, initial error terms can be assumed to be zero ($e_k = 0$ when $k << n-q$). The effect of this assumption is negligible if the recursive process of estimating error terms can start using historical data located far away back from the timepoint $n - q$. The recursive process can be used to estimate future values of error terms after the current point.

The upper and lower limits of the predicted values of the differentiated load time series $\{w_t\}$ can be estimated as follows:

$$[\hat{w}_{n+l,\min}, \hat{w}_{n+l,\max}] = \hat{w}_{n+l} \pm z_{\alpha/2} \sqrt{1 + \sum_{j=1}^{l-1} \pi_j^2} \cdot S_a \tag{3.45}$$

The $z_{\alpha/2}$ is a value such that the integral of the standard normal density function from $z_{\alpha/2}$ to ∞ equals $\alpha/2$ and α is the given significance level. The π_j and S_a are calculated using Equations (3.46) and (3.47), respectively:

$$\pi_j = \phi_1 \pi_{j-1} + \phi_2 \pi_{j-2} + \cdots + \phi_{p+q} \pi_{j-p-q} - \theta_j \qquad (j = 1,\ldots,l-1) \tag{3.46}$$

Equation (3.46) is a recursive formula, and in using it, we have $\pi_0 = 1, \pi_{j-i} = 0$ if $j-i<0$, and $\theta_j = 0$ if $j>q$.

$$S_a = \sqrt{\frac{\sum_{t=k}^{n+l} e_t^2}{n-p-q-1}} \tag{3.47}$$

The numerator in Equation (3.47) is the sum of squares of all error terms before the timepoint $n + l$. The k denotes the startpoint of the error term estimation process.

It is important to appreciate that inverse differentiation operations are needed to transform the predicted values of the differentiated time series $\{w_t\}$ into the forecasted loads for the original load time series $\{y_t\}$. Take the first-order inverse differentiation as an example. By assuming $d = 1$, $D = 1$ and letting $z_t = (1-B^s)y_t$, Equation (3.29) becomes

$$z_t = y_t - y_{t-s} \tag{3.48}$$

$$w_t = z_t - z_{t-1} \tag{3.49}$$

The original load time series $\{y_t\} = \{y_1, y_2, \ldots, y_n\}$ is known, and its differentiated load time series including future predicted values $\{w_t\} = \{w_1, w_2, \ldots, w_n, \hat{w}_{n+1}, \ldots, \hat{w}_{n+l}\}$ has been produced. The forecast load values $\{\hat{y}_{n+1}, \ldots, \hat{y}_{n+l}\}$ can be calculated through a recursion using Equations (3.48) and (3.49) from $t = n + 1$ to $t = n + l$. In the case of $d > 1$ or $D > 1$, the inverse differentiation operations are similar except that these operations are repeated according to the order of d or D.

3.2.3.5 *A Posteriori Test of Load Forecast Accuracy.* An a posteriori test can be used to examine the accuracy of the forecasted loads. The test includes the following steps:

- It is assumed that the load time series $\{y_t\}$ $(t = 1, \ldots, n+L)$ is available, where the first n values, namely, $\{y_t\}$ $(t = 1, \ldots, n)$, are used as historical data to predict the last L loads $\{\hat{y}_t\}$ $(t = n+1, \ldots, n+L)$.
- The differentiated load time series $\{w_t\}$ $(t = 1, \ldots, n+L)$ is calculated from $\{y_t\}$ $(t = 1, \ldots, n+L)$ using Equation (3.29). Similarly, $\{\hat{w}_t\}$ $(t = n+1, \ldots, n+L)$ is calculated from $\{\hat{y}_t\}$ $(t = n+1, \ldots, n+L)$.
- The error time series $\{\delta_l = w_{n+l} - \hat{w}_{n+l}\}$ $(l = 1, \ldots, K)$ is created, where $K \le L$, w_{n+l} $(l = 1, \ldots, K)$ represents the K differentiated actual loads and $\hat{w}_{n+l}$ $(l = 1, \ldots, K)$ represents the K differentiated values of predicted loads corresponding to $\{\hat{y}_t\}$ $(t = n+1, \ldots, n+K)$.
- The sample mean and standard deviation of $\{\delta_l\}$ $(l = 1, \ldots, K)$ are calculated:

$$\bar{\delta} = \frac{1}{K} \sum_{l=1}^{K} \delta_l \tag{3.50}$$

$$\sigma_1 = \sqrt{\frac{1}{K}\sum_{l=1}^{K}(\delta_l - \bar{\delta})^2} \tag{3.51}$$

- The sample mean and standard deviation of $\{w_t\}$ $(t = 1,\ldots,n+K)$ are calculated:

$$\bar{w} = \frac{1}{n+K}\sum_{t=1}^{n+K} w_t \tag{3.52}$$

$$\sigma_2 = \sqrt{\frac{1}{n+K}\sum_{t=1}^{n+K}(w_t - \bar{w})^2} \tag{3.53}$$

If σ_1 and σ_2 are close enough, the load forecast has sufficient accuracy. Otherwise, the forecast model or/and its parameters (p or/and q) should be reconsidered.

3.2.4 Neural Network Forecast

In the regression forecast, an analytical relationship between the load and related variables must be specified. In some cases, this relationship may not be represented by any explicit linear or nonlinear function. In such a situation, a neural network forecast method is a good choice. The neural network can create complex nonlinear mapping relationships between multiple inputs and outputs through a learning process based on historical data.

3.2.4.1 Feedforward Backpropagation Neural Network (FFBPNN). A feedforward backpropagation neural network [15] contains a series of nodes called *neurons* that are located in multiple layers: one input layer, one output layer, and several hidden layers. A simple FFBPNN with the following features is generally sufficient for load forecast:

- One neuron at the output layer, which is the load
- Only one hidden layer with several neurons (less than 10)
- Several neurons at the input layer, which are the variables impacting the load

As shown in Figure 3.1, all the neurons in one layer are connected to all the neurons in the next layer. There is a separate weighting factor on each link between neurons. Signals enter the neurons in the input layer and propagate through the neurons at the hidden layer to the output layer. Each neuron accepts the signals from prior neurons and transmits modified signals to the neurons in the next layer. For each neuron in the input layer, its input and output are the same. For each neuron in the hidden or output layer, as shown in Figure 3.2, its input is the sum of the weighted outputs from neurons in the prior layer, and its output is nonlinearly scaled in the interval [0,1] by a sigmoid function. Mathematically, the input I_j and output O_j for the neuron j in the hidden or output layer is calculated as follows:

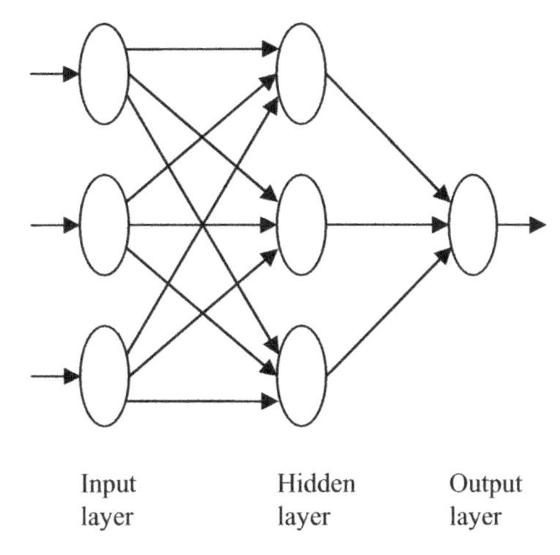

Figure 3.1. Feedforward backpropagation neural network.

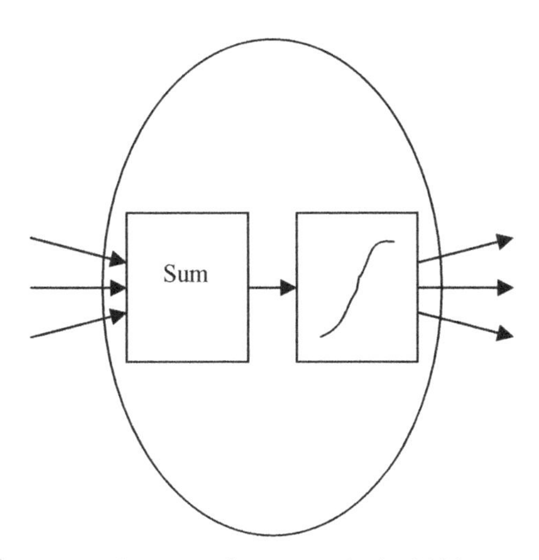

Figure 3.2. Input and output of a neuron in the hidden or output layer.

$$I_j = \sum_i W_{ji} \cdot O_i \tag{3.54}$$

$$O_j = \frac{1}{1 + e^{-(I_j + S_j)}} \tag{3.55}$$

where O_i is the output of the neuron i in the prior layer, W_{ji} is the weighing factor between neurons j and i, and S_j is the threshold of the input of neuron j.

3.2.4.2 Learning Process of FFBPNN. The FFBPNN can be trained using a set of historical data. The inputs are the values of variables impacting the load, and the output is the load. The purpose of a learning process is to determine the values of all weighting factors and thresholds in the neural network, which represent the complex nonlinear relationship between the variables and the load.

The learning process includes the following steps:

1. The initial values of weighting factors W_{ji} and thresholds S_j are given arbitrarily.
2. For each set of historical data, the load is calculated as the output from the FFBPNN using the values of impacting variables as inputs. The following error indicator, which represents the total difference between the calculated loads from the FFBPNN and the actual loads in historical records, is computed:

$$E = \frac{1}{n}\sum_{k=1}^{n}(L_k - O_{lk})^2 \tag{3.56}$$

   Here, n is the number of historical datasets; L_k and O_{lk} are the actual load and the calculated load at the output layer (denoted by subscript l) from the kth set of data. It is worth emphasizing that since the output of FFBPNN must be a value less than 1, the actual load L_k should be a normalized value, which is calculated using the load record divided by a value equal to or larger than the maximum one in all load records.

3. From the output layer to the input layer, the weighting factors W_{ji} and thresholds S_j are backward-updated in order to reduce the error E. The adjusted changes of W_{ji} and S_j are calculated by

$$\Delta W_{ji}^{(p)} = \sum_{k=1}^{n}\alpha \cdot \delta_{jk}^{(p)} O_{ik}^{(p)} + \beta \cdot \Delta W_{ji}^{(p-1)} \tag{3.57}$$

$$\Delta S_{j}^{(p)} = \sum_{k=1}^{n}\alpha \cdot \delta_{jk}^{(p)} + \beta \cdot \Delta S_{j}^{(p-1)} \tag{3.58}$$

   where the subscripts j and i represent current and prior layers, respectively, and j denotes either the output or hidden layer; the subscript k corresponds to the kth set of data; the superscript p represents the iteration count in the learning process; $\Delta W_{ji}^{(p-1)}$ and $\Delta S_{j}^{(p-1)}$ are the adjusted changes of W_{ji} and S_j in the last iteration; O_{ik} is the output at a neuron in the prior layer; the α and β are the learning rate and momentum coefficient, respectively, and their values are selected between 0 and 1.0; and the error term δ_{jk} is calculated as follows. If j is the neuron in the output layer, δ_{jk} is calculated by

$$\delta_{jk} = \delta_{lk} = (L_k - O_{lk}) \cdot O_{lk} \cdot (1 - O_{lk}) \tag{3.59}$$

   where the subscript l denotes the output layer. If j is a neuron in the hidden layer, δ_{jk} is calculated by

$$\delta_{jk} = \delta_{hk} = \delta_{lk} \cdot W_{lh} \cdot O_{hk} \cdot (1 - O_{hk}) \tag{3.60}$$

where the subscript h represents the hidden layer; δ_{lk} has been calculated from Equation (3.59) when the neuron l in the output layer is considered; and W_{lh} is the weighting factor on the link between the neuron l in the output layer and the neuron h in the hidden layer.

If two or more correlated loads are simultaneously forecast using a FFBPNN, multiple neurons in the output layer can be specified. In this case, the total error indicator should be the sum of the error indicators for all output loads, whereas the term $\delta_{lk} \cdot W_{lh}$ in Equation (3.60) should be changed as $\sum_l \delta_{lk} \cdot W_{lh}$.

4. Once $\Delta W_{ji}^{(p)}$ and $\Delta S_j^{(p)}$ are obtained, the weighting factors W_{ji} and thresholds S_j are adjusted by adding $\Delta W_{ji}^{(p)}$ and $\Delta S_j^{(p)}$, respectively. Then the outputted load and error indicator are recalculated. The iteration continues until the convergence criterion $|E^{(p)} - E^{(p-1)}| \le \varepsilon$ is satisfied, where ε is a given permissible error.

3.2.4.3 Load Prediction. After the trained FFBPNN is built, the load forecasting is straightforward. A set of future values of variables impacting the load are input into the FFBPNN, and the output is the predicted load. When new data become available, the learning process should be updated to renew the weighting factors W_{ji} and thresholds S_j. The accuracy of neural network forecast can be estimated using an a posteriori test in which a variance analysis for the differences between forecast and actual loads is required.

3.3 LOAD CLUSTERING

An actual load curve is a chronological one as shown in Figure 3.3. It can be converted into a load duration curve expressed in per unit (a fraction with regard to the peak) as shown in Figure 3.4, in which the chronology is eliminated. Either the chronological or the duration load curve is used in system analysis and reliability assessment, depending on the modeling technique and study purpose. There are two types of load clustering. The first one is to cluster load points in one or more load curves into several load-level groups. This clustering is needed for creating the multistep load model, which is also schematically shown in Figure 3.4. The second one is to cluster different load curves into several curve groups, in each of which load curves have a similar curve shape or time-varying load pattern.

3.3.1 Multistep Load Model

The K-mean clustering technique [10,20,21] can be used to create a multistep load model. It is assumed that a load duration curve is divided into NL load levels. This corresponds to grouping load points of the curve into NL clusters. Each load level is the mean value of those load points in a cluster. The load clustering is performed in the following steps:

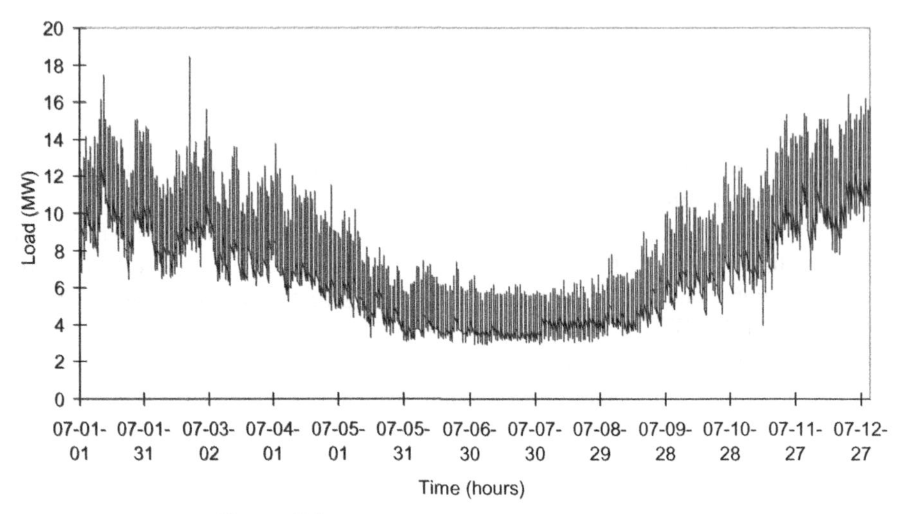

Figure 3.3. A chronological annual load curve.

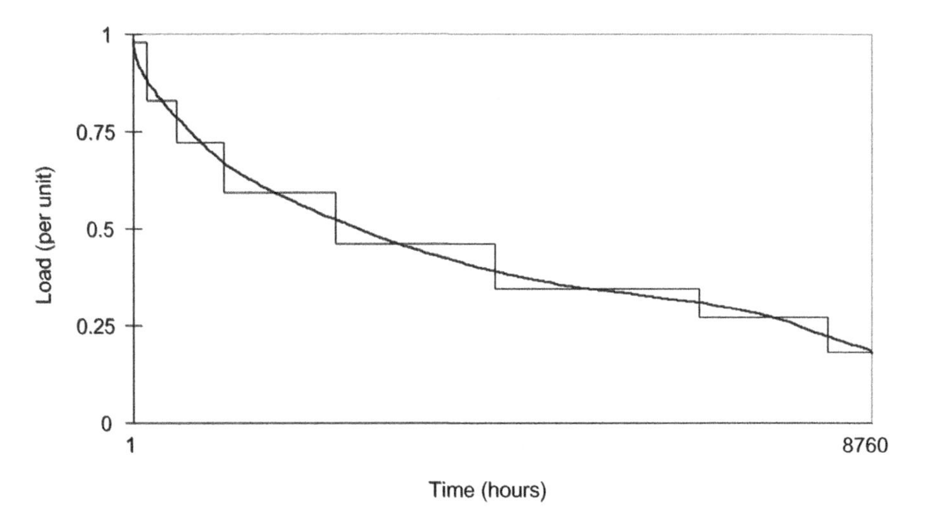

Figure 3.4. Load duration curve and its multistep model.

1. Select initial cluster mean M_i, where i denotes cluster i ($i = 1,\ldots,NL$).
2. Calculate the distance D_{ki} from each load point L_k ($k = 1,\ldots,NP$) (where NP is the total number of load points in the load curve) to the ith cluster mean M_i by

$$D_{ki} = |M_i - L_k| \tag{3.61}$$

3. Group load points by assigning them to the nearest cluster based on D_{ki} and calculate new cluster means by

$$M_i = \frac{\sum_{k \in IC_i} L_k}{NS_i} \tag{3.62}$$

where NS_i is the number of load points in the ith cluster and IC_i denotes the set of the load points in the ith cluster.

4. Repeat steps 2 and 3 until all cluster means remain unchanged between iterations.

The M_i and NS_i obtained are the load level (in MW) and the number of load points for the ith step in the multistep load model, respectively. NS_i also reflects the length of the ith step load level. The multistep load model can be easily expressed in per unit.

The K-mean clustering technique can be extended to the case of multiple load curves where each one represents a curve for a bus or bus group in a region. In this case, the following steps are conducted:

1. Select initial cluster mean M_{ij}, where i denotes cluster i ($i = 1,\ldots,NL$) and j denotes curve j ($j = 1,\ldots,NC$, where NC is the number of load curves).
2. Calculate the Euclidean distance from each load point to each cluster mean by

$$D_{ki} = \left[\sum_{j=1}^{NC} (M_{ij} - L_{kj})^2 \right]^{1/2} \tag{3.63}$$

where D_{ki} denotes the Euclidean distance from the kth load point to the ith cluster mean, and L_{kj} ($k = 1,\ldots,NP, j = 1,\ldots,NC$) is the load at the kth load point on the jth curve.

3. Regroup the load points by assigning them to the nearest cluster based on D_{ki} and calculate the new cluster means by

$$M_{ij} = \frac{\sum_{k \in IC_i} L_{kj}}{NS_i} \qquad (j-1,\ldots,NC) \tag{3.64}$$

Equation (3.64) is the same as Equation (3.62) except that the cluster means are calculated for each of the NC load curves.

4. Repeat steps 2 and 3 until all the cluster means remain unchanged between iterations.

The cluster means are used as the load levels in each cluster for each load curve. Each load level represents NS_i load points in the corresponding cluster in the sense of mean value. It can be seen that the correlation between load levels (cluster means) in each cluster of the load curves is captured in the load clustering technique.

3.3.2 Load Curve Grouping

In system analysis and reliability assessment, it is sometimes necessary to group load curves according to their similarity in the curve shape or time-varying load pattern. A statistic-fuzzy technique can be used for this purpose. Hourly load curves are taken as an example in the following description.

Each load curve of customers or at substations consists of many hourly load points. Its length can be one day, one month, a few months, or one year depending on the purpose and data records available. The statistic-fuzzy technique [22] includes the following steps:

1. The power-consuming behavior varies with time and is represented by the shape of a load curve but not by absolute MW values. For instance, two daily load curves with the same shape, even if their MW values are different, are believed to follow the same pattern (load factor). In order to catch load curve shapes, all load curves for clustering are normalized using their peak values to obtain the load curves represented in per unit with regard to the peak.
2. The similarity of two load curve patterns is described using their nearness degree. Mathematically, a load curve can be regarded as a vector with hourly load points to be its components. The cosine of the angle between two load curve vectors [23] can be used to represent the nearness degree, which is called a *nearness coefficient*. Consider the two load curves expressed in per unit: $\{X_{i1},X_{i2},\ldots,X_{in}\}$ and $\{X_{k1},X_{k2},\ldots,X_{kn}\}$, where X_{il} or X_{kl} $(l = 1,\ldots,n)$ is the lth hourly load. The nearness coefficient between the two load curve vectors is calculated by

$$R_{ik} = \frac{\sum_{l=1}^{n} X_{il}X_{kl}}{\sqrt{\left(\sum_{l=1}^{n} X_{il}^2\right)\cdot\left(\sum_{l=1}^{n} X_{kl}^2\right)}} \tag{3.65}$$

   The value of R_{ik} is between [0,1]. The more the value of R_{ik} approaches 1, the nearer the two load curves are. The nearness coefficients between all load curves in consideration are calculated. Unfortunately, these coefficients cannot be directly used to cluster load curves since they do not have transitivity; A near B, B near C, C near D, and D near E cannot guarantee A near E, particularly when there are many load curves.
3. The nearness of two load curves is actually fuzzy, and therefore nearness relationships of all load curves can form a fuzzy relation matrix, each element of which is the membership function of the relationship between two load curves. Conceptually, the coefficient R_{ik} can be used as the membership function since it represents the nearness degree of two load curves. This membership function matrix is not subjectively assumed but obtained from historical statistics of load data. Apparently, it is a resemblance fuzzy matrix with the following two features:
   - It is symmetric matrix because $R_{ik} = R_{ki}$ (symmetry)
   - Its diagonal element $R_{ii} = 1.0$ (reflexivity)

4. An equivalent fuzzy matrix with transitivity can be obtained through consecutive self-multiplications of a resemblance fuzzy matrix [24]. Let R denotes the fuzzy matrix obtained in step 3. Its self-multiplication is defined as

$$R_m = \underbrace{R \circ R \circ R \cdots\cdots R}_{\text{self-multiplications}} \tag{3.66}$$

$$(R \circ R)_{ij} = \max_{l} \min\{R_{il}, R_{lj}\} \tag{3.67}$$

Equation (3.67) means that the each element in the fuzzy matrix $(R \circ R)$ is calculated using a rule similar to that for multiplication of crisp matrices, except that the product of two elements is replaced by taking the minimum one and the addition of two products by taking the maximum one. It has been proved in fuzzy mathematics that after $m - 1$ self-multiplications of R, where m is a positive integer equal to or less than the number of load curves, we have [see Section B.4.2 (in Appendix B)]:

$$R_{m-1} = R_m = R_{m+1} = R_{m+2} = \cdots \tag{3.68}$$

Namely, transitivity is reached at R_{m-1}.

5. The elements in each column of R_{m-1} are called the *transitive nearness coefficients*. They reflect both direct and indirect nearness relationships between load curves. A non diagonal element in R_{m-1} is selected as the threshold, and the coefficients in the matrix, which correspond to the relation between two load curves, are checked. If a transitive nearness coefficient is equal to or larger than the threshold, then the corresponding load curves are classified in one cluster. A load curve that cannot be classified by this rule, if any, is in a cluster by itself.

Five load curves (in per unit) for a 24-h period with one hour as a point are used to demonstrate the application. These are the typical load curves at five substations in a utility system. The load curves are plotted in Figure 3.5, and their shapes can be directly observed. The *transitive nearness coefficient matrix* of the load curves is obtained using the statistic-fuzzy technique as follows:

$$\begin{bmatrix} 1.000 & 0.999 & 0.983 & 0.983 & 0.997 \\ 0.999 & 1.000 & 0.983 & 0.983 & 0.997 \\ 0.983 & 0.983 & 1.000 & 0.998 & 0.983 \\ 0.983 & 0.983 & 0.998 & 1.000 & 0.983 \\ 0.997 & 0.997 & 0.983 & 0.983 & 1.000 \end{bmatrix}$$

If 0.997 is used as the threshold value, it can be seen from the transitive nearness matrix that load curves 1, 2, and 5 are in one cluster but load curves 3 and 4 in another. This obviously is consistent with the observation from Figure 3.5. If 0.998 is used as the threshold, the load curves are classified into three clusters: (1,2), (3,4), and (5). A larger

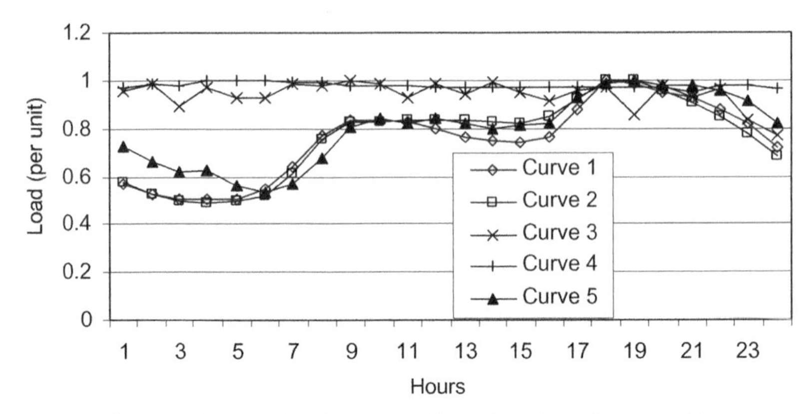

Figure 3.5. Daily load curves at five substations (in per unit).

threshold value results in more clusters, in each of which the load curves are closer. The selection of threshold depends on the preference of users for a specific degree of closeness in one cluster and the data volume of load curves. It should be noted that only relative magnitudes of transitive nearness coefficients are required in clustering. A small difference between the coefficients can still distinguish clusters. This is because the self-multiplication of the fuzzy matrix is a process of taking a minimum or maximum value without any numerical calculations, and therefore no computing error is introduced in creating the transitive nearness coefficients from the resemblance fuzzy matrix.

Apparently, it is impossible to cluster a considerable number of load curves with a long period by direct observation (such as many annual load curves with 8760 hourly points). The statistic-fuzzy technique provides a systematic clustering method for grouping load curves.

3.4 UNCERTAINTY AND CORRELATION OF BUS LOADS

There are always load uncertainty and correlation of bus loads [25]. Bus load uncertainty can be modeled using a normal distribution with the two parameters of mean and standard deviation. The mean can be either the forecast load at the peak (or at any timepoint), or the average load in a given period of a load curve. The standard deviation can also be obtained from a load forecast method as discussed in Section 3.2.

If loads at different buses proportionally increase or decrease, the loads at the buses are completely dependent. This assumption has been used in the practice of many utilities when they prepare power flow cases. Unfortunately, this is not true in real life and can lead to overestimation of line flows in a transmission system. Another extreme assumption is complete independence among all bus loads in a system. Under this second assumption, each bus load can be represented using an independent normal distribution. This is not true, either, and can result in underestimation of line flows. In fact, bus loads are correlated to some extent. A correlated normal distribution load vector can be used to model both uncertainty and correlation of bus loads.

The correlation of bus loads can be expressed using a correlation matrix. It is assumed that load curves at all buses in a given period of interest (such as a month, season, or year) have been obtained. The correlation coefficient between loads at buses i and j (an element of the correlation matrix) is calculated by

$$\rho_{ij} = \frac{(1/n)\sum_{k=1}^{n}(L_{ki} \cdot L_{kj}) - (M_i \cdot M_j)}{(1/n)\left[\sum_{k=1}^{n}(L_{ki} - M_i)^2 \cdot \sum_{k=1}^{n}(L_{kj} - M_j)^2\right]^{1/2}} \tag{3.69}$$

where L_{ki} and L_{kj} are load values at the kth load point on the curves i and j, respectively; n is the number of load points; $M_i = (1/n)\sum_{k=1}^{n} L_{ki}$ and $M_j = (1/n)\sum_{k=1}^{n} L_{kj}$.

In the case of using the multistep load model in Section 3.3.1, the correlation coefficient between the load means in the mth cluster can be calculated by

$$\rho_{ij} = \frac{(1/NS_m)\sum_{k \in IC_m}(L_{ki} \cdot L_{kj}) - (M_{mi} \cdot M_{mj})}{(1/NS_m)\left[\sum_{k \in IC_m}(L_{ki} - M_{mi})^2 \cdot \sum_{k \in IC_m}(L_{kj} - M_{mj})^2\right]^{1/2}} \tag{3.70}$$

where L_{ki} and L_{kj} are load values at the kth load point on the ith and jth load duration curves, respectively; NS_m is the number of load points in the mth cluster; IC_m denotes the set of the load points in the mth cluster; and M_{mi} and M_{mj} are the cluster means in the mth cluster, respectively, for load duration curves i and j. It should be noted that chronology no longer exists in the multistep load model and therefore the correlation given in Equation (3.70) is just the correlation between cluster mean values of load duration curves in each cluster.

The correlation matrix $\boldsymbol{\rho}$ of bus loads is an $N \times N$ matrix where N is the number of buses considered. The elements of matrix $\boldsymbol{\rho}$ are estimated using Equation (3.69) or (3.70). The covariance matrix $\mathbf{C}$ of bus loads is calculated by

$$\mathbf{C} = \begin{bmatrix} S_1^2\rho_{11} & S_1S_2\rho_{12} & \cdots & S_1S_N\rho_{1N} \\ \vdots & \vdots & \vdots & \vdots \\ S_NS_1\rho_{N1} & S_NS_2\rho_{N2} & \cdots & S_N^2\rho_{NN} \end{bmatrix} \tag{3.71}$$

where S_i is the sample standard deviation of the load at the ith bus, which can be obtained with load forecast; and ρ_{ij} is the element in the ith row and jth column of the correlation matrix $\boldsymbol{\rho}$.

The correlated normal distribution load vector that represents both uncertainty and correlation of bus loads can be derived from an independent normal distribution vector. Let $\mathbf{H}$ denote the correlated normal load vector with the mean load vector $\mathbf{B}$ and covariance metric $\mathbf{C}$, where $\mathbf{B}$ is obtained from load forecast or cluster means, and $\mathbf{C}$ from Equation (3.71). Let $\mathbf{G}$ be a standard normal distribution vector whose components are independent of each other, and assume that each component has a mean of zero and a variance of unity. A linear combination of normal distributions is still a normal distribution [26]. There exists therefore a matrix $\mathbf{A}$ that can create the following transformation relationship between and $\mathbf{H}$ and $\mathbf{G}$:

$$\mathbf{H} = \mathbf{AG} + \mathbf{B} \tag{3.72}$$

The mean vector and covariance matrix of **H** can be calculated from Equation (3.72) as follows:

$$E(\mathbf{H}) = \mathbf{A} \cdot E(\mathbf{G}) + \mathbf{B} = \mathbf{A} \cdot \mathbf{0} + \mathbf{B} = \mathbf{B} \tag{3.73}$$

$$E[(\mathbf{H} - \mathbf{B})(\mathbf{H} - \mathbf{B})^T] = E(\mathbf{AGG}^T\mathbf{A}^T) = \mathbf{AA}^T = \mathbf{C} \tag{3.74}$$

Equation (3.74) gives the relation between matrices **A** and **C**. The covariance matrix **C** is a nonnegative definite symmetric matrix and can be triangularized into the unique lower triangular matrix multiplied by its transposed matrix. Consequently, the following recursive formulas can be derived from the relation of $\mathbf{AA}^T = \mathbf{C}$ to calculate the elements of matrix **A** from those of matrix **C**:

$$A_{i1} = \frac{C_{i1}}{\sqrt{C_{11}}} \qquad (i = 1, \ldots, N) \tag{3.75}$$

$$A_{ii} = \sqrt{C_{ii} - \sum_{k=1}^{i-1} A_{ik}^2} \qquad (i = 2, \ldots, N) \tag{3.76}$$

$$A_{ij} = \frac{C_{ij} - \sum_{k=1}^{j-1} A_{ik} A_{jk}}{A_{jj}} \qquad (j = 2, \ldots, N-1; i = j+1, \ldots, N) \tag{3.77}$$

With the matrices **A**, **G**, and **B**, the correlated normal load vector **H** is calculated using Equation (3.72). The bus loads defined by matrix **H** has the mean load vector **B** with the uncertainty and correlation defined by the covariance metric **C** given in Equation (3.71).

3.5 VOLTAGE- AND FREQUENCY-DEPENDENT BUS LOADS

A *bus load* in a transmission system generally refers to an aggregated load seen at a power delivery point (substation). Therefore a bus load representation includes not only power-consuming load devices connected to the bus but also feeders, transformers, reactive power equipment, and even an equivalent subsystem behind the delivery point. A static bus load is voltage- and frequency-dependent and can have the following general representation:

$$P = P_0 \cdot \sum_j \alpha_j \left(\frac{V}{V_0}\right)^{a_j} \cdot [1 + \beta_j (f - f_0)] \tag{3.78}$$

$$Q = Q_0 \cdot \sum_j \eta_j \left(\frac{V}{V_0}\right)^{b_j} \cdot [1 + \theta_j (f - f_0)] \tag{3.79}$$

Here, P and Q are active and reactive powers at a bus, V and f are the voltage and frequency at the bus, V_0 and f_0 are the nominal or initial voltage and frequency, and P_0

and Q_0 are the active and reactive powers consumed at the nominal or initial voltage and frequency. The bus subscript has been omitted for simplicity. The parameters of the model include the coefficients $\alpha_j, \eta_j, \beta_j, \theta_j$ and the exponents a_j and b_j. In the model, the bus power has two portions: one independent of frequency and the other dependent of frequency. Both portions are dependent on voltage. In actual applications, simplified representations are often used for different purposes in the system analysis.

3.5.1 Bus Load Model for Static Analysis

In static system analysis such as power flow, contingency analysis, and static voltage stability, the transition process of system states in the time coordinate is not considered. Two popular bus load models are used in the static analysis.

3.5.1.1 Polynomial Bus Load Model

$$P = P_0 \left[\alpha_1 \left(\frac{V}{V_0} \right)^2 + \alpha_2 \frac{V}{V_0} + \alpha_3 \right] \cdot [1 + \beta(f - f_0)] \tag{3.80}$$

$$Q = Q_0 \left[\eta_1 \left(\frac{V}{V_0} \right)^2 + \eta_2 \frac{V}{V_0} + \eta_3 \right] \cdot [1 + \theta(f - f_0)] \tag{3.81}$$

This model is obtained from the general representation given in Equations (3.78) and (3.79) when $j = 1$, $a_1 = b_1 = 2$; when $j = 2$, $a_2 = b_2 = 1$; and when $j = 3$, $a_3 = b_3 = 0$; and $\beta_1 = \beta_2 = \beta_3 = \beta$; $\theta_1 = \theta_2 = \theta_3 = \theta$.

In most cases of statistic analysis, the portion dependent on frequency is not considered by letting $\beta = 0$ and $\theta = 0$. In this situation, the model becomes only voltage-dependent, which is often called a *ZIP* model since it is composed of constant impedance (*Z*), constant current (*I*), and constant power (*P*) terms. The coefficients a_1, α_2, and α_3 (or η_1, η_2 and η_3) represent the fractions of the three types of load and therefore $a_1 + \alpha_2 + \alpha_3 = 1$ and $\eta_1 + \eta_2 + \eta_3 = 1$. These coefficients can be determined from customer compositions.

3.5.1.2 Exponential Bus Load Model

$$P = P_0 \cdot \left\{ \alpha_1 \left(\frac{V}{V_0} \right)^{a_1} [1 + \beta_1 (f - f_0)] + (1 - \alpha_1) \left(\frac{V}{V_0} \right)^{a_2} \right\} \tag{3.82}$$

$$Q = Q_0 \cdot \left\{ \eta_1 \left(\frac{V}{V_0} \right)^{b_1} \cdot [1 + \theta_1 (f - f_0)] + (1 - \eta_1) \left(\frac{V}{V_0} \right)^{b_2} [1 + \theta_2 (f - f_0)] \right\} \tag{3.83}$$

This model is obtained when the first two terms of the general representation in Equations (3.78) and (3.79) are considered. The frequency coefficient in the second term for the active power load has been assumed to be zero. Equations (3.82) and (3.83) are similar to the static load model used in the LOADSYN program that was developed by the Electric Power Research Institute (EPRI) in the United States [27,28].

3.5.2 Bus Load Model for Dynamic Analysis

Strictly speaking, both generator and loads should be modeled using differential equations in the dynamic system analysis such as transient stability and dynamic voltage stability simulations. In the practice of utilities, however, a static but more complex load model is also often used for dynamic simulation analysis [29,30]. For example, the following bus load model is used in the EPRI extended transient midterm stability program (ETMSP):

$$P = F_p \cdot P_0 \left\{ \alpha_1 \left(\frac{V}{V_0} \right)^2 + \alpha_2 \frac{V}{V_0} + \alpha_3 + \sum_{j=4}^{5} \alpha_j \left(\frac{V}{V_0} \right)^{a_j} \cdot [1 + \beta_j (f - f_0)] \right\} \tag{3.84}$$

$$Q = F_q \cdot Q_0 \left\{ \eta_1 \left(\frac{V}{V_0} \right)^2 + \eta_2 \frac{V}{V_0} + \eta_3 + \sum_{j=4}^{5} \eta_j \left(\frac{V}{V_0} \right)^{b_j} \cdot [1 + \theta_j (f - f_0)] \right\} \tag{3.85}$$

where $\sum_{j=1}^{j=5} \alpha_j = 1$ and $\sum_{j=1}^{j=5} \eta_j = 1$; F_p and F_q represent the fractions of bus load representation using the static model for dynamic performance.

Obviously, this model is a combination of the polynomial and exponential models given in Section 3.5.1. In the model, the first three terms provide a representation of an aggregated load, whereas the fourth and fifth terms can be used to model two types of motors or one motor and one discharging lighting load.

It is important to appreciate that the static load models given by Equations (3.78)–(3.85) may not be realistic at a low voltage and may create a divergence problem. It is a common approach to switch the whole load model to a constant impedance representation when the bus voltage is below a threshold.

In cases where the static bus load model cannot represent a realistic situation, a representation of load dynamics can be used. These cases include, but are not limited to, the following:

- It is necessary to individually model a motor or a motor group in some cases if dynamic induction motor loads represent a large percentage (such as 60%) in the total loads of a system.
- The extinction characteristic of discharging loads at the voltage level of 0.7 per unit should be modeled in simulations if discharging loads reach a relatively large percentage (such as 25%) of a substation load dominated by commercial customers.
- The saturation characteristic of transformers may need to be considered if large transformers are treated as a portion of a load representation.

3.6 CONCLUSIONS

This chapter discusses the basic aspects of load modeling in probabilistic transmission planning. These include load forecast methods, load clustering, bus load uncertainty and correlation, and voltage/frequency-dependent models. The first three aspects are associated with time–space characteristics of power loads, whereas the last one is a technical attribute of power loads.

The three main load forecast techniques have been discussed: regression, probabilistic time series, and neural network. The regression methods include linear and nonlinear regressions. The essential idea in the regression methods is to create an analytical relation between the load and related variables impacting the load using historical samples. This relation is assumed to be applicable for the future. The related variables can be economical, environmental, and social factors. The probabilistic time series technique includes three models of AR, MA, and ARMA. The basic nature of the time series is to treat a load curve as a stochastic process, and the load behavior in historical records is assumed to continue in the future. This technique does not need any information on other variables. The neural network technique can be used to build a complex and implicit relation between the load and other related variables that the regression methods may not be able to capture. In this sense, it may improve accuracy of load forecast in some cases. One important point to be appreciated is that any forecast for the future is always associated with an error in the prediction. Fortunately, the forecast methods presented here can provide not only the predicted load but also its standard deviation or confidence range, which can be used in load uncertainty analysis or sensitivity studies.

The uncertainty and correlation modeling of bus loads is a crucial step toward probabilistic system analysis. The advantage of the model presented that distinguishes it from others is its capacity to represent both uncertainty and correlation of bus loads. The bus load correlation has not received sufficient attention yet in conventional transmission planning studies.

It is neither possible nor necessary to consider all hourly points on a load curve in transmission planning. The multistep model is a common choice. Besides, grouping of multiple load curves is often required to create an aggregated representation of bus load curves with a similar pattern. The load clustering techniques provide the vehicles in simplification of bus load curve modeling.

The voltage- and frequency-dependent models of aggregated bus loads are generally acceptable for both static and dynamic system analyses in transmission planning. However, it should also be kept in mind that specific representations for dynamic load characteristics may be necessary in some cases.

The load modeling presented in this chapter demonstrates the probabilistic features of load attributes. It is the prerequisite for probabilistic system analysis and reliability evaluation, which will be developed in the next two chapters.

4

SYSTEM ANALYSIS TECHNIQUES

4.1 INTRODUCTION

Power system analysis is one of key technical activities in transmission planning. This includes power flow, optimal power flow (OPF), contingency analysis, voltage stability, transient stability, and other technical analyses.

Conventional system analysis techniques are based on deterministic assumptions. These techniques are still of importance in probabilistic transmission planning. Although it has been recognized for a long time that power systems behave probabilistically in real life, not only because the conventional techniques are still necessary tools, but also because probabilistic analysis methods are derived from the conventional techniques by integrating probabilistic characteristics.

This chapter discusses the conventional techniques used in the system analysis and also addresses basic aspects in probabilistic representations and methods. Probabilistic reliability evaluation, including probabilistic voltage and transient stability assessments, will be presented in Chapter 5. There have been many commercial computer programs for the conventional system analysis techniques, and only their fundamental elements are summarized for review purposes. Power flow equations are described in Section 4.2, the probabilistic power flow is presented in Section 4.3, the optimal power

Probabilistic Transmission System Planning, by Wenyuan Li

flow and interior point method are discussed in Section 4.4, two probabilistic search optimization algorithms are introduced in Section 4.5, the contingency analysis and probabilistic contingency ranking are illustrated in Section 4.6, the static voltage stability evaluation is explained in Section 4.7, and the transient stability solution technique is reviewed in Section 4.8.

4.2 POWER FLOW

Power flow is the most basic type of analysis in transmission planning. Commercial power flow computing programs are available. This subsection provides a brief outline.

Power flow equations, in polar coordinate form, are as follows:

$$P_i = V_i \sum_{j=1}^{N} V_j (G_{ij} \cos\delta_{ij} + B_{ij} \sin\delta_{ij}) \qquad (i = 1,\ldots,N) \tag{4.1}$$

$$Q_i = V_i \sum_{j=1}^{N} V_j (G_{ij} \sin\delta_{ij} - B_{ij} \cos\delta_{ij}) \qquad (i = 1,\ldots,N) \tag{4.2}$$

where P_i and Q_i are, respectively, real and reactive power injections at bus i; V_i and δ_i are, respectively, the magnitude and angle of the voltage at bus i; $\delta_{ij} = \delta_i - \delta_j$; G_{ij} and B_{ij} are, respectively, the real and imaginary parts of the element in the bus admittance matrix; and N is the number of system buses.

Each bus has four variables (P_i, Q_i, V_i, δ_i). In order to solve the $4N$-dimensional equations given in Equations (4.1) and (4.2), two of the four variables for each bus have to be pre-specified. Generally, P_i and Q_i at load buses are known and load buses are called *PQ* buses. P_i and V_i at generator buses are specified and generator buses are called *PV* buses. V_i and δ_i of one bus in the system have to be specified to adjust the power balance of the system and this bus is called the *swing bus*.

4.2.1 Newton–Raphson Method

The Newton–Raphson method is a well-known approach to solving a set of nonlinear simultaneous equations. Equations (4.1) and (4.2) are linearized to yield the following matrix equation:

$$\begin{bmatrix} \mathbf{\Delta P} \\ \mathbf{\Delta Q} \end{bmatrix} = \begin{bmatrix} \mathbf{J}_{P\delta} & \mathbf{J}_{PV} \\ \mathbf{J}_{Q\delta} & \mathbf{J}_{QV} \end{bmatrix} \begin{bmatrix} \mathbf{\Delta \delta} \\ \mathbf{\Delta V/V} \end{bmatrix} \tag{4.3}$$

The Jocobian matrix is a $(N + ND - 1)$-dimensional square matrix where N and ND are the numbers of all buses and load buses, respectively. $\mathbf{\Delta V/V}$ represents a vector whose elements are $\Delta V_i/V_i$. The elements of the Jocobian matrix are calculated by

$$(\mathbf{J}_{P\delta})_{ij} = \frac{\partial P_i}{\partial \delta_j} = V_i V_j (G_{ij} \sin\delta_{ij} - B_{ij} \cos\delta_{ij}) \tag{4.4}$$

$$(\mathbf{J}_{P\delta})_{ii} = \frac{\partial P_i}{\partial \delta_i} = -Q_i - B_{ii}V_i^2 \tag{4.5}$$

$$(\mathbf{J}_{PV})_{ij} = \frac{\partial P_i}{\partial V_j} V_j = V_i V_j (G_{ij} \cos \delta_{ij} + B_{ij} \sin \delta_{ij}) \tag{4.6}$$

$$(\mathbf{J}_{PV})_{ii} = \frac{\partial P_i}{\partial V_i} V_i = P_i + G_{ii}V_i^2 \tag{4.7}$$

$$(\mathbf{J}_{Q\delta})_{ij} = \frac{\partial Q_i}{\partial \delta_j} = -(\mathbf{J}_{PV})_{ij} \tag{4.8}$$

$$(\mathbf{J}_{Q\delta})_{ii} = \frac{\partial Q_i}{\partial \delta_i} = P_i - G_{ij}V_i^2 \tag{4.9}$$

$$(\mathbf{J}_{QV})_{ij} = \frac{\partial Q_i}{\partial V_j} V_j = (\mathbf{J}_{P\delta})_{ij} \tag{4.10}$$

$$(\mathbf{J}_{QV})_{ii} = \frac{\partial Q_i}{\partial V_i} V_i = Q_i - B_{ii}V_i^2 \tag{4.11}$$

The resolution is an iterative process. Given the initial values of V_i and δ_i, the Jocobian matrix is constructed, and Equation (4.3) is solved to find $\Delta\delta_i$ and ΔV_i. Then V_i and δ_i are updated. The process is repeated until the minimum power mismatch at all buses is less than a given small error value.

4.2.2 Fast Decoupled Method

In high-voltage transmission systems, the reactance of a branch is normally much larger than the resistance of a branch and the angle difference between two buses is always very small. This leads to the fact that the element values in the matrix blocks $\mathbf{J}_{PV}$ and $\mathbf{J}_{Q\delta}$ are much smaller than those in $\mathbf{J}_{P\delta}$ and $\mathbf{J}_{QV}$. Equation (4.3) can be decoupled by assuming that $\mathbf{J}_{PV} = \mathbf{0}$ and $\mathbf{J}_{Q\delta} = \mathbf{0}$. Considering $|G_{ij} \sin \delta_{ij}| \ll |B_{ij} \cos \delta_{ij}|$ and $|Q_i| \ll |B_{ii} V_i^2|$, the decoupled equations can be further simplified in the following form:

$$\left[\frac{\mathbf{\Delta P}}{\mathbf{V}}\right] = [\mathbf{B}'][\mathbf{V}\Delta\boldsymbol{\delta}] \tag{4.12}$$

$$\left[\frac{\mathbf{\Delta Q}}{\mathbf{V}}\right] = [\mathbf{B}''][\Delta\mathbf{V}] \tag{4.13}$$

where $[\mathbf{\Delta P/V}]$ and $[\mathbf{\Delta Q/V}]$ represent the vectors whose elements are $\Delta P_i/V_i$ and $\Delta Q_i/V_i$ respectively. $[\mathbf{V}\Delta\boldsymbol{\delta}]$ represents the vector whose elements are $V_i\Delta\delta_i$. The elements of the constant matrices $[\mathbf{B}']$ and $[\mathbf{B}'']$ are calculated by

$$B'_{ij} = \frac{-1}{x_{ij}} \tag{4.14}$$

$$B'_{ii} = -\sum_{j \in R_i} B'_{ij} \tag{4.15}$$

$$B''_{ij} = \frac{-x_{ij}}{r_{ij}^2 + x_{ij}^2} \tag{4.16}$$

$$B''_{ii} = -2b_{i0} - \sum_{j \in R_i} B''_{ij} \tag{4.17}$$

where r_{ij} and x_{ij} are the branch resistance and branch reactance, respectively; b_{i0} is the susceptance between bus i and the ground; and R_i is the set of the buses directly connected to bus i.

4.2.3 DC Power Flow

All the notations in this subsection are the same as those defined above, except the ones specifically defined. The DC power flow equations are based on the following four assumptions:

- The branch resistance is much smaller than its reactance, so that the branch susceptance can be approximated by

$$b_{ij} \approx \frac{-1}{x_{ij}} \tag{4.18}$$

- The voltage angle difference between two buses of a branch is small and therefore

$$\begin{aligned} \sin \delta_{ij} &\approx \delta_i - \delta_j \\ \cos \delta_{ij} &\approx 1.0 \end{aligned} \tag{4.19}$$

- The susceptance between a bus and the ground can be neglected:

$$b_{i0} = b_{j0} \approx 0 \tag{4.20}$$

- All bus voltage magnitudes are assumed to be 1.0 per unit.

With these four assumptions, the real power flow in a branch can be calculated by

$$P_{ij} = \frac{\delta_i - \delta_j}{x_{ij}} \tag{4.21}$$

and therefore bus real power injections are

$$P_i = \sum_{j \in R_i} P_{ij} = B'_{ii}\delta_i + \sum_{j \in R_i} B'_{ij}\delta_j \qquad (i = 1, \ldots, N) \tag{4.22}$$

where B'_{ij} and B'_{ii} are given, respectively, by Equations (4.14) and (4.15).

Using a matrix form, Equation (4.22) can be expressed as follows:

$$[\mathbf{P}] = [B'][\boldsymbol{\delta}] \tag{4.23}$$

Obviously, this is a set of simple linear algebraic equations, and its resolution does not need any iterations. By assuming that bus n is the swing bus and letting $\delta_n = 0$, we find that $[\mathbf{B}']$ is a $(N - 1)$-dimensional square matrix. It is exactly the same as the $[\mathbf{B}']$ in Equation (4.12).

By combining Equations (4.23) and (4.21), we obtain the following linear relationship between real power injections at buses and line flows:

$$[\mathbf{T_p}] = [\mathbf{A}][\mathbf{P}] \tag{4.24}$$

Here, $\mathbf{T}_p$ is the line flow vector and its elements are the line flows P_{ij}; $[\mathbf{A}]$ is the relationship matrix between power injections and line flows, and its dimension is $L \times (N - 1)$, where L is the number of lines and N is the number of buses.

The matrix $[\mathbf{A}]$ can be calculated directly from $[\mathbf{B}']$. Assume that the two buses of line k are numbered as i and j. For $k = 1,\ldots,L$, the kth row of matrix $[\mathbf{A}]$ is the solution of the following set of linear equations

$$[\mathbf{B}'][\mathbf{X}] = [\mathbf{C}] \tag{4.25}$$

where

$$\mathbf{C} = \left[0,\ldots,0,\ \underset{\substack{\uparrow \\ i\text{th element}}}{\frac{1}{x_{ij}}},\ 0,\ldots,0,\ \underset{\substack{\uparrow \\ j\text{th element}}}{-\frac{1}{x_{ij}}},\ 0,\ldots,0\right]^T$$

4.3 PROBABILISTIC POWER FLOW

The inputs of power flow (bus loads, generation patterns, and network configurations) have uncertainties and can be represented using probabilistic distributions. The outputs of power flow are therefore random variables following probabilistic distributions. The outputs of power flow include state variables such as bus voltages and network variables such as line flows. The power flow calculation that models the probabilistic characteristics of inputs and outputs is called the *probabilistic power flow*.

The techniques of solving the probabilistic power flow can be classified into two categories of analytical solution and Monte Carlo simulation. Both techniques have merits and demerits. Analytical methods require relatively lower computational burdens but necessitate unavoidable approximations. In contrast, Monte Carlo methods can handle various complex conditions without simplification but require more calculations.

The transmission planning is an offline process, and computing time may not be a major concern.

There are different analytical probabilistic power flow methods such as linearization modeling, cumulant method and point estimation schemes [31,32]. Point estimation and Monte Carlo simulation are discussed in this section.

4.3.1 Point Estimation Method

For a function F that contains N random variables, the point estimation method is based on the following ideas [33]. The function F is evaluated K times for each input variable using K different values on its probability distribution and the mean values of the other remaining input variables. The total number of evaluations is $K \times N$. The raw moments of output function F as a random variable can be calculated using the $K \times N$ values of F. A specific point estimation scheme with $K = 3$, which is proved to have the best performance [32], is described as follows.

A random input variable p_i can be a bus load, a generation power, or a system component parameter (such as line reactance) that is represented using a probability distribution. The three values of p_i can be obtained by

$$p_{ik} = \mu_{p_i} + \xi_{ik}\sigma_{p_i} \qquad (i = 1,\ldots,N; k = 1,2,3) \tag{4.26}$$

where the μ_{p_i} and σ_{p_i} are the mean and standard deviation of p_i and ξ_{ik} is the coefficient representing the kth location on its probability distribution.

The three values of ξ_{ik} and three weighting factors w_{ik} for each random input variable are calculated by

$$\xi_{i1}, \xi_{i2} = \frac{\lambda_{i3}}{2} \pm \sqrt{\lambda_{i4} - \frac{3}{4}\lambda_{i3}^2}, \quad \xi_{i3} = 0 \tag{4.27}$$

$$w_{i1} = \frac{1}{\xi_{i1}^2 - \xi_{i1}\xi_{i2}}, \quad w_{i2} = \frac{1}{\xi_{i2}^2 - \xi_{i1}\xi_{i2}}, \quad w_{i3} = \frac{1}{N} - \frac{1}{\lambda_{i4} - \lambda_{i3}^2} \tag{4.28}$$

where λ_{i3} and λ_{i4} respectively are the skewness and kurtosis of the random input variable p_i, and are calculated by

$$\lambda_{i3} = \frac{\int_{-\infty}^{\infty} (p_i - \mu_{p_i})^3 f_{p_i} dp_i}{\sigma_{p_i}^3}, \quad \lambda_{i4} = \frac{\int_{-\infty}^{\infty} (p_i - \mu_{p_i})^4 f_{p_i} dp_i}{\sigma_{p_i}^4} \tag{4.29}$$

where f_{p_i} is the probability density function of p_i. The weighting factors w_{ik} are used in Equation (4.30).

It has been mathematically proved that $\lambda_{i4} \geq \lambda_{i3}^2 + 1$ [34]. Therefore, the location coefficients ξ_{i1} and ξ_{i2} obtained from Equation (4.27) are definitely real values, and the weighting factors w_{ik} given in Equation (4.28) are well defined.

Once the three values of ξ_{ik} for each random input variable p_i are obtained, the power flows for the three values of ξ_{ik} are calculated while other input variables are specified as their means. This is repeated for each input variable. It can be seen from Equations (4.26) and (4.27) that the third value of p_i is merely its mean since $\xi_{i3} = 0$. Because $\xi_{i3} = 0$ is valid for all random input variables, the power flow case for the third value of each input variable is the same for all input variables, which is the power flow where all input variables are set at their means. In other words, only $2N + 1$ (rather than $3N$) power flows need to be performed in implementation, and therefore this method is called the *2N + 1 point estimation scheme.*

The jth-order raw moment of a random output variable F (such as a bus voltage or a line power flow) can be estimated by

$$E[F^{(j)}] = \sum_{i=1}^{N} \sum_{k=1}^{3} w_{ik} \cdot (F_{ik})^j \tag{4.30}$$

where $E[F^{(j)}]$ denotes the jth-order raw moment of the random output variable F and F_{ik} is the value of F obtained in the corresponding power flow evaluation. With the raw moments, the central moments and cumulants of the random output variable can be calculated. The first raw moment and the second central moment are, respectively, the estimates of the mean and variance of the random output variable. The estimates provide the average values of power flow output variables considering probabilistic distributions of random input variables, and the standard deviations that can be used to indicate their confidence intervals. Moreover, an approximate probability distribution function of any random output variable from power flow can be estimated using either the cumulants and Edgeworth series method or the central moments and Gram–Charlier series method [31].

The demerits of this method include

- The method cannot directly model the uncertainty of network configuration (random failures of transmission components) as component failures are associated with a random change of network topology that cannot be represented using the mean value and standard deviation of an input variable.
- The method cannot handle the correlation between input variables (such as bus loads) since it is based on the assumption that all input variables are independent. Although some other point estimation methods may be able to model the correlation from a mathematical viewpoint, they either do not have a high calculation performance for power flow problems, or cannot deal with asymmetric distribution variables.

4.3.2 Monte Carlo Method

The Monte Carlo method is more flexible, accurate, and straightforward. It can overcome the two demerits of the point estimation method. As mentioned earlier, however, the disadvantage of Monte Carlo simulation is the cost in computing time. This may be acceptable for offline calculations in transmission planning.

The Monte Carlo method for probabilistic power flow is summarized as follows:

1. Create samples of bus loads, which includes the following steps:
   a. Calculate the bus load covariance matrix **C** according to Equation (3.71).
   b. Calculate the matrix **A** using Equations (3.75–3.77).
   c. Create an N-dimensional, normally distributed random vector **G** with each independent component having a mean of zero and a variance of unity using the normal distribution sampling method (see Section A.5.4.2 in Appendix A).
   d. Calculate an N-dimensional, correlative normally distributed random vector **H** using Equation (3.72) in which **B** is the mean vector whose elements are the mean values of bus loads. Obviously, both bus load uncertainty and correlation defined by the matrix **C** are included in the vector **H**.
2. Create samples of system component states by sampling the unavailability of system components (generators, lines, cables, transformers, reactors, capacitors, etc.) [6,10]. Let I_i denote the state of the ith component and U_i be its unavailable probability. A random number R_i uniformly distributed in [0,1] is drawn for the ith component (see Section A.5.2 in Appendix A):

$$I_i = \begin{cases} 0 & \text{(up state)} \qquad \text{if } U_i \geq R_i \\ 1 & \text{(down state)} \qquad \text{if } 0 \leq U_i < R_i \end{cases} \tag{4.31}$$

   If all system components are in the up state, the system is in the normal state. If any one component is in the down state, the system is in a contingency state. It should be appreciated that although component outages may cause greater changes in bus voltages and line flows than random variations of bus loads, the total effect of component outages on the results in probabilistic power flow is generally much lower than that of bus loads as the unavailability of component outages is always very small. The normal system states with load variations have dominant probabilities of occurrence.
3. Create samples of generation power outputs. There are the following two approaches:
   a. The generation patterns are thought to be completely random. Under such an assumption, generation outputs can be treated as negative bus loads and a method similar to that for bus loads can be used. A united covariance matrix **C** for both generation outputs at generator buses and loads at load buses can be created using historical information of chronological generation and load curves at individual buses and the method described in Section 3.4. Then the procedure given in step 1 is used to randomly determine samples of both generation powers and loads at buses.
   b. The generation patterns are thought to be incompletely random. In actual operations, the generation outputs are often determined by operators for a given load level. Under this assumption, one or more generation dispatch rules are applied depending on utilities' practice. For example, an economic dispatch [35] or an optimal power flow (see Section 4.4) can be performed

to calculate generation outputs at each generator bus using the means of bus loads and the system component states obtained in step 2. The calculated outputs are the means of generations. For those generators whose outputs have reached their limits, the outputs can be fixed at the limits. For other generators, a standard deviation, which can be obtained from the historical hourly generation curve, is applied to each of the generators. With the mean and standard deviation, an independent normal distribution is used to randomly sample the output of each generation.

4. Once the samples of bus loads, generator outputs, and system component states are obtained, a power flow evaluation is conducted. Note that a distributed swing concept, in which all dispatchable generators share the effect of a swing bus, should be used in the power flow calculation to meet the power balance. The results of all output variables are recorded.
5. Steps 1–4 are repeated until a stopping rule in the Monte Carlo simulation is met. The average value of each output variable considering the uncertainties of power flow input variables is estimated by

$$\bar{X} = \frac{1}{M}\sum_{i=1}^{M} X_i \tag{4.32}$$

where $\bar{X}$ represents the average value of any output variable (such as a bus voltage or line flow), X_i is the sample value of the output variable obtained in the ith power flow sample, and M is the number of sampled power flows in the Monte Carlo simulation.

The standard deviation of the output variable is estimated by

$$S(\bar{X}) = \sqrt{\frac{1}{M(M-1)}\sum_{i=1}^{M}(X_i - \bar{X})^2} \tag{4.33}$$

It should be noted that $S(\bar{X})$ is not the sample standard deviation. The $S(\bar{X})$ represents a confidence range of the estimated output variable.

6. An experimental probability distribution of an output variable (a bus voltage or line flow) can also be estimated using the information of its sampled values recorded in the Monte Carlo simulation.

4.4 OPTIMAL POWER FLOW (OPF)

There are a considerable number of optimization problems in transmission planning. The optimal power flow (OPF) is the most popular one and includes various implications such as optimal operation simulation, minimization of losses, and maximization of transfer capability. Optimal locations of VAR equipment, optimization of network layouts, and minimal investment cost are examples of other optimization problems.

Different mathematical methods are available for solving the OPF and other optimization problems in transmission planning. These can be classified into deterministic and probabilistic techniques. The deterministic techniques include the Newton method, conventional nonlinear programming methods, and interior point method [36–38]. In general, the interior point method (IPM) is more powerful than other deterministic techniques and appropriate for a large-scale OPF model. This section presents a standard OPF model and its interior point method, whereas two probabilistic search optimization techniques are discussed in Section 4.5.

It is noteworthy that the point estimate and Monte Carlo methods for probabilistic power flow given in Section 4.3 can be also applied to probabilistic optimal power flow in a similar manner. The relevant details are left with readers to figure out and are not repeatedly addressed.

4.4.1 OPF Model

In the power flow model, real powers and voltages at generator buses are prespecified. An essential difference in the optimal power flow (OPF) model is that the real power outputs and voltages at generator buses are calculated from the model with an objective function and constraint conditions. The real power outputs and/or voltages of generators are often used as control variables to be optimized. The OPF model can also include other control variables such as reactive powers of VAR equipment and tap changer turn ratios of transformers. In general, voltages and angles of load buses are state variables. A standard OPF model can be expressed as follows:

$$\min f(\mathbf{P_G}, \mathbf{V_G}, \mathbf{Q_C}, \mathbf{K}) \tag{4.34}$$

$$P_{Gi} - P_{Di} = V_i \sum_{j=1}^{N} V_j (G_{ij} \cos\delta_{ij} + B_{ij} \sin\delta_{ij}) \qquad (i = 1,\ldots,N) \tag{4.35}$$

$$Q_{Gi} + Q_{Ci} - Q_{Di} = V_i \sum_{j=1}^{N} V_j (G_{ij} \sin\delta_{ij} - B_{ij} \cos\delta_{ij}) \qquad (i = 1,\ldots,N) \tag{4.36}$$

$$P_{Gi}^{\min} \le P_{Gi} \le P_{Gi}^{\max} \qquad (i = 1,\ldots,NG) \tag{4.37}$$

$$Q_{Gi}^{\min} \le Q_{Gi} \le Q_{Gi}^{\max} \qquad (i = 1,\ldots,NG) \tag{4.38}$$

$$Q_{Ci}^{\min} \le Q_{Ci} \le Q_{Ci}^{\max} \qquad (i = 1,\ldots,NC) \tag{4.39}$$

$$K_t^{\min} \le K_t \le K_t^{\max} \qquad (t = 1,\ldots,NT) \tag{4.40}$$

$$V_i^{\min} \le V_i \le V_i^{\max} \qquad (i = 1,\ldots,N) \tag{4.41}$$

$$-T_l^{\max} \le T_l \le T_l^{\max} \qquad (l = 1,\ldots,NB) \tag{4.42}$$

where V_i, δ_{ij}, G_{ij}, and B_{ij} are the same as defined in Section 4.2; P_{Gi} and Q_{Gi} are the real and reactive generation power variables at bus i; P_{Di} and Q_{Di} are the real and reactive power loads at bus i; Q_{Ci} is the reactive variable of VAR source equipment at bus i; K_t is the turn ratio variable of transformer t; T_l is the power flow in MVA on branch l; Equations (4.37–4.42) are the upper and lower limits of the corresponding variables;

$T_l^{\max}$ is the rating limit of branch l; and N, NG, NC, NT, and NB are, respectively, the sets of all buses, generator buses, buses with reactive equipment, transformer branches, and all branches in the system. In the objective function, $\boldsymbol{P}_G$, $\boldsymbol{V}_G$, $\boldsymbol{Q}_C$, and $\boldsymbol{K}$ are the control variable vectors whose elements are, respectively, P_{Gi}, V_{Gi}, Q_{Ci}, and K_t, where V_{Gi} is the voltage at generator bus i. Note that each K_t is implicitly included in the elements (G_{ij} and B_{ij}) of admittance matrix and the T_l of transformer branches. The T_l of a line branch is calculated as

$$T_l = \max\{T_{mn}, T_{nm}\} \tag{4.43}$$

where T_{mn} and T_{nm} denote the MVA flows at the two ends of branch l. The m and n are the two bus numbers of branch l. The MVA flow from bus m to bus n is calculated by

$$T_{mn} = \sqrt{P_{mn}^2 + Q_{mn}^2} \tag{4.44}$$

$$P_{mn} = V_m^2(g_{mo} + g_{mn}) - V_m V_n (b_{mn} \sin\delta_{mn} + g_{mn}\cos\delta_{mn}) \tag{4.45}$$

$$Q_{mn} = -V_m^2(b_{mo} + b_{mn}) + V_m V_n (b_{mn}\cos\delta_{mn} - g_{mn}\sin\delta_{mn}) \tag{4.46}$$

where $g_{mn} + jb_{mn}$ is the circuit admittance of branch l and $g_{mo} + jb_{mo}$ is the equivalent admittance of the branch to the ground at the end of bus m. The T_l of a transformer branch has similar calculation formulas, except that the turn ratio K_t is introduced in Equations (4.45) and (4.46), and $g_{mo} + jb_{mo}$ is zero.

The objective function can have different representations depending on the purpose. The most popular and simplest objective function is to minimize the total generation cost, which is used for the purpose of production simulations. In this case

$$f = \sum_{i=1}^{NG} f_i(P_{Gi}) \tag{4.47}$$

where the cost function f_i at each generator can be expressed by the quadratic function

$$f_i = a_i P_{Gi}^2 + b_i P_{Gi} + c_i \tag{4.48}$$

where a_i, b_i, and c_i are the cost coefficients that rely on real power output characteristic of generator i.

Minimization of network losses is another popular objective often used in transmission planning. In this case

$$f = \sum_{i=1}^{NG} P_{Gi} - \sum_{i=1}^{N} P_{Di} \tag{4.49}$$

In many cases, the second term (the sum of all bus loads) is an unchanged constant and plays no role in the resolution unless some bus loads are treated as changeable variables.

The objective function to minimize network losses can also be expressed by the sum of losses on all branches.

Obviously, the two objective functions in Equations (4.47) and (4.49) are the special cases of Equation (4.34), in which $\boldsymbol{Q}_C$ and $\boldsymbol{K}$ are given and $\boldsymbol{V}_G$ becomes a state variable vector. The OPF is a flexible optimization model. Any control variables can be fixed. If the real powers at generator buses except the swing bus are fixed at specified values, the OPF becomes a reactive optimization model, which is useful in reactive source planning. If the objective is still to minimize the network losses in this case, the objective function is just the real power output of the swing bus. The control variables in a reactive OPF can be further limited to include only tap changer turn ratios of transformers (or reactive power outputs of reactive equipment) to investigate the effect of tap changers (or reactive equipment) if necessary. It should be noted that the tap changer turn ratios and the reactive power outputs that rely on capacitor or reactor banks are both discrete integer variables. Mathematically, this is an integer programming problem. However, they are often approximately treated as continuous variables for simplifying calculations. The results of the variables are rounded to the closest discrete values at the end. This treatment leads to a suboptimal solution.

The OPF model given in Equations (4.34–4.42) is just a representative example. There are different OPF models depending on the problem to be solved. The concept of OPF can be also extended to other optimization problems in transmission planning in which the power flow equations are still included as equality constraints but a different objective function and more constraints are imposed. In any case, the mathematical form in an optimization model and the solution method are similar. Therefore, the interior point method will be described below in a general form.

4.4.2 Interior Point Method (IPM)

The primal–dual interior point method [39] has a good convergence feature for a large-scale nonlinear optimization problem. The OPF model given in Equations (4.34–4.42) or any other optimization model can be expressed in the following compact form:

$$\left.\begin{array}{ll} \min & f(\mathbf{x}) \\ \text{subject to} & \\ & \mathbf{g}(\mathbf{x}) = \mathbf{0} \\ & \underline{\mathbf{h}} \le \mathbf{h}(\mathbf{x}) \le \overline{\mathbf{h}} \end{array}\right\} \tag{4.50}$$

Here, f is a scalar function; $\mathbf{x}$ is a control variable vector; and $\mathbf{g}$ and $\mathbf{h}$ represent the two vector equations, respectively, with a set of equality and inequality constraints. Note that the inequality $\underline{\mathbf{h}} \le \mathbf{h}(\mathbf{x}) \le \overline{\mathbf{h}}$ has a general implication and also includes the inequality constraint of the control variable vector itself: $\underline{\mathbf{x}} \le \mathbf{x} \le \overline{\mathbf{x}}$.

4.4.2.1 Optimality and Feasibility Conditions. The inequality constraints in Equation (4.50) can be converted into the equalities by adding nonnegative slack vectors $\mathbf{y}$ and $\mathbf{z}$ as follows:

$$\left.\begin{array}{ll} \min & f(\mathbf{x}) \\ \text{subject to} & \\ & \mathbf{g}(\mathbf{x}) = \mathbf{0} \\ & \mathbf{h}(\mathbf{x}) - \mathbf{y} - \underline{\mathbf{h}} = \mathbf{0} \\ & -\mathbf{h}(\mathbf{x}) - \mathbf{z} + \overline{\mathbf{h}} = \mathbf{0} \\ & \mathbf{y} \geq \mathbf{0}, \quad \mathbf{z} \geq \mathbf{0} \end{array}\right\} \tag{4.51}$$

The nonnegative condition on the slack vectors **y** and **z** can be guaranteed by introducing them into logarithmic barrier terms in the objective function so that Equation (4.51) becomes

$$\left.\begin{array}{ll} \min & f(\mathbf{x}) - \mu^k \sum_{i=1}^{m} (\ln y_i + \ln z_i) \\ \text{subject to} & \\ & \mathbf{g}(\mathbf{x}) = \mathbf{0} \\ & \mathbf{h}(\mathbf{x}) - \mathbf{y} - \underline{\mathbf{h}} = \mathbf{0} \\ & -\mathbf{h}(\mathbf{x}) - \mathbf{z} + \overline{\mathbf{h}} = \mathbf{0} \end{array}\right\} \tag{4.52}$$

In Equation (4.52), the logarithmic terms impose strictly positive conditions on the slack variables so that it is not necessary to explicitly express their nonnegativity constraints. μ^k is called a *barrier parameter*, where k denotes an iteration count in the resolution process discussed below. The equality-constrained optimization problem in Equation (4.52) can be solved using the Lagrange multiplier approach. The Lagrangian function $L_\mu(\mathbf{w})$ is constructed as

$$L_\mu(\mathbf{w}) = f(\mathbf{x}) - \mu^k \sum_{i=1}^{m} (\ln y_i + \ln z_i) - \boldsymbol{\lambda}^T \mathbf{g}(\mathbf{x}) - \boldsymbol{\gamma}^T (\mathbf{h}(\mathbf{x}) - \mathbf{y} - \underline{\mathbf{h}}) - \boldsymbol{\pi}^T (-\mathbf{h}(\mathbf{x}) - \mathbf{z} + \overline{\mathbf{h}}) \tag{4.53}$$

where $\mathbf{w} = \{\mathbf{x}, \mathbf{y}, \mathbf{z}, \boldsymbol{\lambda}, \boldsymbol{\gamma}, \boldsymbol{\pi}\}$; $\boldsymbol{\lambda}$, $\boldsymbol{\gamma}$, and $\boldsymbol{\pi}$ are called the *dual-variable vectors*, and **x**, **y**, and **z** are called the *primal-variable vectors*; **x** and $\boldsymbol{\lambda}$ are the n-dimensional vectors, and the other variables are the m-dimensional vectors.

According to the Kuhn–Tucker optimality condition, the local minimum point of the Lagrangian function is reached when its gradient equals zero:

$$\frac{\partial L_\mu(\mathbf{w})}{\partial \mathbf{w}} = \begin{bmatrix} \left[\frac{\partial f(\mathbf{x})}{\partial \mathbf{x}}\right] - \left[\frac{\partial \mathbf{g}(\mathbf{x})}{\partial \mathbf{x}}\right]^T \boldsymbol{\lambda} - \left[\frac{\partial \mathbf{h}(\mathbf{x})}{\partial \mathbf{x}}\right]^T \boldsymbol{\gamma} + \left[\frac{\partial \mathbf{h}(\mathbf{x})}{\partial \mathbf{x}}\right]^T \boldsymbol{\pi} \\ \boldsymbol{\gamma} - \mu^k \mathbf{Y}^{-1} \mathbf{u} \\ \boldsymbol{\pi} - \mu^k \mathbf{Z}^{-1} \mathbf{u} \\ -\mathbf{g}(\mathbf{x}) \\ -\mathbf{h}(\mathbf{x}) + \mathbf{y} + \underline{\mathbf{h}} \\ \mathbf{h}(\mathbf{x}) + \mathbf{z} - \overline{\mathbf{h}} \end{bmatrix} = [\mathbf{0}] \tag{4.54}$$

where $\boldsymbol{Y} = \text{diag}(y_1,y_2,\ldots,y_m)$, $\mathbf{Z} = \text{diag}(z_1,z_2,\ldots,z_m)$, and $\mathbf{u} = (1,1,\ldots,1)^T$. By left-multiplying the second and third terms in (4.54) by $\boldsymbol{Y}$ and $\boldsymbol{Z}$, respectively, we obtain

$$\frac{\partial L_\mu(\mathbf{w})}{\partial \mathbf{w}} = \begin{bmatrix} \left[\frac{\partial f(\mathbf{x})}{\partial \mathbf{x}}\right] - \left[\frac{\partial \mathbf{g}(\mathbf{x})}{\partial \mathbf{x}}\right]^T \boldsymbol{\lambda} - \left[\frac{\partial \mathbf{h}(\mathbf{x})}{\partial \mathbf{x}}\right]^T \boldsymbol{\gamma} + \left[\frac{\partial \mathbf{h}(\mathbf{x})}{\partial \mathbf{x}}\right]^T \boldsymbol{\pi} \\ \mathbf{Y}\boldsymbol{\gamma} - \mu^k \mathbf{u} \\ \mathbf{Z}\boldsymbol{\pi} - \mu^k \mathbf{u} \\ -\mathbf{g}(\mathbf{x}) \\ -\mathbf{h}(\mathbf{x}) + \mathbf{y} + \underline{\mathbf{h}} \\ \mathbf{h}(\mathbf{x}) + \mathbf{z} - \overline{\mathbf{h}} \end{bmatrix} = [\mathbf{0}] \tag{4.55}$$

In Equations (4.55), the first term, along with $\boldsymbol{\gamma} \geq 0$ and $\boldsymbol{\pi} \geq 0$, ensures the dual feasibility; the fourth, fifth, and sixth terms, together with $\mathbf{y} \geq 0$ and $\mathbf{z} \geq 0$, ensures the primal feasibility; and the second and third terms are called the *complementarity conditions*.

The primal–dual interior point method for solving Equation (4.50) is an iteration process. Given an initial value of μ^0 and a starting point $\mathbf{w}^0$, the set of nonlinear equations in (4.55) is solved, a step length in the correction direction is calculated, and then the variable vector $\mathbf{w}$ is updated. The process is continued with the reduced barrier parameter μ^k. During the iteration, the nonnegativity of the slack variables and multipliers must be satisfied at every point. The iteration terminates when the violations of primal feasibility and dual feasibility and the complementarity gap are smaller than the prespecified tolerances. The resolution process gradually reaches optimality and feasibility as μ^k approaches zero.

4.4.2.2 Procedure of IPM. The primal–dual interior point method includes the following steps:

1. Specify an initial value of μ^0 and select the starting point $\mathbf{w}^0$ that strictly meets the positive conditions.
2. Solve the correction equation of Equation (4.55) at the current point to obtain a correction direction. By applying the Newton–Raphson method, we can build up the following correction equation:

$$\begin{bmatrix} \left[\frac{\partial^2 L_\mu}{\partial \mathbf{x}^2}\right] & 0 & 0 & -\left[\frac{\partial \mathbf{g}}{\partial \mathbf{x}}\right]^T & -\left[\frac{\partial \mathbf{h}}{\partial \mathbf{x}}\right]^T & \left[\frac{\partial \mathbf{h}}{\partial \mathbf{x}}\right]^T \\ 0 & \boldsymbol{\Gamma} & 0 & 0 & \mathbf{Y} & 0 \\ 0 & 0 & \boldsymbol{\Pi} & 0 & 0 & \mathbf{Z} \\ -\left[\frac{\partial \mathbf{g}}{\partial \mathbf{x}}\right] & 0 & 0 & 0 & 0 & 0 \\ -\left[\frac{\partial \mathbf{h}}{\partial \mathbf{x}}\right] & \mathbf{I} & 0 & 0 & 0 & 0 \\ \left[\frac{\partial \mathbf{h}}{\partial \mathbf{x}}\right] & 0 & \mathbf{I} & 0 & 0 & 0 \end{bmatrix} \begin{bmatrix} \Delta\mathbf{x} \\ \Delta\mathbf{y} \\ \Delta\mathbf{z} \\ \Delta\boldsymbol{\lambda} \\ \Delta\boldsymbol{\gamma} \\ \Delta\boldsymbol{\pi} \end{bmatrix} = \begin{bmatrix} \mathbf{b}_x \\ \mathbf{b}_y \\ \mathbf{b}_z \\ \mathbf{b}_\lambda \\ \mathbf{b}_\gamma \\ \mathbf{b}_\pi \end{bmatrix} \tag{4.56}$$

Here, $\boldsymbol{\Gamma} = \text{diag}(\gamma_1,\gamma_2,\ldots,\gamma_m)$, $\boldsymbol{\Pi} = \text{diag}(\pi_1,\pi_2,\ldots,\pi_m)$, and $\mathbf{I} = \text{diag}(1,1,\ldots,1)$.

$$\left[\frac{\partial^2 L_\mu}{\partial \mathbf{x}^2}\right] = \left[\frac{\partial f^2(\mathbf{x})}{\partial \mathbf{x}^2}\right] - \left[\frac{\partial \mathbf{g}^2(\mathbf{x})}{\partial \mathbf{x}^2}\right]^T \lambda - \left[\frac{\partial \mathbf{h}^2(\mathbf{x})}{\partial \mathbf{x}^2}\right]^T \gamma + \left[\frac{\partial \mathbf{h}^2(\mathbf{x})}{\partial \mathbf{x}^2}\right]^T \pi \tag{4.57}$$

$$\begin{bmatrix} \mathbf{b}_x \\ \mathbf{b}_y \\ \mathbf{b}_z \\ \mathbf{b}_\lambda \\ \mathbf{b}_\gamma \\ \mathbf{b}_\pi \end{bmatrix} = \begin{bmatrix} -\left[\frac{\partial f(\mathbf{x})}{\partial \mathbf{x}}\right] + \left[\frac{\partial \mathbf{g}(\mathbf{x})}{\partial \mathbf{x}}\right]^T \lambda + \left[\frac{\partial \mathbf{h}(\mathbf{x})}{\partial \mathbf{x}}\right]^T \gamma - \left[\frac{\partial \mathbf{h}(\mathbf{x})}{\partial \mathbf{x}}\right]^T \pi \\ -\mathbf{Y}\gamma + \mu^k \mathbf{u} \\ -\mathbf{Z}\pi + \mu^k \mathbf{u} \\ \mathbf{g}(\mathbf{x}) \\ \mathbf{h}(\mathbf{x}) - \mathbf{y} - \underline{\mathbf{h}} \\ -\mathbf{h}(\mathbf{x}) - \mathbf{z} + \overline{\mathbf{h}} \end{bmatrix} \tag{4.58}$$

It should be noted that the coefficient matrix and right-side vector in Equation (4.56) are the calculated values at the current point in the iteration. The superscript denoting iteration has been omitted for simplicity.

3. Update the primal and dual variables in the correction direction using the following formulas:

$$\begin{bmatrix} \mathbf{x} \\ \mathbf{y} \\ \mathbf{z} \end{bmatrix}^{(k+1)} = \begin{bmatrix} \mathbf{x} \\ \mathbf{y} \\ \mathbf{z} \end{bmatrix}^{(k)} + \eta \alpha_p^k \begin{bmatrix} \Delta\mathbf{x} \\ \Delta\mathbf{y} \\ \Delta\mathbf{z} \end{bmatrix}^{(k)} \tag{4.59}$$

$$\begin{bmatrix} \lambda \\ \gamma \\ \pi \end{bmatrix}^{(k+1)} = \begin{bmatrix} \lambda \\ \gamma \\ \pi \end{bmatrix}^{(k)} + \eta \alpha_d^k \begin{bmatrix} \Delta\lambda \\ \Delta\gamma \\ \Delta\pi \end{bmatrix}^{(k)} \tag{4.60}$$

Here, η is a scalar between (0,1) to ensure the positive conditions in the next point, and its value is often given as 0.9995; α_p^k and α_d^k are the step lengths for the primal and dual variables, respectively, and they are selected by

$$\alpha_p^k = \min\left\{1, \min_{\Delta y_i \le -\delta}\left\{\frac{y_i}{|\Delta y_i|}\right\}, \min_{\Delta z_i \le -\delta}\left\{\frac{z_i}{|\Delta z_i|}\right\}\right\} \tag{4.61}$$

$$\alpha_d^k = \min\left\{1, \min_{\Delta \gamma_i \le -\delta}\left\{\frac{\gamma_i}{|\Delta \gamma_i|}\right\}, \min_{\Delta \pi_i \le -\delta}\left\{\frac{\pi_i}{|\Delta \pi_i|}\right\}\right\} \tag{4.62}$$

where δ is a given tolerance.

4. Check whether the following convergence criteria are met: (a) the barrier parameter μ^k is sufficiently small, (b) the equality constraints are satisfied, and (c) the changes in the objective function and variables between two iterations are negligible. The criteria are mathematically expressed as follows:

$$\left.\begin{aligned} \mu_k &\le \varepsilon_0 \\ \|\mathbf{g}(\mathbf{x})\| &\le \varepsilon_1 \\ \|\Delta\mathbf{x}\| &\le \varepsilon_2 \\ \frac{|f(\mathbf{x}^k)-f(\mathbf{x}^{k-1})|}{|f(\mathbf{x}^k)|} &\le \varepsilon_3 \end{aligned}\right\} \tag{4.63}$$

If the convergence criteria are satisfied at the new point, the iteration ends. Otherwise, go to step 5.

5. Update the barrier parameter using Equation (4.64) and return to step 2:

$$\mu^{k+1} = \tau^k \frac{(\mathbf{y}^k)^T \boldsymbol{\gamma}^k + (\mathbf{z}^k)^T \boldsymbol{\pi}^k}{2m} \tag{4.64}$$

The τ^k here is a positive parameter smaller than 1.0. It is often chosen as $\tau^k = \max\{0.99\tau^{k-1}, 0.1\}$, with $\tau^0 = 0.2\text{~}0.3$. The m is the dimension of $\mathbf{y}$ or $\mathbf{z}$.

4.5 PROBABILISTIC SEARCH OPTIMIZATION ALGORITHMS

Probabilistic search optimization techniques include evolutionary, particle swarm, and annealing algorithms [40–43]. These techniques are much slower in convergence compared to the interior point method (IPM) and often produce only a suboptimal solution due to the limitation of computation time. However, they can overcome the difficulties encountered with deterministic techniques such as IPM in dealing with discrete variables, nonconvex objective functions, and local optimums. In these cases, the application of probabilistic search techniques is a good option for an optimization problem in transmission planning, particularly when a relatively small size is considered. The genetic algorithm and the particle swarm optimization technique are discussed in this section.

4.5.1 Genetic Algorithm (GA)

The genetic algorithm is the most popular one in the family of evolutionary algorithms that mimic the process of biological evolution. The genetic algorithm is based on the principle of the fittest offspring surviving to produce better generations. In using the genetic algorithm, a fitness function is derived from the objective function in the original optimization model. A solution (a set of control variable values) that meets all constraints is an individual that is often called a chromosome. A population is composed of multiple individuals. At each generation, some individuals are randomly selected for mating according to the level of fitness. The operators used to produce offspring are based on the concept in natural genetics. The process includes selection, recombination (crossover), mutation and reinsertion [44]. A probabilistic search for population of individuals (multiple solutions) instead of finding a single solution is the most essential feature of the genetic algorithm.

4.5.1.1 Fitness Function. The genetic algorithm searches the optimal solution by maximizing the fitness function and therefore the objective function in the original optimization model such as an OPF problem must be converted into an appropriate fitness function to be maximized. In general, the original optimization or OPF model is used to minimize the objective function $f(\mathbf{x})$, where $\mathbf{x}$ is the control variable vector [see Equation (4.34) or (4.50)]. In this case, the fitness function F can be selected as either of the following two equations:

$$F = A - f(\mathbf{x}) \tag{4.65}$$

$$F = \frac{B}{f(\mathbf{x})} \tag{4.66}$$

Here, A is a prespecified value that is estimated to be larger than the maximum of $f(\mathbf{x})$; B is such a value that all possible values of F will fall in a range more suitable for calculations. It should be emphasized that a frequently used approach to inequality constraints in the original optimization model is to introduce them into the objective function as additional penalty function terms.

4.5.1.2 Selection. *Selection* refers to choosing of individuals for mating (recombination). There are different selection schemes. Roulette-wheel selection is the simplest one, which is based on a stochastic sampling technique.

A value of control variable vector $\mathbf{x}$ in the original optimization or OPF model is an individual, and each of its components is a control variable. Assume that a population contains N individuals (i.e., N feasible solutions) and M individuals need to be selected for mating, where $M < N$. An appropriate N can be determined according to the scale of optimization problem and the amount of calculations required. Roulette-wheel selection includes the following steps:

1. Calculate the fitness functions for all the N individuals.
2. Calculate the probabilities of all the N individuals by

$$P_{si} = \frac{F_i(\mathbf{x})}{\sum_{i=1}^{N} F_i(\mathbf{x})} \qquad (i = 1,\ldots,N) \tag{4.67}$$

   where F_i is the value of the fitness function for the ith individual. Apparently, the probability P_{si} for the ith individual reflects the relative level of fitness. A larger probability value represents a higher relative level of fitness.

3. Determine M individuals for mating. The N probability values P_{si} $(i = 1,\ldots,N)$ are placed in the interval [0,1], and M random numbers R_k $(k = 1,\ldots,M)$ in the interval [0,1] are created. The individuals corresponding to the places of the M random numbers are selected for mating. An example of $N = 9$ and $M = 6$ is shown in Figure 4.1, where in the given nine individuals, the randomly selected six individuals are 1, 2, 3, 4, 6, and 9.

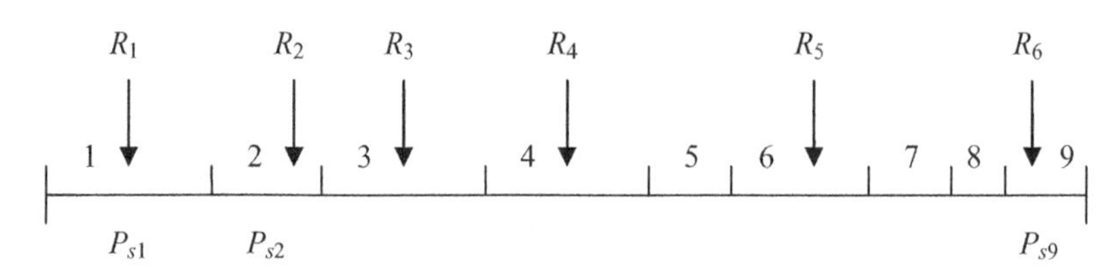

Figure 4.1. Determination of individuals for mating.

4.5.1.3 Recombination. *Recombination* refers to production of offspring individuals by combining the information contained in the selected parent individuals for mating. Binary-variable-based recombination was also called *crossover*. Real-variable-based recombination methods have been developed. However, binary-variable-based recombination is still popular since real numbers, integers, and other discrete variables can all be expressed using binary values.

1. *Crossover Using Binary Values.* Each variable in each individual is coded into a binary value. There are different crossover approaches. The simplest approach is the single-point crossover. For example, we have the following binary expressions for the kth variable (i.e., a component of control variable vector) of two parent individuals P_i and P_j:

$$x_k(P_i) = 101|001 \qquad x_k(P_j) = 001|110$$

A crossover position is randomly determined using random sampling. For instance, it is located right after the third place as shown by "|." The kth variable of two new offspring individuals are obtained by exchanging the binary numbers of the parents after "|":

$$x_k(O_i) = 101110 \qquad x_k(O_j) = 001001$$

Another approach is uniform crossover, which can be conducted as follows. A random string of 0 and 1 whose length is the same as or longer than that of the binary values of parent individuals is created using random sampling. If the number at the mth position on the string is 1, the number at the mth position in the offspring $x_k(O_i)$ inherits the number at the same position from the parent $x_k(P_i)$; otherwise, it inherits the number at the same position from the parent $x_k(P_j)$. Similarly, if the number at the mth position on the string is 0, the number at the mth position in the offspring $x_k(O_j)$ inherits the number at the same position from the parent $x_k(P_i)$; otherwise, it inherits the number at the same position from the parent $x_k(P_j)$. For instance, if the string created by random sampling is 110011 for the kth variable of the two parents P_i and P_j, then the kth variable of two new offspring individuals are obtained from the uniform crossover:

$$x_k(O_i) = 101101 \qquad x_k(O_j) = 001010$$

2. *Recombination Using Real Values.* There are also different recombination approaches for real variables. The intermediate recombination is the most popular one. The variable values of offspring individuals are produced somewhere around the variable values of parents using the rule

$$x_k(O) = x_k(P_i) \cdot \alpha_k + x_k(P_j) \cdot (1 - \alpha_k) \tag{4.68}$$

where $x_k(O)$, $x_k(P_i)$, and $x_k(P_j)$ are the kth variables of an offspring individual and the parents P_i and P_j, respectively; α_k is a random number uniformly distributed between $[-d, 1 + d]$ for the kth variable, where $d = 0$~0.25. In general, two α_k are randomly created for each variable of each parent pair so that two parents are recombined to produce two offspring individuals.

4.5.1.4 Mutation. After recombination, every offspring individual may experience mutation with a low probability. This probability can be prespecified with a small value, such as 0.001–0.1 depending on the problem. Another approach is to use a simple rule to determine the probability. For example, the probability of mutating a variable can be inversely proportional to the number of variables in an individual. A uniform random number is created for every variable of an individual. If the random number is less than the given probability, the mutation occurs. Otherwise, the mutation for this variable does not take place. The purpose of mutation is to create a new individual from the existing one through small perturbations on the values of its variables. The mutation operation may possibly bring one or more individuals out of a local optimum.

1. *Binary Mutation.* For a binary-valued individual, the mutation of its variable can be conducted by changing the number from 1 to 0 or from 0 to 1 at a randomly selected position. Take the $x_k(O_i) = 101101$ as an example. If the randomly selected position is the third place, this variable is mutated and becomes $x_k(O_i) = 100101$.
2. *Real-Valued Mutation.* Real-valued mutation is based on the following operator:

$$x_k(M) = x_k + s_k \cdot r \cdot (x_{k\max} - x_{k\min}) \cdot 2^{-u\beta} \tag{4.69}$$

The x_k here is the kth variable of an individual before mutation, and $x_k(M)$ represents the variable after mutation; s_k is a random sign number that has an equal probability of being only +1 or −1; r is the mutation range that is often specified in the interval [0.001,0.1]; $x_{k\max}$ and $x_{k\min}$ are the upper and lower bounds of x_k; u is a uniformly distributed random number over [0,1]; and β is the mutation precision factor that can be an integer between 4 and 10. The appropriate values of r and β are varied for different problems and should be determined by testing.

4.5.1.5 Reinsertion. The offspring individuals produced by selection, recombination, and mutation must be inserted into the population. A simple principle is to maintain the same size of population at each generation. The approach of combining elitism and fitness is applied for reinsertion. At each generation, N_p best-fit parents are kept and the rest is replaced by best-fit offspring, where "best fit" is identified in terms of the fitness function. The number N_o of inserted offspring individuals is therefore equal to $N - N_p$, where N_p can be prespecified as about 50% of N initially and then adjusted according to the average fitness function values of the best-fit parents and the best-fit offspring. This approach implements truncating selection of offspring individuals before inserting them into the population. At the same time, the best-fit individuals can live for many generations. It is noted that the best-fit parents and the best-fit offspring are determined as two separate groups using the fitness function values. Whether the retained parents are better or worse than the inserted offspring is not inspected. It is possible that some parents may be replaced by offspring with lower fitness at some generation, resulting in decreased average fitness of population. However, a relatively bad offspring will be automatically eliminated in the evolution from generation to generation.

4.5.1.6 Procedure of Genetic Algorithm. The genetic algorithm as a probabilistic search optimization technique has merits and demerits compared to the deterministic interior point method. On one hand, it can deal with discrete variables and nonconvex or discrete objective functions, and avoid a possible local optimum. On the other hand, it requires much more computational time. In general, the genetic algorithm is appropriate in the following cases:

- A relatively small system size
- Discrete control variables, such as a reactive optimization problem using banks of capacitors and reactors, or/and transformer tap changer positions as optimized variables
- A nonconvex or discrete objective function, such as a network topology optimization problem

The control variables in the standard OPF model given in Equations (4.34–4.42) are used as an example to explain the procedure of the genetic algorithm. Although it may not be the most appropriate algorithm for this particular OPF problem, the procedure of applying the genetic algorithm to other optimization problems in transmission planning is similar. The procedure includes the following steps:

1. The control variables $\mathbf{x} = (\mathbf{P_G}, \mathbf{V_G}, \mathbf{Q_C}, \mathbf{K})$ in the OPF model are the variables of an individual. A fitness function given in Equation (4.65) or (4.66) is used. Note that the inequality constrains of state or network variables may be transformed as penalty terms in the objective function before it is converted to the fitness function. The population size (the number of individuals) N and the number M of individuals for mating at each generation are specified. If binary values are used for variables, a coding process is needed.
2. N initial individuals (i.e., N feasible but not optimal solutions) are randomly created. Each of the N individuals is produced as follows:

a. Each control variable is decided by

$$x_k = r_k(x_{k\max} - x_{k\min}) + x_{k\min} \tag{4.70}$$

where x_k is the kth variable; $x_{k\max}$ and $x_{k\min}$ are the upper and lower bounds of x_k; r_k is a uniformly distributed random number over [0,1] for the kth variable. Each variable has an independent r_k. If a variable is an integer, the term $r_k(x_{k\max} - x_{k\min})$ should be rounded to an integer.

b. A power flow is calculated using the created control variable vector **x**.

c. If the power flow is converged and the values of all state and network variables are within their limits, the obtained **x** is an initial individual. Otherwise, go to step 2a and resample r_k for all variables.

3. *M* individuals for mating are selected using the approach described in Section 4.5.1.2.

4. The genetic operations (recombination, mutation, and reinsertion) are performed to create *N* new individuals for the next generation using the approaches suggested in Sections 4.5.1.3–4.5.1.5. The power flows for all individuals are calculated. The feasibility of offspring individuals must be guaranteed. This includes the following:

   a. If a control variable in an offspring individual created by the genetic operation is not within its limits, it should be fixed at either its upper or lower limit.

   b. If a power flow diverges, or a state or network variable obtained from a power flow is beyond its limits, the corresponding individual is abandoned. If the inequality constrains of state or network variables have been treated as penalty terms in the objective function, violations of the limits are unlikely to occur.

5. If the fitness function values do not show significant change in a few consecutive generations, or the iteration process reaches the given maximum number of generations, go to step 6. Otherwise go to step 3.

6. Binary values of variables are decoded, and the optimized results are obtained.

4.5.2 Particle Swarm Optimization (PSO)

Particle swarm optimization is based on simulation of social behavior of a swarm such as fish schooling or bird flocking. In particle swarm optimization, each particle tries to search its best position (solution) in a multidimensional space (dimension of control variables). During swimming or flying, each particle adjusts its position according to its current velocity, its own previous experience, and the experience of other particles. The process gradually achieves the best position (optimal solution). A particle is somewhat similar to an individual in the genetic algorithm, and a particle swarm is similar to a population in the GA. Particle swarm optimization is also a probabilistic search technique, which moves positions from a set of solutions to another set of solutions.

4.5.2.1 Inertia Weight Approach. The inertia weight approach is based on the following update rule [45]:

$$V_{ik}^{(t+1)} = wV_{ik}^{(t)} + \varphi_1 \cdot \alpha_1^{(t)} \cdot (Pb_{ik}^{(t)} - x_{ik}^{(t)}) + \varphi_2 \cdot \alpha_2^{(t)} \cdot (Gb_k^{(t)} - x_{ik}^{(t)}) \quad (4.71)$$

$$x_{ik}^{(t+1)} = x_{ik}^{(t)} + V_{ik}^{(t+1)} \quad (4.72)$$

Here, the V_{ik} and x_{ik} are the velocity and position of the kth variable of the ith particle, respectively; Pb_{ik} is the best position of the kth variable of the ith particle, where "best" implies that this position leads to the best objective function value; Gb_k is the best global position of the kth variable in all particles; φ_1 and φ_2 are two acceleration factors; α_1 and α_2 are two uniformly distributed random numbers over [0,1]; w is an inertia weighting factor that reflects the effect of velocity in the immediately previous iteration; and the superscript (t) or $(t + 1)$ represents the iteration count.

It can be seen that the essence of particle swarm optimization is to calculate the new velocity of each variable of all particles considered using its current velocity, the distance between its current position and its own previous best position, and the distance between its current position and the previous best position of the same variable among other particles in the swarm. The two distances are randomly weighted in updating the velocity. Once the new velocity is calculated, the new position is updated by moving a step using the velocity.

The convergence speed is impacted by the three parameters w, φ_1, and φ_2. If the values of φ_1 and φ_2 are too small, the optimization process will be slow because a relatively large number of iterations are required. However, excessively large values of φ_1 and φ_2 may cause instability in numerical calculations. The appropriate values of these two parameters are varied for different problems, and there is no firm rule for selecting them. A trial is needed to determine their values. The inertia weighting factor w is often specified as a real value in the interval [0.0,1.0] and can be proportionally decreased with the iteration

$$w = w_{\max} - \frac{t}{t_{\max}}(w_{\max} - w_{\min}) \quad (4.73)$$

where $w_{\max}$ and $w_{\min}$ are the maximum and minimum weighting values; t and $t_{\max}$ are the current and maximum counts of iterations, respectively.

4.5.2.2 Constriction Factor Approach. On the basis of the concept of the constriction factor [46,47], the velocity updating equation is modified as follows:

$$V_{ik}^{(t+1)} = K \cdot [V_{ik}^{(t)} + \varphi_1 \cdot \alpha_1^{(t)} \cdot (Pb_{ik}^{(t)} - x_{ik}^{(t)}) + \varphi_2 \cdot \alpha_2^{(t)} \cdot (Gb_k^{(t)} - x_{ik}^{(t)})] \quad (4.74)$$

$$K = \frac{2}{\left|2 - \beta - \sqrt{\beta^2 - 4\beta}\right|} \quad (4.75)$$

where $\beta = \varphi_1 + \varphi_2$. K is called a *constriction factor*. Compared to Equation (4.71), in which the inertia weighting factor w impacts only $V_{ik}^{(t)}$, the K in Equation (4.74) has the same effect on all three terms. It has been found that this modification improves the convergence performance. Obviously, β must be larger than 4. Very often, φ_1 is selected to be equal to φ_2.

An alternative modified approach is based on the following rule for velocity updating [47]:

$$V_{ik}^{(t+1)} = K \cdot \left[V_{ik}^{(t)} + \sum_{j=1}^{N} \varphi_j \cdot \alpha_j^{(t)} \cdot (Pb_{jk}^{(t)} - x_{ik}^{(t)}) \right] \tag{4.76}$$

In this equation, the new velocity of a particle variable is based on its current velocity and the distances between its current position and the previous best positions of the same variable of each particle in the swarm. Each particle has its own independent random number α_j and acceleration factor φ_j. N is the number of particles in the swarm.

4.5.2.3 Procedure of PSO. The procedure of using the particle swarm search technique to solve an optimization problem or OPF model in transmission planning is somewhat similar to that of using the genetic algorithm. It includes the following steps:

1. The control variables in the original optimization problem or OPF model are the variables of a particle. For the OPF model given in Equations (4.34–4.42), for instance, the control variables are $\mathbf{x} = (\mathbf{P_G}, \mathbf{V_G}, \mathbf{Q_C}, \mathbf{K})$.
2. N initial particles (i.e., N feasible but not optimal solutions) are randomly created using the same approach given in step 2 of Section 4.5.1.6, including power flow calculations and feasibility checking. N velocities are also randomly created using a similar approach.
3. The objective function values obtained from the update of each variable in each particle are calculated. The "best position" refers to the one leading to the minimum objective function. The initial best position Pb_{ik} of each variable in each particle is its initial position in the first iteration. The best global position Gb_{ik} of each variable is the one corresponding to the minimum in the objective function values obtained from the update of the same variable in the N particles.
4. The velocities of each variable in each particle are updated using Equation (4.71), (4.74), or (4.76).
5. The positions of each variable in each particle are updated using Equation (4.72). The power flows for the current positions of all the particles are calculated. The feasibility checking for inequality constraints is similar to that as described in step 4 of Section 4.5.1.6.
6. The best position of each variable in each particle and the best global positions of each variable in all the particles are updated.
7. If the objective function values do not have a significant change in a few consecutive iterations, or if the iteration process reaches the given maximum number of iterations, the optimization terminates. Otherwise go to step 4.

4.6 CONTINGENCY ANALYSIS AND RANKING

This section discusses the methods for steady-state contingency analysis and ranking. The task of steady-state contingency analysis is to calculate power flows in outage states in which one or more system components are out of service. A transmission system must satisfy security criteria in both normal and outage states. Conceptually, the contingency analysis can be performed by running full power flow studies for all specified outages. However, this will impose a large computational burden, particularly when system outage states with multicomponent failures are considered in the probabilistic reliability evaluation that will be illustrated in the next chapter. Fast contingency analysis methods are used to reduce computational efforts.

Contingency ranking has a twofold purpose: (1) contingencies in the front of a ranking list can be selected for detailed investigations, and (2) a ranking list provides information on importance of contingencies. Conventional contingency ranking is based on deterministic performance indices. A probabilistic risk-based ranking method is presented in this section.

4.6.1 Contingency Analysis Methods

The essential feature in fast contingency analysis methods is avoidance of full steps in power flow calculations for contingency states. A popular idea is to calculate power flows in a post-outage state using the information from the preoutage state. Described below are two basic contingency analysis methods.

4.6.1.1 AC Power-Flow-Based Method. Consider a case of branch i–j outage in a transmission system. Assume that the flows on the two ends of the branch i–j before its outage are $P_{ij}+jQ_{ij}$ and $P_{ji}+jQ_{ji}$, and imagine that there are two additional power injections at buses i and j in the preoutage state, which are denoted by $\Delta P_i + j\Delta Q_i$ and $\Delta P_j + j\Delta Q_j$. If the additional power injections can produce power flow increments so that the power flows on the system are the same as those in the postoutage state, the effect of the additional power injections is completely equivalent to the outage of branch i–j. It can be proved that the power flows on branch i–j and the additional power injections in the preoutage state have the following relationship [10,48]:

$$\begin{bmatrix} P_{ij} \\ Q_{ij} \\ P_{ji} \\ Q_{ji} \end{bmatrix} = \left[\begin{bmatrix} 1 & 0 & 0 & 0 \\ 0 & 1 & 0 & 0 \\ 0 & 0 & 1 & 0 \\ 0 & 0 & 0 & 1 \end{bmatrix} - \begin{bmatrix} \frac{\partial P_{ij}}{\partial P_i} & \frac{\partial P_{ij}}{\partial Q_i} & \frac{\partial P_{ij}}{\partial P_j} & \frac{\partial P_{ij}}{\partial Q_j} \\ \frac{\partial Q_{ij}}{\partial P_i} & \frac{\partial Q_{ij}}{\partial Q_i} & \frac{\partial Q_{ij}}{\partial P_j} & \frac{\partial Q_{ij}}{\partial Q_j} \\ \frac{\partial P_{ji}}{\partial P_i} & \frac{\partial P_{ji}}{\partial Q_i} & \frac{\partial P_{ji}}{\partial P_j} & \frac{\partial P_{ji}}{\partial Q_j} \\ \frac{\partial Q_{ji}}{\partial P_i} & \frac{\partial Q_{ji}}{\partial Q_i} & \frac{\partial Q_{ji}}{\partial P_j} & \frac{\partial Q_{ji}}{\partial Q_j} \end{bmatrix} \right] \begin{bmatrix} \Delta P_i \\ \Delta Q_i \\ \Delta P_j \\ \Delta Q_j \end{bmatrix} \tag{4.77}$$

The sensitivity matrix in Equation (4.77) can be calculated from the explicit expression of the branch power as a function of two terminal bus voltages and the Jacobian matrix of Newton–Raphson power flow equations in the preoutage system state. The additional power injections at buses i and j can be found by solving Equation (4.77). Then the increments of bus voltage magnitudes and angles due to the branch i–j outage can be obtained by solving the following equation:

$$[\mathbf{J}]\begin{bmatrix}\boldsymbol{\Delta\delta} \\ \mathbf{\Delta V/V}\end{bmatrix} = [\mathbf{\Delta I}] \tag{4.78}$$

Here, $[\mathbf{J}]$ is the Jacobian matrix of the power flow equations in the preoutage state, $[\mathbf{\Delta V/V}]$ is the voltage magnitude increment subvector whose elements are $\Delta V_i/V_i$, $[\boldsymbol{\Delta\delta}]$ is the voltage angle increment subvector whose elements are δ_i, and $[\mathbf{\Delta I}]$ is defined as follows:

$$[\mathbf{\Delta I}] = [0,\ldots,0,\Delta P_i,0,\ldots,0,\Delta P_j,0,\ldots,0,\Delta Q_i,0,\ldots,0,\Delta Q_j,0,\ldots,0]^T$$

The bus voltages in the postoutage state are calculated by adding the voltage increments obtained from Equation (4.78). Once bus voltages are updated, the branch power flows following the branch i–j outage can be calculated. Note that no iteration is needed in solving Equation (4.78), although it seems that it was in the same format as Equation (4.3).

In the case of outage of one or more generators or reactive equipment (capacitors or reactors), a redispatch of outputs of other generators or reactive equipment is needed in order to maintain system power balance. A rule of redispatch such as a proportional change principle or other rules should be specified. The changes in outputs of generators or reactive equipment are the corresponding components in $[\mathbf{\Delta I}]$, and Equation (4.78) is directly solved. In the case of relatively large changes in bus generation or reactive power injection due to a generator or reactive equipment outage, a few iterations may be needed to enhance accuracy in the contingency analysis.

A similar procedure can be applied to multiple branch outages or mixed outages of branches and generators/reactive equipment. The concept can also be applied using the fast decoupled power flow model, in which equivalent real and reactive power injections are calculated separately.

4.6.1.2 DC Power-Flow-Based Method. DC power-flow-based contingency analysis provides faster and sufficiently accurate real power flow solutions following branch or generator outages. This will considerably reduce computing time when a huge number of outage events are considered, particularly in probabilistic reliability evaluation.

The bus impedance matrices after multiple branch outages can be calculated directly from the bus impedance matrix before the outages [10]:

$$\mathbf{Z}(S) = \mathbf{Z}(0) + \mathbf{Z}(0)\mathbf{M}\mathbf{Q}\mathbf{M}^T\mathbf{Z}(0) \tag{4.79}$$

$$\mathbf{Q} = [\mathbf{W} - \mathbf{M}^T\mathbf{Z}(0)\mathbf{M}]^{-1} \tag{4.80}$$

Here, $\mathbf{Z}(0)$ and $\mathbf{Z}(S)$ are the bus impedance matrices before and after the branch outages, in which the resistance of each branch is neglected. The (0) and (S) denote the preoutage and postoutage system states, respectively. $\mathbf{W}$ is the diagonal matrix, each of whose diagonal element is the reactance of an outage branch. $\mathbf{M}$ is the submatrix of the bus–branch incidence matrix, which is composed of the columns corresponding to the outage branches. The superscript T denotes transposition of a matrix.

The branch flows after the branch outages can be calculated by

$$\mathbf{T}(S) = \mathbf{A}(S)(\mathbf{PG} - \mathbf{PD}) \tag{4.81}$$

where $\boldsymbol{T}(S)$ is the real power flow vector in the outage state; $\mathbf{PG}$ and $\mathbf{PD}$ are the generation output and load power vectors, respectively. $\mathbf{A}(S)$ is the relation matrix between real power flows and power injections in the outage state S. The mth row of $\mathbf{A}(S)$ can be calculated as

$$\mathbf{A}_m(S) = \frac{\mathbf{Z}_r(S) - \mathbf{Z}_q(S)}{X_m} \tag{4.82}$$

where X_m is the reactance of the mth branch; the subscripts r and q denote the two bus numbers of the mth branch; and $\mathbf{Z}_r(S)$ and $\mathbf{Z}_q(S)$ are the rth and qth rows of $\mathbf{Z}(S)$, respectively.

For a contingency associated with a generator outage, the redispatch of other generators is needed to maintain system power balance. The changes in generators' outputs are used to modify the corresponding components in the $\mathbf{PG}$ vector of Equation (4.81).

If there are only branch outages, the power injections at all buses remain unchanged in the outage state. In this case, the simpler formula can be derived to calculate the branch flows in the outage state directly from the branch flows in the preoutage state. For example, the following equations can be used for single branch outage states [35]:

$$T_m(S) = T_m(0) + \rho_{mk} \cdot T_k(0) \tag{4.83}$$

$$\rho_{mk} = \frac{-B_m \cdot D_{mk} \cdot \Delta b_k}{(1 + \Delta b_k D_{kk}) B_k} \qquad (m \neq k) \tag{4.84}$$

$$\rho_{kk} = \frac{(1 - B_k \cdot D_{kk}) \cdot \Delta b_k}{(1 + \Delta b_k D_{kk}) B_k} \tag{4.85}$$

$$D_{mk} = (\mathbf{M}^m)^T \mathbf{Z}(0) \mathbf{M}^k \tag{4.86}$$

$$D_{kk} = (\mathbf{M}^k)^T \mathbf{Z}(0) \mathbf{M}^k \tag{4.87}$$

Here, k denotes any outage branch and m, any branch in the system; $T_m(0)$, $T_k(0)$, and $T_m(S)$ are the power flows on branches m and k in the preoutage state (0) and single branch outage state (S), respectively; B_m and B_k are the mutual admittances of branches m and k, respectively; Δb_k is the changed value of B_k after removal of one or several circuits in the branch k; and $\mathbf{M}^m$ and $\mathbf{M}^k$ are the column vectors of $\mathbf{M}$ corresponding to, respectively, branch m and branch k.

It can be seen that, similar to the AC power-flow-based method, the DC power-flow-based method also calculates branch flows in the outage states using the information in the preoutage state. Besides, the likelihood of single-component outages is much greater than that of multiple component outages. Therefore, Equations (4.83–4.87) can lead to a considerable decrease in computing efforts in probabilistic reliability evaluation, in which component outages are randomly sampled.

4.6.2 Contingency Ranking

4.6.2.1 Ranking Based on Performance Index. The conventional contingency ranking is based on the following performance indices [49,50]:

- *Branch Power-Based Index.* This is expressed as

$$PI_s = \sum_{i=1}^{NL} w_{si} \left(\frac{S_i}{S_i^{\max}} \right)^{2m_s} \tag{4.88}$$

  where S_i is the apparent power on branch i, $S_i^{\max}$ is the apparent power limit of branch i, w_{si} is the weighting factor for branch i, NL is the number of branches in the system, and m_s is the integer exponent for the performance index PI_s.
- *Bus Voltage-Based Index.* This is expressed as

$$PI_v = \sum_{i=1}^{ND} w_{vi} \left(\frac{V_i - V_i^{\text{sp}}}{\Delta V_i^{\max}} \right)^{2m_v} \tag{4.89}$$

  where V_i is the voltage magnitude at bus i, V_i^{sp} is the specified voltage magnitude at bus i, $\Delta V_i^{\max}$ is the permissible voltage deviation at bus i, w_{vi} is the weighting factor for bus i, ND is the number of PQ buses in the system, and m_v is the integer exponent for the performance index PI_v.

The weighting factor w_{si} or w_{vi} reflects the relative importance of each branch or bus and can be determined on the basis of an engineering judgment. The exponents (m_s and m_v) play a role of reducing mask effects.

4.6.2.2 Ranking Based on Probabilistic Risk Index. The performance indices given in Equations (4.88) and (4.89) have been widely used for contingency ranking. A disadvantage of these indices is that only the consequence of an outage is reflected but the probability of occurrence of the outage is missed. In fact, even a relatively severe contingency may not create a high average risk if it has a very low probability of occurrence. The true risk due to a contingency is a combination of consequence and probability. The following indices are presented for contingency risk ranking:

$$RI_s = \sum_{i=1}^{NL} P_c \cdot w_{si} \left(\frac{S_i}{S_i^{\max}} \right)^{2m_s} \tag{4.90}$$

$$RI_v = \sum_{i=1}^{ND} P_c \cdot w_{vi} \left(\frac{V_i - V_i^{\text{sp}}}{\Delta V_i^{\max}} \right)^{2m_v} \tag{4.91}$$

Here, P_c is the probability of a system contingency state with component outage(s), and all other symbols are the same as defined in Equations (4.88) and (4.89). It is important to note that P_c is not the outage probability of component(s). P_c can be calculated by

$$P_c = \prod_{j=1}^{n_d} U_j \cdot \prod_{j=1}^{n-n_d} (1 - U_j) \tag{4.92}$$

where U_j is the unavailability or outage probability of the jth component, n is the total number of components in the system, and n_d is the number of outage components in a contingency event. If only branches are considered, n is the number of branches; if both branches and generators are considered, n is the total number of branches and generators. For single-component contingency events, n_d is 1.

4.7 VOLTAGE STABILITY EVALUATION

Voltage stability is a crucial issue in many transmission systems. Both dynamic and static approaches can be applied to voltage stability analysis. A set of mixed differential and algebraic equations is used to model the system behavior in the dynamic analysis approach, whereas snapshots of various system conditions are considered in the static analysis method. In actual applications of utilities, the static analysis method is more popular. This section illustrates the continuation power flow and reduced Jacobian matrix techniques, which have been widely used in commercial computer programs.

4.7.1 Continuation Power Flow Technique

The continuation power flow technique [51] is a robust power flow method that can ensure convergence at the voltage collapse point (nose point) and unstable equilibrium points on the lower portion of a *P–V* (or *Q–V*) curve. Therefore, this technique can be used to recognize voltage instability by identifying the collapse point.

A system can approach the collapse point on a *P–V* curve and lose voltage stability as the load level is gradually increased. A load parameter variable λ is introduced into the conventional power flow Equations (4.1) and (4.2) to obtain

$$P_{Gio}(1 + \lambda \cdot K_{Gi}) - P_{Lio}(1 + \lambda \cdot K_{Li}) = V_i \sum_{j=1}^{N} V_j (G_{ij} \cos \delta_{ij} + B_{ij} \sin \delta_{ij}) \qquad (i = 1, \ldots, N) \tag{4.93}$$

$$Q_{Gio} - Q_{Lio}(1 + \lambda \cdot K_{Li}) = V_i \sum_{j=1}^{N} V_j (G_{ij} \sin \delta_{ij} - B_{ij} \cos \delta_{ij}) \qquad (i = 1, \ldots, N) \tag{4.94}$$

where P_{Lio} and Q_{Lio} are the real and reactive base loads, P_{Gio} and Q_{Gio} are real and reactive base generation powers, λ is the parameter variable representing an increased load percentage, and K_{Li} and K_{Gi} are the constant multipliers to designate the change rates of load and generation at bus i. When $\lambda = 0$, it corresponds to the base load and when λ is increased to λ_{nose}, it corresponds to the critical load condition (the collapse point). K_{Li} and K_{Gi} should be selected such that the increased total load and generation in the system are balanced.

The continuation power flow technique includes two steps: prediction and correction.

4.7.1.1 Prediction Step.

Equations (4.93) and (4.94) can be reexpressed in a more compact form as follows:

$$\mathbf{F}(\boldsymbol{\delta}, \mathbf{V}, \lambda) = \mathbf{0} \tag{4.95}$$

Taking the partial derivatives of Equation (4.95) yields

$$\begin{bmatrix} \mathbf{F}_\delta & \mathbf{F}_V & \mathbf{F}_\lambda \end{bmatrix} \begin{bmatrix} \mathbf{d\delta} \\ \mathbf{dV} \\ d\lambda \end{bmatrix} = [\mathbf{0}] \tag{4.96}$$

where $[\mathbf{F}_\delta \quad \mathbf{F}_V \quad \mathbf{F}_\lambda]$ is termed the *partial derivative vector* and $[d\boldsymbol{\delta} \quad d\mathbf{V} \quad d\lambda]^T$ is called the *tangent vector*. Since the additional unknown variable λ is added, one more equation is needed to solve Equation (4.96) for the tangent vector. This can be done by choosing one component of the tangent vector as +1 or −1. The state variable corresponding to this component is called the *continuation parameter*. Equation (4.96) becomes

$$\begin{bmatrix} \mathbf{F}_\delta & \mathbf{F}_V & \mathbf{F}_\lambda \\ & \mathbf{e}_k & \end{bmatrix} \begin{bmatrix} \mathbf{d\delta} \\ \mathbf{dV} \\ d\lambda \end{bmatrix} = [\mathbf{e}'_k] \tag{4.97}$$

where $\mathbf{e}_k$ is a row vector in which all elements are zeros except for the kth element (corresponding to the continuation parameter), which is equal to 1; $\mathbf{e}'_k$ is a column vector in which all elements are zeros, except that the kth element can be +1 or −1.

Introduction of the continuation component ensures a nonzero norm of the tangent vector and guarantees that Equation (4.97) always has a solution even at and beyond the collapse point. Initially, λ is chosen as the continuation parameter, and the sign of the corresponding component in the tangent vector is set to be positive (+1). In subsequent iterations, the continuation component is determined at the beginning of the correction step and is chosen as the state variable whose tangent component has the greatest absolute value, which is obtained in the previous prediction step. Its sign is positive (+1) if the state variable is increasing and negative (−1) if the state variable is decreasing.

Once the tangent vector is found by solving Equation (4.97), the prediction solution of the state variable vector is obtained as

$$\begin{bmatrix} \boldsymbol{\delta}^* \\ \mathbf{V}^* \\ \lambda^* \end{bmatrix} = \begin{bmatrix} \boldsymbol{\delta}^0 \\ \mathbf{V}^0 \\ \lambda^0 \end{bmatrix} + \sigma \begin{bmatrix} \mathbf{d}\boldsymbol{\delta} \\ \mathbf{dV} \\ d\lambda \end{bmatrix} \tag{4.98}$$

where the asterisk superscript * denotes the predicted values and the superscript 0 denotes the values before the current prediction step starts. The scale step length σ is selected such that a power flow solution exists in the correction step. Otherwise, the step length is reduced and the correction step is repeated until a solution is found.

4.7.1.2 Correction Step.

In the correction step, Equation (4.95) is solved by specifying the state variable corresponding to the current continuation parameter to be equal to its predicted value obtained in the prediction step, and thus the new set of equations is

$$\begin{bmatrix} \mathbf{F}(\boldsymbol{\delta}, \mathbf{V}, \lambda) \\ x_k - x_k^* \end{bmatrix} = [\mathbf{0}] \tag{4.99}$$

where x_k is the state variable corresponding to the current continuation parameter and x_k^* is its predicted value.

The tangent component $d\lambda$ is positive before the power flow solution reaches the collapse point, zero at the collapse point, and negative beyond the collapse point. Therefore, the sign of $d\lambda$ provides an indicator of whether the collapse point has been reached or passed.

4.7.1.3 Identification of Voltage Collapse Point.

Although the continuation power flow method can guarantee convergence at and beyond the collapse point, it is very time-consuming and should be applied only when the operation condition is close to the collapse point. To identify the voltage collapse point, the conventional power flow and continuation power flow methods should be used in a coordinated manner. This is illustrated in Figure 4.2. Further discussion of the continuation power flow method can be found in References 52 and 53.

4.7.2 Reduced Jacobian Matrix Analysis

The incremental form of power flow equations can be expressed as follows:

$$\begin{bmatrix} \Delta\mathbf{P} \\ \Delta\mathbf{Q} \end{bmatrix} = \begin{bmatrix} \mathbf{J}_{\mathrm{P}\delta} & \mathbf{J}_{\mathrm{PV}} \\ \mathbf{J}_{\mathrm{Q}\delta} & \mathbf{J}_{\mathrm{QV}} \end{bmatrix} \begin{bmatrix} \Delta\boldsymbol{\delta} \\ \Delta\mathbf{V} \end{bmatrix} \tag{4.100}$$

where $\Delta\mathbf{P}$ and $\Delta\mathbf{Q}$ are the incremental change vector in bus real and reactive power injections, respectively; $\Delta\boldsymbol{\delta}$ and $\Delta\mathbf{V}$ are the incremental change vector in bus voltage angle and magnitude, respectively.

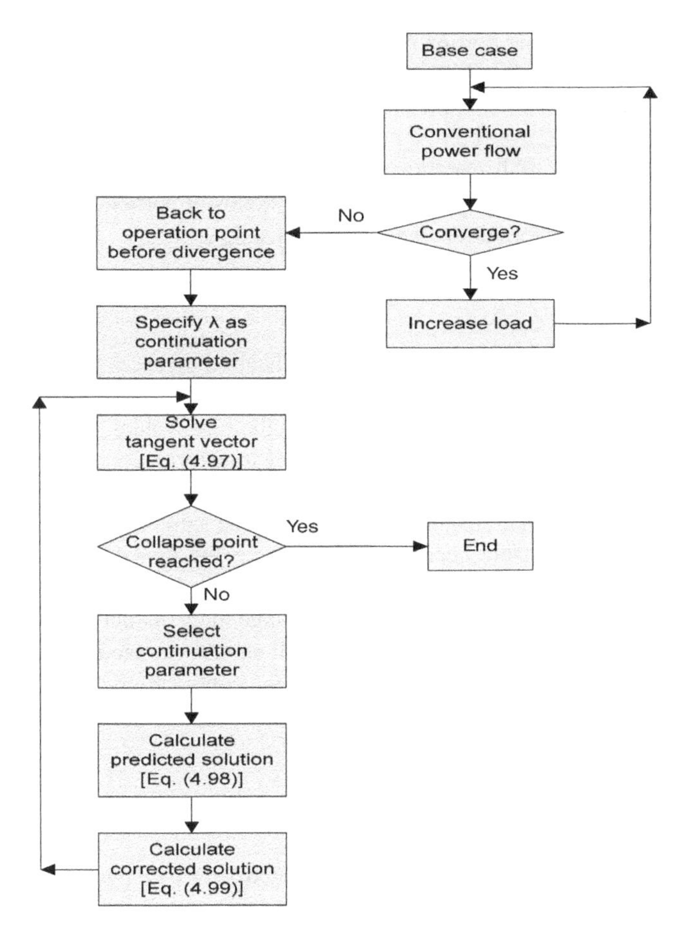

Figure 4.2. Identification of voltage collapse point using continuation power flow.

Voltage stability can be evaluated by only considering the incremental relationship between reactive powers (Q) and voltages (V). In other words, $\mathbf{\Delta P}$ can be assumed to be zero at each operation point in the Q–V analysis. By letting $\mathbf{\Delta P} = \mathbf{0}$, we can derive the following equation from Equation (4.100) [54]:

$$\Delta \mathbf{Q} = \mathbf{J}_R \Delta \mathbf{V} \quad (4.101)$$

where

$$\mathbf{J}_R = \mathbf{J}_{QV} - \mathbf{J}_{Q\delta} (\mathbf{J}_{P\delta})^{-1} \mathbf{J}_{PV} \quad (4.102)$$

Here, $\mathbf{J}_R$ is called the *reduced Jacobian matrix* of the system. The singularity of $\mathbf{J}_R$ is equivalent to the singularity of the full Jacobian. If $\mathbf{J}_R$ is singular at an operation point, the system voltage collapses. Mathematically, a power flow can have a solution on either the lower or upper portion of a *Q*–*V* curve. However, a solution on the lower portion results in voltage instability because it implies that the voltage at a bus will drop further as the reactive support at the bus increases, although $\mathbf{J}_R$ at this solution point is nonsingular.

The modal analysis of $\mathbf{J}_R$ can be used to identify both the voltage collapse and the voltage instability corresponding to a solution on the lower portion of a *Q*–*V* curve. By applying similarity transformation, we can factorize $\mathbf{J}_R$ as

$$\mathbf{J}_R = \xi \Lambda \eta \tag{4.103}$$

where ξ and $\eta = \xi^{-1}$ are the right and left multiplication matrices for similarity transformation; Λ is the diagonal matrix whose elements λ_i are the eigenvalues of $\mathbf{J}_R$.

There are a lot of algorithms for calculating the eigenvalues. The *QR* algorithm, which is based on an iterative orthogonal similarity transformation, is the most famous one. If all $\lambda_i > 0$, the system is voltage-stable since each pair of modal voltage and modal reactive power variations are oriented along the same direction. If any $\lambda_i < 0$, the system is voltage-instable because a variation in the modal reactive power will create an opposite variation in the corresponding bus voltage. If $\lambda_i = 0$, the modal voltage collapses, as this means that any small change in the modal reactive power will result in an infinite change in the modal voltage. It should be noted that the concept of λ_i here is completely different from that of $d\lambda$ in the continuation power flow method, although their signs provide a similar indication on voltage stability or instability.

4.8 TRANSIENT STABILITY SOLUTION

Transient stability is a major topic and is associated with dynamic representations of system components, including generators, synchronous condensers, excitation systems, motors, protection relays, HVDC converters, FACTS, and other control devices. The intent of this section is to provide a brief summary of the popular solution techniques of transient stability that have been used in commercial computer programs, rather than to repeat the topics that have been discussed in other textbooks. Transient stability analysis can be conducted using time-domain simulation or the energy function method. Time-domain simulation provides details of dynamic process and is a main tool used in the utility industry. Mathematically, transient stability analysis is a resolution of a set of differential and algebraic equations. Two approaches can be applied for solving the differential equations: explicit and implicit integrations. Considerable calculations indicate that the explicit integration approach has poor numerical stability whereas the trapezoidal-rule-based implicit integration approach is numerically stable. This section focuses on the latter.

4.8.1 Transient Stability Equations

All dynamic devices are represented using a set of differential and algebraic equations:

$$\frac{\mathbf{dx}_d}{dt} = \mathbf{f}_d(\mathbf{x}_d, \mathbf{V}_d) \tag{4.104}$$

$$\mathbf{I}_d = \mathbf{g}_d(\mathbf{x}_d, \mathbf{V}_d) \tag{4.105}$$

where $\mathbf{x}_d$ is the state vector of individual devices, $\mathbf{I}_d$ is the current injection vector from the devices into the network, and $\mathbf{V}_d$ is the voltage vector of the buses to which the devices are connected.

The network can be represented using a set of algebraic equations. The transient stability equations of overall system can be expressed in the following general form

$$\frac{\mathbf{dx}}{dt} = \mathbf{f}(\mathbf{x}, \mathbf{V}) \tag{4.106}$$

$$\mathbf{I}(\mathbf{x}, \mathbf{V}) = \mathbf{Y} \cdot \mathbf{V} \tag{4.107}$$

where $\mathbf{x}$ is the state vector of the system, $\mathbf{I}$ is the current injection vector (including real and imaginary components), $\mathbf{V}$ is the bus voltage vector (including real and imaginary components), and $\mathbf{Y}$ is the admittance matrix of the network. The initial conditions ($\mathbf{x}_0$,$\mathbf{V}_0$) at $t = 0$ are known.

4.8.2 Simultaneous Solution Technique

The simultaneous solution technique [55] is summarized as follows. By applying the trapezoidal rule for implicit integration, we can rewrite Equation (4.106) as

$$\mathbf{x}_{n+1} = \mathbf{x}_n + \frac{\Delta t}{2}[\mathbf{f}(\mathbf{x}_{n+1}, \mathbf{V}_{n+1}) + \mathbf{f}(\mathbf{x}_n, \mathbf{V}_n)] \tag{4.108}$$

where ($\mathbf{x}_n$, $\mathbf{V}_n$) is the solution at $t = t_n$ and ($\mathbf{x}_{n+1}$, $\mathbf{V}_{n+1}$) is the solution at $t = t_{n+1} = t_n + \Delta t$. Obviously, the original differential equations are converted into the algebraic equations through the trapezoidal rule. The solution of the unknown variable vectors ($\mathbf{x}_{n+1}$, $\mathbf{V}_{n+1}$) at $t = t_{n+1}$ must also satisfy the following system equation:

$$\mathbf{I}(\mathbf{x}_{n+1}, \mathbf{V}_{n+1}) = \mathbf{Y} \cdot \mathbf{V}_{n+1} \tag{4.109}$$

A solution of the unknown variable vectors ($\mathbf{x}_{n+1}$, $\mathbf{V}_{n+1}$) can be obtained by solving the following set of simultaneous nonlinear algebraic equations:

$$\mathbf{F}(\mathbf{x}_{n+1}, \mathbf{V}_{n+1}) = \mathbf{x}_{n+1} - \mathbf{x}_n - \frac{\Delta t}{2}[\mathbf{f}(\mathbf{x}_{n+1}, \mathbf{V}_{n+1}) + \mathbf{f}(\mathbf{x}_n, \mathbf{V}_n)] = \mathbf{0} \tag{4.110}$$

$$\mathbf{G}(\mathbf{x}_{n+1}, \mathbf{V}_{n+1}) = \mathbf{Y} \cdot \mathbf{V}_{n+1} - \mathbf{I}(\mathbf{x}_{n+1}, \mathbf{V}_{n+1}) = \mathbf{0} \tag{4.111}$$

This set of equations can be solved using the Newton–Raphson method. The iterative correction equations can be built as follows

$$\begin{bmatrix} -\mathbf{F}(\mathbf{x}_{n+1}^k, \mathbf{V}_{n+1}^k) \\ -\mathbf{G}(\mathbf{x}_{n+1}^k, \mathbf{V}_{n+1}^k) \end{bmatrix} = \begin{bmatrix} \mathbf{J}_1 & \mathbf{J}_2 \\ \mathbf{J}_3 & \mathbf{J}_4 \end{bmatrix} \begin{bmatrix} \Delta\mathbf{x}_{n+1}^k \\ \Delta\mathbf{V}_{n+1}^k \end{bmatrix} \tag{4.112}$$

where the superscript k denotes the iteration count. The Jacobian matrix is calculated at the point $(\mathbf{x}_{n+1}^k, \mathbf{V}_{n+1}^k)$ as follows (where $\mathbf{U}$ is the unit matrix):

$$\mathbf{J}_1 = \mathbf{U} - \frac{\Delta t}{2} \frac{\partial \mathbf{f}}{\partial \mathbf{x}} \tag{4.113}$$

$$\mathbf{J}_2 = -\frac{\Delta t}{2} \frac{\partial \mathbf{f}}{\partial \mathbf{V}} \tag{4.114}$$

$$\mathbf{J}_3 = -\frac{\partial \mathbf{I}}{\partial \mathbf{x}} \tag{4.115}$$

$$\mathbf{J}_4 = \mathbf{Y} - \frac{\partial \mathbf{I}}{\partial \mathbf{V}} \tag{4.116}$$

Equation (4.112) is solved in an iterative manner. In the iteration, the unknown vectors are updated at each step by

$$\begin{bmatrix} \mathbf{x}_{n+1}^{k+1} \\ \mathbf{V}_{n+1}^{k+1} \end{bmatrix} = \begin{bmatrix} \mathbf{x}_{n+1}^{k} \\ \mathbf{V}_{n+1}^{k} \end{bmatrix} + \begin{bmatrix} \Delta\mathbf{x}_{n+1}^k \\ \Delta\mathbf{V}_{n+1}^k \end{bmatrix} \tag{4.117}$$

A modified resolution scheme is used to conduct a partitioned iteration. From Equation (4.112), the $\Delta\mathbf{x}_{n+1}$ and $\Delta\mathbf{V}_{n+1}$ can be expressed as follows:

$$\Delta\mathbf{x}_{n+1}^k = -\mathbf{J}_1^{-1}[\mathbf{F}(\mathbf{x}_{n+1}^k, \mathbf{V}_{n+1}^k) + \mathbf{J}_2 \Delta\mathbf{V}_{n+1}^k] \tag{4.118}$$

$$\Delta\mathbf{V}_{n+1}^k = (\mathbf{J}_4 - \mathbf{J}_3\mathbf{J}_1^{-1}\mathbf{J}_2)^{-1}[-\mathbf{G}(\mathbf{x}_{n+1}^k, \mathbf{V}_{n+1}^k) + \mathbf{J}_3\mathbf{J}_1^{-1}\mathbf{F}(\mathbf{x}_{n+1}^k, \mathbf{V}_{n+1}^k)] \tag{4.119}$$

Equation (4.119) is used to calculate $\Delta\mathbf{V}_{n+1}$ first, and then Equation (4.118) is used to calculate $\Delta\mathbf{x}_{n+1}$. In the Newton iteration, the block matrices $\mathbf{J}_1$, $\mathbf{J}_2$, and $\mathbf{J}_3$ are updated at each step, whereas $(\mathbf{J}_4 - \mathbf{J}_3\mathbf{J}_1^{-1}\mathbf{J}_2)$ as a whole can be updated only once in several steps or only when a change in the network topology occurs [56].

4.8.3 Alternate Solution Technique

Equations (4.106) and (4.107) can be solved alternately. The current injection vector $\mathbf{I}$ is the interface variable vector between alternate iterations. At initial point $t = 0$, the value of state variable vector $\mathbf{x}$ does not change instantly and is known. The solution technique includes the following steps:

1. The initial estimates of $\mathbf{x}_0$ and $\mathbf{V}_0$ at the timepoint $n = 0$ are given.
2. With the current $\mathbf{x}_n$ and $\mathbf{V}_n$, $\mathbf{I}_d$ is calculated from Equation (4.105) first, and then Equation (4.107) is solved to update the network voltage vector $\mathbf{V}_n$. Note that $\mathbf{x}_d$ and $\mathbf{V}_d$ in Equation (4.105) are the subvectors of $\mathbf{x}$ and $\mathbf{V}$, respectively.

3. With the updated $\mathbf{V}_n$, Equation (4.106) is solved to update the state variables $\mathbf{x}_n$.
4. If a given converge criterion is met, go to the next time point $n + 1$. Otherwise, go back to step 2.

It should be appreciated that the Jacobian matrices used in the alternative and simultaneous solution techniques are different.

4.9 CONCLUSIONS

This chapter reviewed the basic system analysis techniques in transmission planning. These include the power flow equations and solution methods, the optimal power flow model and its interior point method, contingency analysis for postoutage states and contingency ranking, continuation power flow and reduced Jacobian matrix for voltage stability, and implicit integration-based solution techniques for transient stability. All these techniques have been widely applied in both operation and planning of transmission systems. Understanding the fundamental elements of the techniques will help transmission planners make better use of the computing programs that have utilized the techniques and interpret results obtained from the programs. More importantly, these analysis techniques are the basis of the probabilistic evaluation methods that have been described in this chapter and will be extended in the next chapter.

The basic concepts associated with probabilistic system assessments are presented using probabilistic power flow as an example. The similar concepts can be applied or extended to probabilistic OPF and contingency analysis. While the probabilistic evaluation methods will be further developed in the following chapters, the probabilistic power flow itself is a powerful computing tool to capture uncertain factors in transmission planning. The probabilistic contingency ranking indices proposed in this chapter are based on a combination of consequence and probability of contingencies, and are the useful complements to the conventional performance indices for contingency ranking.

The genetic algorithm and particle swarm optimization are discussed as examples of probabilistic search optimization techniques that are capable of handling discrete variables and nonconvex objective functions and avoiding local optimums. However, it should be noted that the two algorithms are very slow in convergence speed for a large-scale system problem. In general, the interior point method should be selected first unless discrete variables or/and nonconvex objective functions are a crucial concern in an OPF or other optimization model for transmission planning.

5

PROBABILISTIC RELIABILITY EVALUATION

5.1 INTRODUCTION

Probabilistic reliability evaluation is the most important activity in probabilistic transmission planning. In reliability evaluation, the load models in Chapter 3 and the system analysis techniques in Chapter 4 are combined with probabilistic system state selection methods to create reliability indices that truly represent the system reliability level. Some indices can be converted into monetary values so that the reliability worth based on the indices and the economic assessment of investment and operation costs could be performed on a unified valuation basis.

Reliability evaluation is divided into two areas of the system adequacy and system security: *Adequacy* relates to the existence of sufficient facilities within the system to satisfy the load demand and system operational constraints, and is therefore associated with static conditions. *Security* relates to the ability of the system to respond to dynamic disturbances arising within the system and is therefore associated with the dynamic conditions that may cause transient or voltage instability.

Probabilistic reliability evaluation includes four aspects:

- Reliability data
- Reliability indices

Probabilistic Transmission System Planning, by Wenyuan Li

- Reliability worth assessment
- Reliability evaluation models and methods

Reliability data will be discussed with other data for transmission planning in Chapter 7. The definitions of reliability indices are discussed in Section 5.2 and the basic concepts of reliability worth assessment, in Section 5.3. Adequacy evaluation techniques for substation configurations and composite systems are described in Sections 5.4 and 5.5, respectively. Security assessments, which are divided into probabilistic voltage stability and probabilistic transient stability, are presented in Sections 5.6 and 5.7, respectively.

5.2 RELIABILITY INDICES

Reliability indices can be classified into two categories:

1. Indices used to estimate the reliability level of a future system with various changes, including load growth, change in operation conditions, modification or enhancement in network configuration, and addition or retirement of facilities in the system.
2. Indices used to measure the historical performance of the existing system or equipment

The reliability indices discussed in this section pertain to category 1. Category 2 indices will be addressed in Chapter 7 as part of the data required in probabilistic transmission planning.

There are many possible indices that can be used to measure transmission system reliability. Most reliability indices are the expected values of random variables, although a probability distribution can be calculated in some cases. It is important to appreciate that an expected value is not a deterministic quantity. It is the long time average of the phenomenon under study. The expected indices provide the average outcomes due to various factors such as system component availability or failure probability, component capacities, load characteristics and uncertainty, system topological configurations, and operational conditions.

5.2.1 Adequacy Indices

The six adequacy indices are as follows:

1. *Probability of Load Curtailments (PLC).* This is expressed as

$$\mathrm{PLC} = \sum_{i \in S} p_i \tag{5.1}$$

where p_i is the probability of the system state i and S is the set of all system states associated with load curtailments.

2. *Expected Frequency of Load Curtailments (EFLC).* This is expressed (in outages per year) as

$$\text{EFLC} = \sum_{i \in S} (F_i - f_i) \tag{5.2}$$

where F_i is the frequency of entering (or departing) the load curtailment state i and f_i is the portion of F_i that corresponds to not passing through the boundary wall between the load curtailment state set and non-load-curtailment state set. In other words, f_i is the frequency of transition from other load curtailment states to the load curtailment state i. In using a state enumeration or state sampling (nonsequential Monte Carlo simulation) technique, it is difficult or extremely time-consuming to identify all transitions between load curtailment states. The ENLC (expected number of load curtailments) index is often used to approximate the EFLC:

$$\text{ENLC} = \sum_{i \in S} F_i \tag{5.3}$$

It can be seen that f_i is ignored in calculating the ENLC and therefore the ENLC is an upper-bound estimate of the EFLC. This approximation is generally acceptable in transmission reliability evaluation because the probability of transitions from one load curtailment state to another is small. In most cases, the system transits from a load curtailment state to the normal operation (non-load-curtailment) state by a repairing or recovering process. The system state frequency F_i can be calculated by

$$F_i = p_i \sum_{k \in N_i} \lambda_k \tag{5.4}$$

where λ_k is the kth departure rate of components in system state i and N_i is the set of all possible departure rates in the system state i. For example, if a component in system state i is in the normal state, the departure rate is its outage rate, whereas if it is in an outage state, the departure rate is its repair or recovery rate. A component may be represented by a multistate model such as the one including not only up and down states but also one or more derated states. In this case, a component can have multiple departure rates.

3. *Expected Duration of Load Curtailments (EDLC).* This is expressed (in hours per year) as

$$\text{EDLC} = \text{PLC} \times T \tag{5.5}$$

where T is the time length considered (in hours). One year is often considered in the reliability evaluation for transmission planning and thus $T = 8760$ hours.

4. *Average Duration of Load Curtailments (ADLC).* This is expressed (in hours per outage) as

$$\text{ADLC} = \frac{\text{PLC} \times T}{\text{EFLC}} \tag{5.6}$$

5. *Expected Demand Not Supplied (EDNS).* This is expressed (in MW) as

$$\text{EDNS} = \sum_{i \in S} p_i C_i \tag{5.7}$$

   where C_i is the load curtailment (in MW) in the system state i.
6. *Expected Energy Not Supplied (EENS).* This is expressed (in MWh/year) as

$$\text{EENS} = \sum_{i \in S} C_i F_i D_i = \sum_{i \in S} p_i C_i T \tag{5.8}$$

   where D_i is the duration (in hours) of the system state i. It can be seen from Equation (5.8) that the frequency (F_i, in occurrences per year), duration (D_i, in hours per occurrence), and probability p_i of a system state satisfy the following relationship:

$$p_i \cdot T = F_i \cdot D_i \tag{5.9}$$

   This is a general relationship, which applies to any system state in the Markov state space regardless of whether if it is a load curtailment or non-load-curtailment state.

It is noted that these six indices are defined from a viewpoint of the overall system. The indices can be similarly defined for individual buses.

5.2.2 Reliability Worth Indices

It is almost impossible to directly measure reliability worth; therefore system reliability worth is usually assessed by considering unreliability costs. The damage costs due to outages are used as a surrogate of reliability worth. The two reliability worth indices are as follows:

1. *Expected Damage Cost (EDC).* This is expressed (in k$/year) as

$$\text{EDC} = \sum_{i \in S} C_i \cdot F_i \cdot W(D_i) \tag{5.10}$$

   where C_i is the load curtailment (in MW) in the system state i; F_i and D_i are the frequency (outages/year) and duration (hours/failure) of system state i; $W(D_i)$ is the customer damage function (in $/kW), which is a function of the duration; and S is the set of all system states associated with load curtailments.
2. *Unit Interruption Cost (UIC).* This is expressed (in $/kWh) as

$$\text{UIC} = \text{EDC/EENS} \tag{5.11}$$

In practice applications, the UIC is often estimated first using different methods, including the direct use of statistical data of customers' interruption costs (see Section 5.3), and then the EDC can be calculated using the EENS and UIC. It is noteworthy that UIC is also called *interruption energy assessment rate* (IEAR) in other literature. The term of UIC is used in this book.

Both EDC and UIC can be calculated either for the entire system or at individual buses. UIC or $W(D_i)$ is a composite one if there are multiple customer sectors at a bus. It is a single UIC or customer damage function if there is just one customer sector at a bus. For the entire system or a region, there are always mixed customers, and a composite UIC or customer damage function is used.

5.2.3 Security Indices

In security evaluation, probabilistic voltage stability or transient stability is assessed. There are two outcomes for any contingency event: either system stable or system unstable. If the system loses stability, the consequence is disastrous (such as a massive blackout). Various emergency control measures, which are called *special protection systems* (SPSs) or *remedial-action schemes* (RASs), are applied to prevent system instability. If the measures are taken in time, system instability can be avoided but the system still suffers consequences such as load shedding, generation rejection, line tripping, or export reduction, which may create economic losses. The following two indices are used in probabilistic security assessment:

1. *Probability of System Instability (PSI).* This is expressed as

$$\mathrm{PSI} = \sum_{i \in SS} p_i \tag{5.12}$$

where p_i is the probability of system instability state i and SS is the set of all system instability states. A system instability state is composed of a contingence (or fault) event that causes system instability and the system operation condition when the contingence occurs, and thus p_i can be expressed as the product of two probabilities of p_{si} and p_{ci}, where p_{si} is the probability of the system operation condition and p_{ci} is the probability of the contingency (or fault) leading to instability. The system operation condition can include the network configuration, load level, generation pattern, and availability of noncontingence components. Therefore p_{si} can be further expressed as the product of multiple factors, each of which represents the probability of an individual component. The PSI index could be calculated before and after a remedial control measure is taken. The difference between the two cases reflects the effect of the control measure.

2. *Risk Index.* This is expressed as

$$\mathrm{RI} = \sum_{i \in SD} p_i R_i \tag{5.13}$$

where p_i is the probability of the system state i; SD is the set of all system states, each of which is composed of a contingence (or fault) event and the system operation condition when the contingence occurs; and R_i represents the consequence due to the contingence (or fault), which can be measured by load curtailments or cost damages. Note that the set SD in Equation (5.13) is different from the set SS in Equation (5.12) in the sense that the set SD includes not only instable states but also the system states that become stable after remedial control actions (such as load shedding or generation rejection) but lead to losses or damages. Specific expressions for RI for probabilistic voltage and transient instability will be given in Sections 5.6.4 and 5.7.5, respectively.

5.3 RELIABILITY WORTH ASSESSMENT

5.3.1 Methods of Estimating Unit Interruption Cost

Estimation of the unit interruption cost ($/kWh) is the key to evaluating the EDC using Equation (5.11). The following four methods can be used to estimate the unit interruption cost [57]:

1. *Method Based on Customer Damage Functions.* This method provides the average social damage cost due to electricity supply interruptions. It is important to recognize that the unit interruption cost is customer-specific. In general, a utility should use the unit interruption cost that is based on its own customer survey and system analysis. This method includes various complex factors that may not be easily considered using other methods. The customer damage function will be discussed in detail in Section 5.3.2.
2. *Method Based on Capital Investments.* Generally, a capital investment for system reinforcement can lead to an improvement in system reliability. There is a quantifiable relationship between the average annual capital investment ($/year) and the incremental reduction in the system EENS index (MWh/year). Therefore an average unit interruption cost in $/kWh can be estimated from previous planning projects that required capital investments and created variations in the EENS index. The detail of this method can be found in Reference 6.
3. *Method Based on Gross Domestic Product (GDP).* The GDP for a province, state, or country divided by the total annual electric energy consumption of the province, state, or country results in a dollar value per kilowatt-hour. This number reflects the average economic damage cost due to 1 kWh of electric energy loss in that province, state, or country. This method does not directly account for some damage components such as equipment damage costs due to power interruptions. However, the method is very simple and can be used in many cases, particularly for a government-owned utility.
4. *Method Based on Revenue Lost to a Utility Due To Power Outages.* In this method, the electricity price per kWh is directly used as UIC. Obviously, this

method only takes revenue losses of the utility into consideration and can be used in cases where the utility's own benefits are focused.

5.3.2 Customer Damage Functions (CDFs)

5.3.2.1 Customer Survey Approach. Customer damage functions (CDFs) can be obtained by customer surveys. There three basic customer survey approaches:

- *Contingent Valuation Approach.* A monetary value of power interruption cost is quantified through either the consumer's willingness to pay (WTP) to avoid an interruption, or the willingness to accept (WTA) compensation for an interruption. In general, WTP values are significantly less than WTA values since customers normally object to providing further money for a service for which they consider already paid through an electricity rate. The WTP and WTA values may serve as upper and lower bounds.
- *Direct Costing Approach.* In this approach, respondents are asked to identify impacts and evaluate the costs associated with particular outage scenarios. The results obtained by this approach are consistent with situations where most losses tend to be tangible, directly identifiable, and quantifiable. The approach is most applicable for industrial and other larger electricity users.
- *Indirect Costing Approach.* In this approach, respondents are asked questions that relate to the content of their experience. These can include the cost of hypothetical insurance to compensate for possible interruption effects, preparatory actions that the respondent may take in the event of interruptions, or considerations in ranking a set of reliability/rate alternatives. Evaluations of financial burden that customers would like to bear to alleviate the interruption effects can be obtained through responses to the questions. This approach is most appropriate to cases where customers may lack experience in rating reliability worth, such as residential customers.

5.3.2.2 Establishment of CDF. The establishment of customer damage functions includes the following steps:

1. Prepare questionnaires for different customer sectors.
2. Perform a customer survey.
3. Process the data obtained from the customer survey, including filtering bad and invalid information using statistical methods.
4. Calculate customer damage functions for individual customer sectors.
5. Calculate composite customer damage functions at different substations, in different regions, and for the whole system.

The customer damage function portrays the damage cost as a function of interruption duration. The CDF can be expressed in \$/kW or \$/kWh. Tables 5.1 and 5.2 provide examples of customer damage functions in \$/kW and \$/kWh, respectively. More

TABLE 5.1. Customer Damage Functions in $/kW

Duration (min)	Residential	Commercial	Industrial	Unknown Mix	Composite
0–19	0.20	11.40	5.50	1.90	3.76
20–59	0.60	26.40	8.60	4.00	7.55
60–119	2.80	40.10	19.60	8.50	14.41
120–239	5.00	72.60	33.60	15.10	25.49
240–480	7.20	147.60	52.10	26.50	45.41

TABLE 5.2. Customer Damage Functions in $/kWh

Duration (min)	Residential	Commercial	Industrial	Unknown Mix	Composite
10	1.20	68.40	33.00	11.40	22.58
40	0.90	39.60	12.90	6.00	11.32
90	1.90	26.70	13.10	5.70	9.63
180	1.70	24.20	11.20	5.00	8.52
360	1.20	24.60	8.60	4.40	7.54
Average	1.38	36.70	15.76	6.50	11.92

statistical data of damage costs for North America can be found in References 58–60. Composite CDFs each in the last column of the two tables are based on the percentage composition of residential 52.36%, commercial 17.61%, and industrial 30.03%. The following observations can be made:

- The damage cost in $/kW, which corresponds to an interruption duration range, increases as the interruption duration increases.
- The damage cost in $/kWh decreases as the interruption duration increases except for residential customers. Note that the value in $/kWh is the average damage cost for energy loss of 1 kWh regardless of duration. The trends for commercial and industrial customers in the table indicate that the hourly energy damage is higher at the beginning of interruption and decreases as the interruption period is longer. However, residential customers do not have such a trend.
- Commercial customers have the highest damage costs, followed by industrial customers. Residential customers have relatively low damage costs.

Note that the term "unknown mix" in the tables refers to unknown customer sectors and the CDF of this category is completely different from the composite CDF that is based on the percentage composition of customer sectors. The values in the last line of Table 5.2 are the average damage dollars per kWh. These values are useful when one single number of $/kWh for each customer category is used.

5.3.3 Application of Reliability Worth Assessment

Reliability and economy are two conflicting factors in system planning. One purpose of assessing reliability worth is to represent system reliability in a dollar value so that

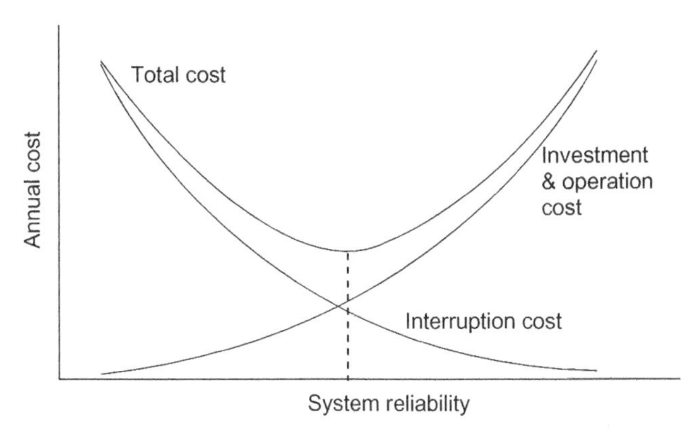

Figure 5.1. Investment & operation, interruption, and total costs versus system reliability.

it can be assessed together with investment and operation costs on the same valuation basis. The basic concept of the probabilistic cost criteria has been described in Section 2.2 and can be further illustrated using the cost–reliability curves in Figure 5.1. The damage cost due to interruptions decreases as system reliability improves while the investment and operation cost increases. The total cost to utility or society is the sum of these two individual costs. The total cost exhibits a minimum point at which an optimum or target level of reliability is achieved. This is one of most important concepts in probabilistic planning and will be applied to actual examples in Chapters 9–12. The unreliability cost will be considered as an additional cost component in the general economic assessment method in Chapter 6.

5.4 SUBSTATION ADEQUACY EVALUATION

Substation reliability can be evaluated using either Monte Carlo simulation or analytical methods [6]. In general, the analytical methods are often more efficient and preferable for substation reliability evaluation since the failure probabilities of substation components are usually very low and the network size of a substation is relatively small. There are two popular enumeration techniques for reliability evaluation of substation configurations: minimum cutset enumeration [61,62] and network state enumeration [63]. In the cutset technique, the minimum cutsets that lead to a network failure are identified and then disjoint operations are conducted for evaluating the failure probability of the substation network. A minimum cutset contains only failed components; therefore the minimum cutsets are not mutually exclusive. This is the reason why the subsequent disjoint calculations are required. The disjoint operations are generally complex when dependent failures, multiple failure modes, and switching actions are considered. In the network state technique, network states are directly enumerated by considering network component states, and then a network connectivity approach is used to identify failure states. A network state is defined by both failed and nonfailed components and is always disjoint with other network states. Because of this feature, the network state technique surpasses the minimum cutset technique in dealing with dependent failures between components; multiple failure modes, including short-circuit faults and breaker stuck conditions; and incorporation of breaker switching and protection coordination.

The network state technique is summarized in this section. A simplified minimum cutset technique will be introduced using an actual planning example in Chapter 11. More materials for substation reliability evaluation can be found in References 6, 10, and 11.

5.4.1 Outage Modes of Components

A substation component has two forced failure modes: passive and active. *Passive failure* refers to a failure that does not cause any operation of protection devices and therefore has no impact on the remaining healthy (nonfailed) components. Service is restored after the failed component is repaired. An open circuit is an example of passive failures. *Active failure* is a failure that causes an operation of the primary protection zone around the failed component and therefore results in outage of other healthy components. The actively failed component is isolated by switching actions leading to service restored to some or all load points. The failed component itself enters a repairing state and is restored only after the repair. A short fault is an example of active failures. A maintenance outage can be considered as the third outage mode. A stuck condition is the fourth failure mode only for a breaker or switch, which creates the operation of a wider protection zone and outages of more healthy components.

The typical four-state model for a substation component is given in Figure 5.2. This model includes forced passive failure, forced active failure, and maintenance outage but not breaker stuck conditions. It should be noted that the switching state is caused by an active failure and thus is associated with multiple component outages. In the figure, λ denotes a failure or outage rate and μ denotes a repair, or recovery, or switching rate. The subscripts a, p, r, m, and sw represent active failure, passive failure, repair, maintenance, and switching, respectively.

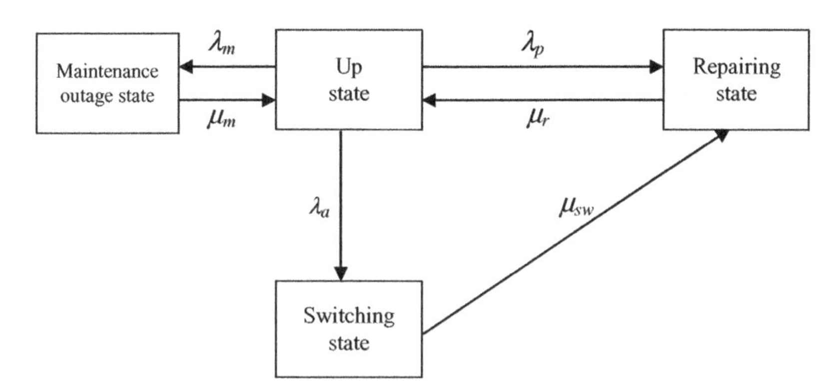

Figure 5.2. Four-state model of a substation component.

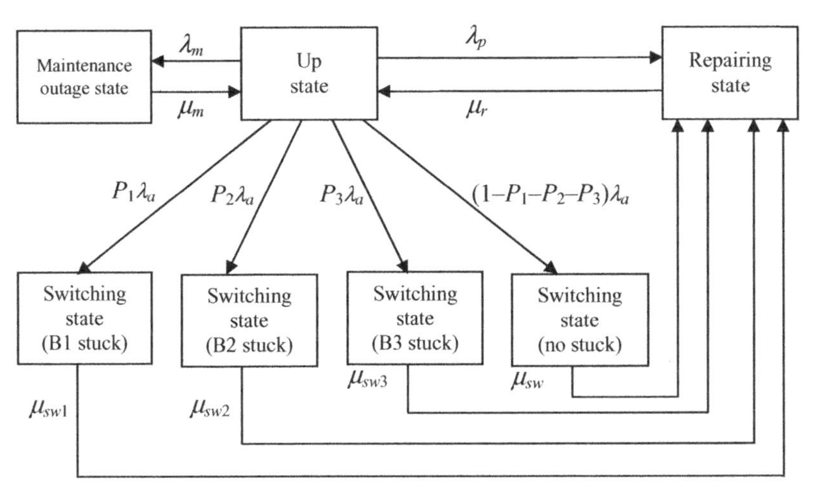

Figure 5.3. State model considering breaker stuck conditions.

When stuck conditions of breakers are considered, the switching state in Figure 5.2 is divided into more substates. Each substate corresponds to a breaker stuck condition. This is represented by the model shown in Figure 5.3. It is assumed that an active failure requires three breakers to switch (open or close). Each of the three breakers may be stuck. P_1 is the probability that the first breaker is stuck but other two are not stuck. P_2 and P_3 have the similar definition. Note that each stuck breaker causes outages of more and different healthy components depending on the configuration and protection logic in a substation.

The state probabilities in the models shown in Figures 5.2 and 5.3 can be calculated using the Markov equation method (see Section C.2.3 in Appendix C).

5.4.2 State Enumeration Technique

Once the state probabilities in the multistate model of each component are obtained, the enumeration process is straightforward by means of considering combinations of component states. The probability $p(s)$ of a substation network state is calculated by

$$P(s)=\prod_{i\in n1}P_{ri}\prod_{i\in n2}P_{si}\prod_{i\in n3}P_{mi}\prod_{i\in n4}(1-P_{ri}-P_{si}-P_{mi}) \tag{5.14}$$

where n_1, n_2, n_3, and n_4 are, respectively, the sets of components in the repairing, switching, maintenance, and success states for the network state s; P_{ri}, P_{si}, and P_{mi} are the probabilities of the ith component being in its three outage states (repairing, switching, and maintenance states), respectively. Note that the switching state of a component can be associated with outages of other healthy components but the outages of healthy components are caused only by a switching action, and their own failure probabilities make no contribution to the calculation.

The network states are enumerated. The enumeration process stops at a given threshold probability for network states. In an actual substation network reliability evaluation, it is generally sufficient to enumerate up to only the second-order failure level, at which two components are in the failure state. This is because the probability of network states associated with simultaneous failures (failed or maintained) of more than two components is extremely low, and neglecting it will not create an effective error from an engineering viewpoint. Note that the outages of healthy components caused by switching actions are not accounted for a failure in the enumeration.

It is obvious that substation network states enumerated using Equation (5.14) are mutually exclusive. Therefore the total failure probability P_f is the direct sum of probabilities of all network failure states

$$P_f = \sum_{s \in G} P(s) \tag{5.15}$$

where G is the set of all substation network failure states. G relies on the failure criterion associated with loss of load at substation buses. A substation generally includes multiple bus sections, and each load may be supplied from one single or multiple power sources. The failure criterion can be "at least one load bus isolated from power supply," "any two load buses isolated," or "all load buses isolated," and so on. This depends on the purpose of reliability evaluation. An isolated bus means that the load at this bus has to be curtailed. The next section presents a labeled bus set approach to identify whether a network state belongs to G.

5.4.3 Labeled Bus Set Approach

The primary components in a substation include transformers, breakers, switches, and bus sections. A bus section is a bus, and other equipment is a branch with two buses at its ends. All buses in the substation configuration are numbered. The following labeled bus set approach can be used to identify connectivity between any pair of power source and load bus in a substation network with multiple sources and/or multiple load buses. For each enumerated network state with equipment outage, the connectivity between each source bus and all load buses is examined. The procedure includes the following steps:

1. All branches, which are expressed using pairs of bus numbers, are put in a full branch set.
2. The branches that are in the outage state, including the outages of both unhealthy (failed and maintained) and healthy (out-of-service by switching actions) components, are crossed out from the above mentioned full branch set to form a residual branch set.
3. Each source bus is checked. If a source bus does not remain in the residual branch set, this indicates that no connection exists between the source bus and all load buses.
4. If a source bus remains in the residual branch set, the source bus is first labeled. All the buses that are directly connected to the labeled buses through branches

can be further labeled using the link information shown in the residual branch set. Starting from the first labeled bus, the labeling process is continued until it can no longer proceed or until all buses are labeled. All the labeled buses form a bus set. If a load bus is not found in the labeled bus set, it does not have a connection to the source bus. Otherwise, there is a connection between them. This labeling process is repeated for all the source buses in the residual branch set.

5. The usual failure criterion for a substation configuration is that if a load bus has been disconnected from all the source buses, this load bus has a failure (load curtailment). A substation network state with at least one load bus failure belongs to the set G.

The labeled bus set approach is running very quickly on a computer since the process of building a residual branch set, labeling buses starting from any source bus, and creating a labeled bus set can be automatically coded in programming. The effects of a protection scheme or switching action, such as the information on which breakers are opened after a short-circuit fault and which breakers are reclosed after isolation of a failed component, can be either automatically identified in a computer program or predefined in a data file because this is the known information in the protection design of a substation. In other words, the impacts of protection logic and switching can be easily incorporated into the state enumeration technique.

5.4.4 Procedure of Adequacy Evaluation

The reliability evaluation method for a substation configuration includes the following steps:

1. The four-state model shown in Figure 5.2 can be applied to all substation components. If the maintenance outage is not considered for a component, the model becomes a three-state representation by deleting the state for maintenance outage. The switching state for each component is individually examined to determine the following:
   a. Which healthy components are out of service because of an active failure
   b. What switching actions are performed to isolate the failed component in order to transit to the repairing state
   c. Which breaker/switch has a possible stuck condition with a probability of occurrence
2. The state space equations of all components are solved using the Markov equation method to obtain the probabilities of each state of the components (see Section C.2.3 in Appendix C).
3. The state enumeration technique described in Section 5.4.2 is used to select system states. In most cases, considering only first- and second-order failure events is sufficient.
4. If the breaker stuck conditions are considered, a subenumeration process is performed. A system state associated with breaker switching is divided into

several substates. Each sub-state corresponds to one breaker stuck or no breaker stuck condition as shown in Figure 5.3. The probabilities of the substates are calculated by

$$P_j(sb) = \begin{cases} P(s)\left[P_{bj}\prod_{\substack{k=1\\k\neq j}}^{m}(1-P_{bk})\right] & (j=1,\cdots,m;\ \text{breaker } j \text{ stuck}) \\ P(s)\left\{1-\sum_{i=1}^{m}\left[\left(P_{bi}\prod_{\substack{k=1\\k\neq i}}^{m}(1-P_{bk})\right)\right]\right\} & (j=0;\ \text{no breaker stuck}) \end{cases} \tag{5.16}$$

where $P(s)$ is the probability of the network state that is associated with m switching breakers. This probability has been obtained in step 3. $P_j(sb)$ is the probability of the substate corresponding to the jth breaker stuck or no breaker stuck ($j = 0$) condition. P_{bj} or P_{bk} or P_{bi} is the probability of breaker j or breaker k or breaker i stuck. It should be noted that the substates of two or more breakers stuck simultaneously have been assumed not to occur, and thus their probabilities have been included in the probability of no breaker stuck. For this reason, the probability of no breaker stuck in Equation (5.16) is expressed as 1.0 minus the sum of probabilities of all the individual breaker stuck substates.

5. The connectivity between source points and load buses is examined for each system state or its substate using the labeled bus set approach described in Section 5.4.3. If a load bus is disconnected from all the sources, this system state or substate is identified as a failure state for the load bus and the curtailed load is recorded.
6. The reliability indices at each load bus are evaluated using the following equations:

   a. *Probability of Load Curtailments (PLC).* This is expressed as

$$\mathrm{PLC}_k = \sum_{i=1}^{N_k} P_{ik} \tag{5.17}$$

   where P_{ik} is the probability of the ith network failure state or substate associated with the load bus k; N_k is the number of the failure states or substates in which the load at the load bus k has to be curtailed.

   b. *Expected Energy Not Supplied (EENS).* This is expressed (in MWh/year) as

$$\mathrm{EENS}_k = \sum_{i=1}^{N_k} P_{ik} L_k T \tag{5.18}$$

   where L_k is the average load (in MW) at the load bus k during the period T (in hours) considered, which is often one year in planning.

c. *Expected Frequency of Load Curtailments (EFLC).* This is expressed in the number of failures per year. For each failure state of a load bus, the state frequency can be obtained using the relationship between the state probability and departure rates:

$$F_{ik} = P_{ik} \sum_{j=1}^{M_i} \lambda_j \tag{5.19}$$

Here, F_{ik} is the frequency of the ith failure state associated with the load bus k; λ_j is the departure rate of the jth component in state i, and it can be a failure, repair, switching, maintenance, or recovery rate depending on the status of the component in state i; and M_i is the number of the departure rates in state i. The EFLC index is the cumulative failure frequency, which can be approximately estimated by

$$\text{EFLC}_k = \sum_{i=1}^{N_k} F_{ik} \tag{5.20}$$

The transition frequencies between failure states should have been deducted from the sum of all failure state frequencies to obtain the accurate EFLC_k index. However, these transition frequencies cannot be directly calculated using the state enumeration technique. Ignorance of this deduction in the evaluation is acceptable from an engineering viewpoint as a transition between failure states almost does not take place in a real operation or maintenance process of substations.

d. *Average Duration of Load Curtailments (ADLC).* This is expressed (in hours per failure) as

$$\text{ADLC}_k = \frac{\text{PLC}_k \cdot T}{\text{EFLC}_k} \tag{5.21}$$

7. The system reliability indices for a whole substation can be evaluated using the similar equations. Note that the system EENS index is the sum of all bus EENS indices but the system PLC or EFLC index is not the sum of bus PLC or EFLC indices because a network failure state may be associated with simultaneous load curtailments at several buses.

It should be recognized that this enumeration process considered the average load at a load bus for a given time period. If the load curve is modeled using the multistep model, the total index is the sum of products of the index and probability for each individual load level.

5.5 COMPOSITE SYSTEM ADEQUACY EVALUATION

Composite generation and transmission system adequacy evaluation is one of the most important technical activities in probabilistic transmission planning. This is a more complex task compared to substation reliability evaluation. It is associated with the two aspects of system analysis and practical considerations in selecting system states. System analysis is not simple connectivity identification but requires power flow calculations, contingency analysis, rescheduling of generations, elimination of system limit violations, and load shedding. Selection of system states involves different considerations and various outage models, such as derated states, common cause outages, dependent outages, weather-related failures of transmission lines, bus load uncertainty and correlation, reservoir dispatches, and operating conditions [6,10,25,64–76].

Composite system adequacy evaluation was briefly summarized in Section 2.3.1. This section provides more details.

5.5.1 Probabilistic Load Models

The load models used in the composite system reliability evaluation include the following three aspects:

- Load curves
- Load uncertainty
- Load correlation

5.5.1.1 Load Curve Models. When the sequential Monte Carlo method is used, a chronological load curve is directly employed. For the state enumeration or state sampling method, a nonchronological load duration curve is utilized. The following three situations are encountered in practical applications:

- A single-system load curve is considered, and loads at all buses are scaled proportionally to follow the shape of the given system load curve. In this case, a multistep model that is similar to that in Figure 3.4 can be created to represent the load duration curve.
- Loads at some buses may remain flat all the time, such as an industrial customer with the same power demand every hour. It is easy to model the constant loads at certain buses while other bus loads follow a load curve.
- Bus loads are classed into multiple bus groups that follow different load curves. In this case, several load duration curves have to be created, with each curve representing a bus group. The multistep load models have to capture the correlation between all the load curves.

The K-mean clustering technique employed to create the multistep load models has been described in Section 3.3.1.

5.5.1.2 Load Uncertainty Model. The load uncertainty at each load level can be represented using a normally distributed random variable and can be modeled as

$$M_{\sigma ij} = z_{ij}\sigma_{ij} + M_{ij} \tag{5.22}$$

where M_{ij} is the ith load level on the jth curve in the multistep load model obtained using the K-mean technique described in Section 3.3.1, $M_{\sigma ij}$ is a random sample value around M_{ij}, z_{ij} is the standard normal distribution random variable corresponding to the ith load level on the jth curve, and σ_{ij} is the standard deviation representing the uncertainty of M_{ij}.

In the AC power-flow-based system analysis, both real and reactive power loads at buses are needed. The sample values of bus reactive loads can be calculated from sample values of real loads and power factors.

5.5.1.3 Load Correlation Model. If the correlation between bus loads is considered, the bus load correlation model described in Section 3.4 can be applied.

5.5.2 Component Outage Models

In general, substation layouts are not modeled in composite system reliability evaluation since substation reliability evaluation is performed separately. The main components in composite system reliability are transmission lines, cables, transformers, capacitors, reactors, and generating units.

5.5.2.1 Basic Two-State Model. The forced repairable failures of system components are represented using a two-state model as shown in Figure 5.4. The parameters in the model have the following relationships:

$$U = \frac{\lambda}{\lambda + \mu} = \frac{\text{MTTR}}{\text{MTTF} + \text{MTTR}} = \frac{f * \text{MTTR}}{8760} \tag{5.23}$$

where U is the forced unavailability, λ is the failure rate (failures/year), μ is the repair rate (repairs/year), MTTR is the mean time to repair (hours/repair), MTTF is the mean time to failure (hours/failure), and f is the average failure frequency (failures/year).

A planned outage of a system component can be represented using a similar two-state model, in which all the parameters (transition rates or mean times) correspond to the planned outage. The separate modeling for forced and planned outages will not create an effective error [6]. Similarly, a common cause outage can also be represented using an independent two-state model, in which the down state includes simultaneous multiple component outages and the failure and repair rates correspond to a group of components in the common-cause outage [69].

5.5.2.2 Multistate Model. One or more derated states of generating units or HVDC links often need to be considered. In this case, a multistate model can be applied.

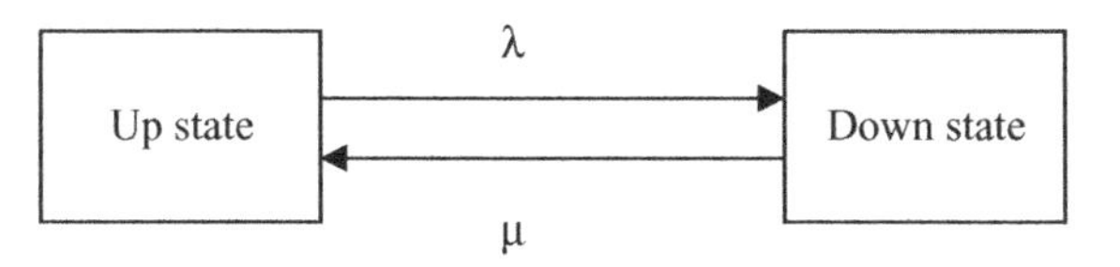

Figure 5.4. Two-state model of a repairable component.

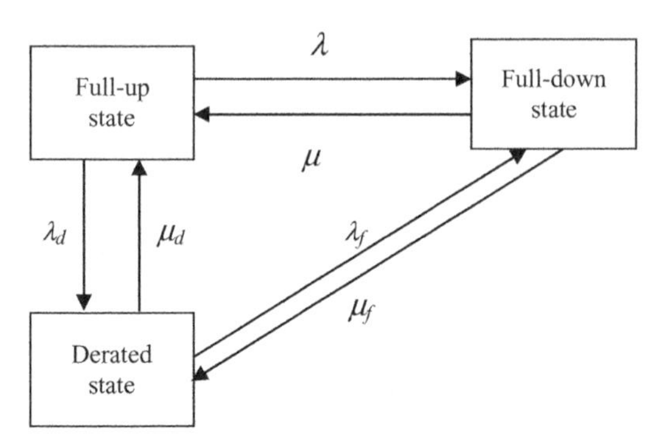

Figure 5.5. Three-state model with a derated state.

A three-state model with one derated state is shown in Figure 5.5, in which λ and μ denote outage and repair rates, respectively, and the subscript d represents the transition between up and derated states. In most situations, the transition rates (represented by the subscript f) between derated and full-down states can be ignored. More derated states can be modeled in a similar manner. The probabilities of each state in a multistate model can be calculated using the Markov equation method (see Section C.2.3 in Appendix C).

5.5.3 Selection of System Outage States

Similar to substation reliability evaluation, composite system reliability evaluation can be performed using either the state enumeration or Monte Carlo simulation techniques. Both techniques have the same requirements in system analysis but are different in selection of system states. The enumeration method of system states is similar to that described in Section 5.4.2.

In general, Monte Carlo methods are more flexible and efficient for incorporating various outage models (particularly for multistate models) and complex operation considerations. There are two Monte Carlo methods: nonsequential and sequential sampling.

5.5.3.1 Nonsequential Sampling. Nonsequential sampling is also called the *component state sampling*. Take the three-state model shown in Figure 5.5 as an example. Each component has the up, down, and derated states. A random number R_i uniformly distributed between [0,1] is created for the ith component:

$$I_i = \begin{cases} 0 \quad \text{(up)} & \text{if} \quad R_i > PP_i + PF_i \\ 1 \quad \text{(down)} & \text{if} \quad PP_i < R_i \le PP_i + PF_i \\ 2 \quad \text{(derated)} & \text{if} \quad 0 \le R_i \le PP_i \end{cases} \tag{5.24}$$

Here, I_k is the indicator variable of the sampled state for the ith component; PF_i and PP_i are the probabilities of the forced outage and derated states for the ith component, respectively. For a two-state model without the derated state, only PF_i is needed and PP_i is zero. Planned and common-cause outages can be also sampled using separate random numbers. All sampled component states form a system state for the system analysis.

5.5.3.2 Sequential Sampling. The sequential sampling can be classified as either state duration sampling or system state transition sampling [10,77]. State duration sampling consists of the following steps:

1. All components are assumed to be in the up state initially.
2. The duration of each component residing in its present state is sampled. The probability distribution of the state duration should be assumed. For example, the sampling value of the state duration following an exponential distribution is given by

$$D_i = \frac{1}{\lambda_i} \ln R_i \tag{5.25}$$

   where R_i is a uniformly distributed random number between [0,1] corresponding to the ith component. If the present state is the up state, λ_i is the outage rate of the ith component; if the present state is the down state, λ_i is its repair rate. The methods of generating random variables following different probability distributions can be found in Sections A.5.3 and A.5.4 of Appendix A.
3. Step 2 is repeated for all components in the timespan considered (years). The chronological state transition processes of each component in the given timespan are obtained.
4. The chronological system state transition cycle is obtained by combining the state transition processes of all components.

5.5.4 System Analysis

A system analysis is conducted for each sampled system state. This includes contingency analysis and load curtailment evaluation. Load curtailment occurs in relatively few system states. The set of system states G_0 can be divided into four subsets as follows:

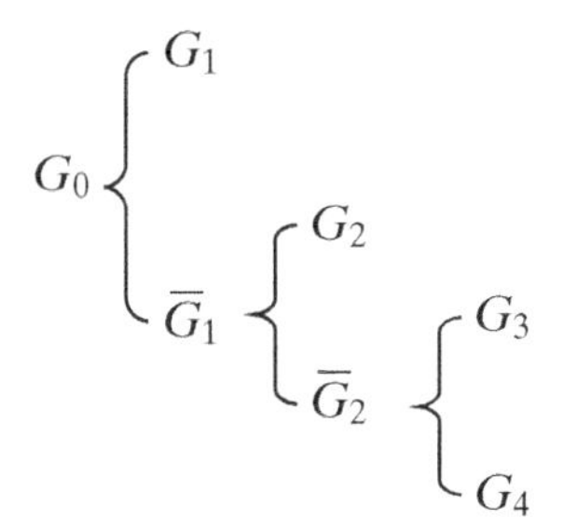

G_1 is the normal state subset without contingency, $\bar{G}_1$ is the contingency state subset, G_2 is the subset of the contingency states that lead to no system problem, $\bar{G}_2$ is the subset of the contingency states that create system problems, G_3 is the subset of the contingency states in which system problems can disappear by remedial actions without load curtailment, and G_4 is the subset of the contingency states in which system problems can be resolved only by load curtailments. The subset G_4 here is similar to the set G of substation network failure states in Sections 5.4.2 and 5.4.3.

For a system state selected by either state enumeration or Monte Carlo simulation method, it is necessary to judge whether it belongs to $\bar{G}_1$. If so, it is necessary to determine by means of contingency analysis techniques whether it belongs to $\bar{G}_2$. If so, a remedial action is taken. Only the system states belonging to the subset G_4 make contributions to reliability indices.

The contingency analysis includes generating unit outages and transmission component outages. The analysis for generating unit contingencies is straightforward. If the remaining generation capacity at each generator bus can compensate for the unavailable capacity due to loss of one or more generators at the same bus, no load curtailment is required. Otherwise, generation rescheduling should be performed using the optimization model that is discussed in the next section.

The transmission contingency analysis is more complex, The purpose is to calculate line flows and bus voltages following one or more component outages and identify whether there is any overloading, or voltage violation, or isolated bus, or split island. The transmission contingency analysis methods have been discussed in Section 4.6.1. If the contingency analysis indicates the existence of any system problem, the optimization model presented in the next section is applied.

5.5.5 Minimum Load Curtailment Model

In any contingency state belonging to $\bar{G}_2$, either generation outputs at some buses cannot be maintained because of generating unit contingencies or there exist system problems such as branch overloading or voltage violations or system splits due to transmission component outages. For the contingency state, redispatching outputs of generators and reactive power sources and changing transformer turn ratios are needed to maintain the power balance and eliminate the system problems, and at the same time to avoid load curtailments if possible or to minimize the total load curtailment if unavoidable. The following minimization model of load curtailment can be applied for this purpose:

$$\min \sum_{i=1}^{ND} w_i C_i \tag{5.26}$$

$$P_{Gi} - P_{Di} + C_i = V_i \sum_{j=1}^{N} V_j (G_{ij} \cos\delta_{ij} + B_{ij} \sin\delta_{ij}) \qquad (i = 1, \ldots, N) \tag{5.27}$$

$$Q_{Gi} + Q_{Ci} - Q_{Di} + C_i \frac{Q_{Di}}{P_{Di}} = V_i \sum_{j=1}^{N} V_j (G_{ij} \sin\delta_{ij} - B_{ij} \cos\delta_{ij}) \qquad (i = 1, \ldots, N) \tag{5.28}$$

$$P_{Gi}^{\min} \le P_{Gi} \le P_{Gi}^{\max} \quad (i = 1,\ldots,NG) \tag{5.29}$$

$$Q_{Gi}^{\min} \le Q_{Gi} \le Q_{Gi}^{\max} \quad (i = 1,\ldots,NG) \tag{5.30}$$

$$Q_{Ci}^{\min} \le Q_{Ci} \le Q_{Ci}^{\max} \quad (i = 1,\ldots,NC) \tag{5.31}$$

$$K_t^{\min} \le K_t \le K_t^{\max} \quad (t = 1,\ldots,NT) \tag{5.32}$$

$$V_i^{\min} \le V_i \le V_i^{\max} \quad (i = 1,\ldots,N) \tag{5.33}$$

$$-T_l^{\max} \le T_l \le T_l^{\max} \quad (l = 1,\ldots,NB) \tag{5.34}$$

$$0 \le C_i \le P_{Di} \quad (i = 1,\ldots,ND) \tag{5.35}$$

In these equations P_{Gi}, Q_{Gi}, P_{Di}, Q_{Di}, Q_{Ci}, K_t, N, NG, NC, NT, and NB are the same as defined in the OPF model in Section 4.4.1; C_i is the load curtailment variable at load bus i and ND is the number of load buses; and w_i is the weighting factor reflecting importance of loads.

It can be seen that there are the following differences between this minimization model and the OPF model in Section 4.4.1:

- The bus load curtailment variables are introduced. The reactive loads are curtailed proportionally in terms of power factors at each bus. The introduction of load curtailment variables guarantees that the optimization model always has a solution in any outage state.
- The objective function is to minimize the total load curtailment. The values of load curtailment variables are parts of the solution. Nonzero curtailments provide the information needed to calculate reliability indices.
- The weighting factors provide a load curtailment order, which can be automatically conducted in the resolution. A greater weighting factor represents a more important bus load.

The minimization model can be solved using the interior point method presented in Section 4.4.2.

5.5.6 Procedure of Adequacy Evaluation

The adequacy evaluation method for a composite system using the nonsequential Monte Carlo simulation includes the following steps:

1. The state probabilities of system components are calculated using the component outage models in Section 5.5.2. If a component is represented by a two-state model, the probabilities of up and down states can be obtained directly from the statistical data. If a component is represented by a multistate model, the state space equation of the model is solved using the Markov equation method to obtain its state probabilities.
2. The multistep load model is created using the method described in Section 3.3.1. The following state sampling process is performed for each load level.

3. A system state is selected using Monte Carlo simulations. This is associated with random determination of bus loads and component states:
   a. Load uncertainty and correlation are modeled using the approaches described in Section 5.5.1.
   b. Component states (up, down, or derated) are sampled using the approach described in Section 5.5.3.1. Note that the forced, planned, and common-cause outages are modeled using independent and separate random variables.
4. A contingency analysis is performed using the method summarized in Section 5.5.4 and detailed in Section 4.6.1.
5. An optimal power flow analysis is conducted using the minimization load curtailment model in Section 5.5.5. If the load curtailment is not zero, the selected state is a failure one. The load curtailments in system failure states are recorded.
6. Steps 3–5 are repeated for all system states sampled in the Monte Carlo simulations and for all load levels in the multistep load model.
7. The reliability indices for the system or at each bus are calculated using the following equations:
   a. *Probability of Load Curtailments (PLC).* This is expressed as

$$\mathrm{PLC} = \sum_{i=1}^{NL}\left(\sum_{s\in G_{4i}} \frac{n(s)}{N_i}\right)\frac{T_i}{T} \tag{5.36}$$

   where $n(s)$ is the number of states s occurring in the sampling; N_i is the total number of samples at the ith load level in the multistep load model; G_{4i} is the set of all system failure states at the ith load level, where the definition of G_4 has been given in Section 5.5.4; T_i is the time length (in hours) of the ith load level; T is the total hours of the load curve considered, which is often one year; and NL is the number of load levels.

   b. *Expected Energy Not Supplied (EENS).* This is expressed (in MWh/year) as

$$\mathrm{EENS} = \sum_{i=1}^{NL}\left(\sum_{s\in G_{4i}} \frac{n(s)C(s)}{N_i}\right)\cdot T_i \tag{5.37}$$

   where $C(s)$ is the load curtailment (in MW) in state s.

   c. *Expected Frequency of Load Curtailments (EFLC).* This is expressed (in failures per year) as

$$\mathrm{EFLC} = \sum_{i=1}^{NL}\sum_{s\in G_{4i}}\left(\frac{n(s)}{N_i}\sum_{j=1}^{m(s)}\lambda_j\right)\frac{T_i}{T} \tag{5.38}$$

   where λ_j is the jth departure rate of the components in state s; and $m(s)$ is the total number of transition rates departing from state s. As in Equation

(5.20), the transition frequencies between failure states have been ignored in Equation (5.38), leading to an approximate estimation of the frequency index.

d. *Average Duration of Load Curtailments (ADLC).* This is expressed (in hours per failure) as

$$\mathrm{ADLC} = \frac{PLC \cdot T}{EFLC} \tag{5.39}$$

There are similar procedures when the state enumeration or sequential sampling techniques are used. The system analysis for selected system states is the same for all the techniques; the difference is associated only with selection of system states. If the sequential sampling technique is utilized, the frequency index can be exactly estimated but requires much more efforts in computations. A compromise between the computational cost and the accuracy just for the frequency index should be carefully considered when a simulation method is selected. The formulas for index calculations have varied expressions for the different techniques. More detailed information can be found in References 6 and 10.

5.6 PROBABILISTIC VOLTAGE STABILITY ASSESSMENT

The purpose of probabilistic voltage stability assessment is to evaluate the average voltage instability risk under various contingencies. The voltage stability analysis for a huge number of system contingency states has to be conducted in the assessment. The continuation power flow method is very slow and thus inappropriate for this purpose. A new technique, which includes an optimization model, the modal analysis of reduced Jacobian matrix and Monte Carlo simulations, is presented to assess the average voltage instability risk of a transmission system under various outage conditions [78].

The basic idea is as follows. For each contingency state, the optimization model is used to identify insolvability of power flow. If there is no power flow solution, this may be caused by either system voltage collapse (system instability) or numerical computation instability. Singularity analysis of the reduced Jacobian matrix can be used to differentiate the two cases. If the power flow is solvable, the solution obtained from the optimization model may correspond to either the upper portion (stable) or lower portion (instable) of a *Q–V* curve. The signs of eigenvalues of the reduced Jacobian matrix can be used to distinguish the two cases. Therefore, system voltage instability can be recognized by combining the optimization model and the reduced Jacobian matrix method. Introduction of the optimization model avoids considerable power flow calculations in the continuation power flow method, which are required to gradually reach the collapse point. The Monte Carlo simulation is used to select system contingency states, from which the average voltage instability risk indices can be assessed.

5.6.1 Optimization Model of Recognizing Power Flow Insolvability

In this optimization model, real and reactive powers of generators, reactive power outputs of VAR equipments, turn ratios of transformers, and real power curtailments at load buses are control variables; real and imaginary parts of bus voltages are state variables; and the minimization of total load curtailment is the objective function.

$$\min \sum_{i=1}^{ND} w_i C_i \tag{5.40}$$

subject to

$$P_{Gi} - P_{Di} + C_i - \sum_{ij \in S_{Li}} P_{Lij} - \sum_{ij \in S_{Ti}} P_{Tij} = 0 \qquad (i = 1, \ldots, N) \tag{5.41}$$

$$Q_{Gi} + Q_{Ci} - Q_{Di} + C_i \frac{Q_{Di}}{P_{Di}} - \sum_{ij \in S_{Li}} Q_{Lij} - \sum_{ij \in S_{Ti}} Q_{Tij} = 0 \qquad (i = 1, \ldots, N) \tag{5.42}$$

$$e_i = K_t e_m \qquad (t = 1, \ldots, NT) \tag{5.43}$$

$$f_i = K_t f_m \qquad (t = 1, \ldots, NT) \tag{5.44}$$

$$P_{Gi}^{\min} \le P_{Gi} \le P_{Gi}^{\max} \qquad (i = 1, \ldots, NG) \tag{5.45}$$

$$Q_{Gi}^{\min} \le Q_{Gi} \le Q_{Gi}^{\max} \qquad (i = 1, \ldots, NG) \tag{5.46}$$

$$Q_{Ci}^{\min} \le Q_{Ci} \le Q_{Ci}^{\max} \qquad (i = 1, \ldots, NC) \tag{5.47}$$

$$K_t^{\min} \le K_t \le K_t^{\max} \qquad (t = 1, \ldots, NT) \tag{5.48}$$

$$0 \le C_i \le P_{Di} \qquad (i = 1, \ldots, ND) \tag{5.49}$$

where P_{Gi}, Q_{Gi}, P_{Di}, Q_{Di}, Q_{Ci}, K_t, N, NG, NC, and NT are the same as defined for the minimum load curtailment model in Section 5.5.5; e_i and f_i are the real and imaginary part variables of voltage at bus i; C_i is load curtailment variable at load bus i and ND is the number of load buses; w_i is the weighting factor reflecting importance of load; P_{Tij} and Q_{Tij} are real and reactive flows on branch i–j of a transformer with an on-load tap changer (OLTC) and S_{Ti} is the set of OLTC transformer branches connected to bus i; and P_{Lij} and Q_{Lij} are real and reactive flows on non-OLTC branches (lines and transformers without OLTC) and S_{Li} is the set of non-OLTC branches connected to bus i.

The OLTC transformer branches are modeled using an ideal transformer with a turn ratio as a control variable in series with a regular branch representation. A dummy bus m is added in the middle of each transformer branch t with its two original buses i and j, as shown in Figure 5.6. In Equations (5.43) and (5.44), the subscripts i and m are the two end buses of the ideal transformer branch t. If the bus i is located at the high-voltage side, P_{Tij} and Q_{Tij} are calculated using Equations (5.50) and (5.51); if the bus i is located at the low-voltage side, P_{Tji} and Q_{Tji} are calculated using Equations (5.52) and (5.53).

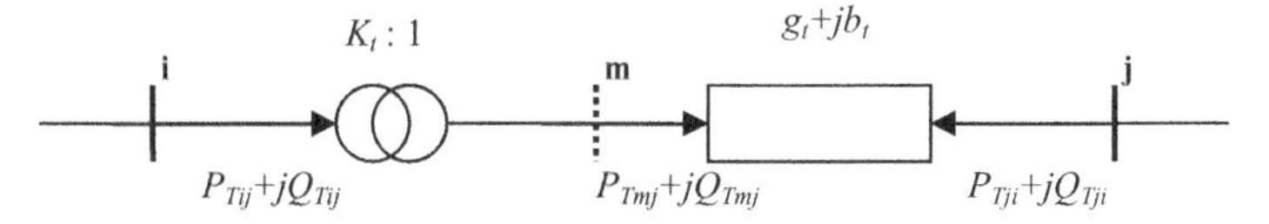

Figure 5.6. OLTC transformer branch representation.

$$P_{Tij} = (e_m^2 + f_m^2 - e_m e_j - f_m f_j)g_t + (e_m f_j - e_j f_m)b_t \tag{5.50}$$

$$Q_{Tij} = -(e_m^2 + f_m^2 - e_m e_j - f_m f_j)b_t + (e_m f_j - e_j f_m)g_t \tag{5.51}$$

$$P_{Tji} = (e_j^2 + f_j^2 - e_m e_j - f_m f_j)g_t + (e_j f_m - e_m f_j)b_t \tag{5.52}$$

$$Q_{Tji} = -(e_j^2 + f_j^2 - e_m e_j - f_m f_j)b_t + (e_j f_m - e_m f_j)g_t \tag{5.53}$$

where g_t+jb_t is the admittance of branch t.

The sum of P_{Lij} or Q_{Lij} for non-OLTC branches at bus i can be calculated using the modified conventional power flow equations in the rectangular coordinate form as follows:

$$\sum_{ij \in S_{Li}} P_{Lij} = \sum_{j=1}^{N} [G'_{ij}(e_i e_j + f_i f_j) + B'_{ij}(f_i e_j - e_i f_j)] \tag{5.54}$$

$$\sum_{ij \in S_{Li}} Q_{Lij} = \sum_{j=1}^{N} [G'_{ij}(f_i e_j - e_i f_j) - B'_{ij}(e_i e_j + f_i f_j)] \tag{5.55}$$

In Equations (5.54) and (5.55), G'_{ij} and B'_{ij} are the elements of a modified admittance matrix, in which the contributions of OLTC transformer branches are excluded since these branches have been explicitly represented in the model.

Comparison of this model with the optimization model in Section 5.5.5 leads to the following observations:

- The bus load curtailment variables are introduced in the both models. As mentioned in Section 5.5.5, the introduction of load curtailment variables guarantees that the optimization model always has a solution in any outage state.
- The constraints of this optimization model are all in a linear or quadratic form because the power flow equations are represented in the rectangular coordinate using real and imaginary parts of bus voltages, and the OLTC transformers are represented by inserting a dummy bus. This results in a constant Hessian matrix that does not need to be updated in the iteration process when the interior point method (see Section 4.4.2) is used to solve the model. Thus the computing time can be greatly reduced.
- Only the constraints of control variables are included in this model. The constraints of state variables (upper and lower limits of bus voltages) and network variables (rating limits of branches) are removed. The model fully corresponds to solvability of power flow. In other words, load curtailments cannot originate from the constraints of state and network variables.

The optimization model described above has the following features:

- When the solution of the model does not lead to load curtailment, the corresponding power flow is solvable and the power flow solution has been obtained. When the solution of the model results in any load curtailment, this indicates that the corresponding power flow has no solution but the model recovers solvability and provides a critical solution with minimum load curtailment.
- Unlike the continuation power flow method, in which many consecutive power flow solutions are required to gradually approach the voltage collapse point, the optimization model needs to be solved only once for judging the solvability of power flow.
- The model can flexibly handle the constraints of control variables. Either all control variables are taken into consideration to enhance system voltage stability, or some control variables can be fixed if necessary. This depends on the operational requirements to be considered.

5.6.2 Method for Identifying Voltage Instability

When the optimization model leads to a solution without load curtailment, it is a solution of corresponding power flow equations. In most cases, this solution is voltage-stable. Mathematically, however, it is possible that the solution can be located in the lower portion of a *Q*–*V* (or *P*–*V*) curve, which represents voltage instability (uncontrollability). When the optimization model has a solution with minimum load curtailment, this indicates that the corresponding power flow does not have a solution since loads have to be curtailed to bring solvability back. The minimum load curtailment ensures that the solution obtained from the optimization model is a critical solution. It is generally accepted in the utility's practice that if there is no power flow solution, the system is thought to be voltage-instable. The optimization model creates a solution point by only one power flow calculation step, at which system voltage instability can be judged.

As discussed in Section 4.7.2, the signs of eigenvalues of the reduced Jacobian matrix $\mathbf{J}_R$ can be used to recognize voltage instability. The optimization model directly provides a solution point to which the reduced Jacobian matrix technique can be applied. The method of identifying voltage instability for any contingency state includes the following three steps:

1. The optimization model is solved.
2. If the solution requires load curtailments, the system is thought to have reached the voltage collapse point.
3. If the solution does not require any load curtailment, the eigenvalues of the reduced Jacobian matrix $\mathbf{J}_R$ at the solution point are calculated. If the minimum eigenvalue is larger than a given positive threshold (a small value close to 0.0), system voltage stability is confirmed. Otherwise, if at least one eigenvalue is smaller than the threshold, the system is thought to be voltage-unstable.

In applications, the following modifications in step 2 or/and step 3 can be performed:

- Conceptually, a critical solution with minimum load curtailment from the optimization model corresponds to the voltage collapse point only when there is no numerical computation instability problem. Unfortunately, the numerical instability problem, which leads to a false load curtailment case, although rare, may occur. This depends on not only the algorithm but also the coding quality of a computer program. To exclude such a falsehood from the set of voltage instability cases, step 2 can be modified as follows. The eigenvalues of the reduced Jacobian matrix $\mathbf{J}_R$ at the critical solution with minimum load curtailment are calculated. If the absolute value of at least one eigenvalue is smaller than the threshold, the voltage collapse is confirmed. Otherwise, if the minimum eigenvalue is larger than the threshold, the false load curtailment is caused by numerical instability but not by voltage collapse. This modification can improve the accuracy of results. However, it should be pointed out that in general, the error due to numerical instability is small as long as the computer program is sufficiently robust, and this modification may not be needed in many practical cases in order to reduce computational efforts.
- Considerable calculations indicate that generally, if a power flow solution is located in the lower portion of a *Q–V* curve, there must be at least one bus whose voltage is lower than the normally permissible operational voltage limit. Step 3 can be modified as follows. If a solution without load curtailment is obtained from the optimization model, all bus voltages at this solution are checked. If all the bus voltages are within the normal operational limits, the system voltage stability at the solution point is confirmed. Otherwise, if there is at least one bus whose voltage is lower than the normal voltage level, the reduced Jacobian matrix method described in step 3 is applied. This modification can significantly reduce computational efforts since finding the eigenvalues of $\mathbf{J}_R$ is a time-consuming process and the majority of contingency cases will not cause voltage instability. Strictly speaking, this alternative approach may introduce a very small risk of misjudgment and should be used with caution for real-time voltage stability assessment. However, it is sufficiently accurate and acceptable for calculating the probabilistic voltage instability risk indices for planning purposes.

5.6.3 Determination of Contingency System States

In the probabilistic voltage stability risk assessment, various contingency system states are randomly sampled. Determination of contingency system states includes

- Selection of precontingency system states (network topology, generation capacities, and bus loads)
- Selection of contingencies

5.6.3.1 Selection of Precontingency System States. The system components (transmission components and generating units) are divided into two groups:

crucial component set and noncrucial component set. A failure event of components in the crucial set may cause voltage instability, whereas a failure event of components in the noncrucial set does not cause voltage instability. This division can be determined from the knowledge and experience of engineers on an actual system. A conservative division is always used. If we are uncertain about a component, it is assigned to the crucial set. The unavailability due to forced failures of the components in the crucial set is not taken into consideration in selecting a precontingency system state since their forced failures will be considered as random contingencies later. The unavailability due to both forced failures and planned outages of components in the non-crucial set and the unavailability due to only planned outages of components in the crucial set make contributions to the probability of a precontingency system state. The precontingency state must be a stable one. A Monte Carlo sampling method similar to that described in Section 5.5.3.1, or an enumeration technique is used to select precontingency system states.

Voltage stability is sensitive to load profiles. The real power load at a bus can be assumed to follow a normal distribution, whereas its reactive load is assumed to have a change proportional to the random variation of the real load with a constant power factor. A standard normally distributed random number x_i for the real load at the ith bus is created using the method in Section A.5.4.2 of Appendix A. The bus loads are calculated by

$$P_{Di} = x_i \sigma_i + P_{Di}^{\text{mean}} \tag{5.56}$$

$$Q_{Di} = P_{Di} \frac{Q_{Di}^{\text{mean}}}{P_{Di}^{\text{mean}}} \tag{5.57}$$

where P_{Di}^{mean} and Q_{Di}^{mean} are the mean values of real and reactive loads at the ith bus; P_{Di} and Q_{Di} are their sampled values, which are used in the optimization model; x_i is a standard normal distribution random variable; and σ_i is the standard deviation of P_{Di}^{mean}.

Obviously, the load model is the same as that in Section 5.5.1.2. Besides, the methods for load curve modeling in Section 5.5.1.1 and load correlation modeling in Section 5.5.1.3 can be also applied if necessary.

5.6.3.2 Selection of Contingencies. A forced failure (fault event) of components in the crucial set leads to a contingency. Such a forced failure can occur when these components are in service (not in a planned outage). The status as to whether a component is in planned outage has been randomly determined in selecting the precontingency system state. Selection of contingencies depends on the forced failure probability of the components. It should be noted that this probability is not the forced unavailability of components given in Equation (5.23), which is calculated by both failure and repair rates. In the probabilistic voltage stability assessment, we are concerned about only occurrence of a forced failure event but not a subsequent repair (or recovery) process because as long as it occurs, the system may lose voltage stability in a very short time, which has nothing to do with repairing. The probability of occurrence of a contingency event can be estimated using the Poisson distribution with a constant occurrence rate. Using the Poisson distribution formula, the probability of no contingency occurring in the time period t is given by

$$P_{no} = \frac{e^{-\lambda_o t}(\lambda_o t)^0}{0!} = e^{-\lambda_o t} \tag{5.58}$$

where P_{no} is the probability of no contingency occurrence, λ_o is the average occurrence rate of a contingency, and t is the duration considered.

The probability of a failure or contingency occurring in t is

$$P_o = 1 - e^{-\lambda_o t} \tag{5.59}$$

Obviously, Equation (5.59) is simply an occurrence probability following the exponential distribution. In fact, the Poisson and exponential distributions are essentially consistent since both are based on the constant rate assumption. Equation (5.59) is applied to all the components in the crucial set. Each component has a different average contingency occurrence rate, which can be calculated from historical fault records.

5.6.4 Assessing Average Voltage Instability Risk

The procedure of assessing voltage instability risk using the Monte Carlo simulation includes the following steps:

1. Precontingency system states are selected using the Monte Carlo simulation methods described in Section 5.6.3.1. As mentioned earlier, the forced unavailability of components in the crucial set is excluded in selection of precontingency system states.
2. Contingencies (forced failures) of the components in the crucial set that are not in a planned outage status during the selection of precontingency system states are randomly determined. A uniformly distributed random number R in the interval [0,1] is created for each component or component group (simultaneous multiple component failures) in the crucial set. If $R < P_o$, the contingency occurs; otherwise the contingency does not occur.
3. The optimization model in Section 5.6.1 is used to create a solution point for the selected contingency, which is either a solution without load curtailment or a critical solution with minimum load curtailment.
4. The reduced Jacobian matrix method outlined in Section 5.6.2 and detailed in Section 4.7.2 is used to judge whether each of the sampled contingency states is voltage-stable.
5. The following two voltage instability risk indices are calculated:

$$\text{PVI} = \frac{m}{M} \tag{5.60}$$

$$\text{ELCAVI} = \sum_{j=1}^{M} \sum_{i=1}^{N_j} \frac{C_{ij}}{M} \tag{5.61}$$

where PVI is the probability of voltage instability, ELCAVI is the expected load curtailment to avoid voltage instability, m is the number of system states that are voltage-unstable, M is the total number of sampled system states, C_{ij} is the load curtailment at the ith bus in the jth sampled system state, and N_j is the number of buses in the jth system state. Note that the number of buses may be changed since isolated buses may occur as a result of branch outages in a sampled system state. The PVI index represents the average likelihood of voltage instability occurring in various contingencies. The ELCAVI index represents the average load curtailments required to prevent voltage instability in various contingencies. The load curtailments at isolated buses are not included in the ELCAVI index, as these are not considered as consequences to avoid system voltage instability.

6. Steps 1–5 are repeated until a variance coefficient of the PVI or ELCAVI index is smaller than a specified threshold.

The voltage instability risk indices can be used as system performance indicators in transmission planning along with other adequacy indices. For instance, an acceptable PVI index target may be established after considerable studies for a system are conducted. If the PVI index for a future year in planning is higher than the target, this indicates that the voltage stability performance of the system has deteriorated and some reinforcement is needed. The difference in the PVI or ELCAVI index before and after a reinforcement project represents the improvement in system voltage stability due to the project.

5.7 PROBABILISTIC TRANSIENT STABILITY ASSESSMENT

The purpose of probabilistic transient stability assessment in transmission planning is to evaluate the average transient instability risk under various fault events. Similar to the probabilistic voltage stability assessment, the probabilistic transient stability assessment needs to evaluate both probabilities and consequences of a huge number of fault events. The probability and consequence of a fault depend on multiple factors, including prefault system conditions, fault location and types, protection schemes, and disturbance sequences [79,80]. The consequence evaluation requires the simulation of transient stability and impact analysis. The Monte Carlo simulation method for probabilistic transient stability assessment is discussed in this section.

5.7.1 Selection of Prefault System States

Selection of prefault system states in the probabilistic transient stability assessment is similar to that in the probabilistic voltage stability assessment. The state sampling method for generation and transmission components in Section 5.5.3.1 is used to determine network topology and available generation capacities in prefault system states. As mentioned in Section 5.6.3.1, it is important to appreciate that the unavailability of forced failures of the components in the crucial set is excluded in calculating the probability of a prefault system state since these failures are the fault events but not a portion of a prefault state. The modeling methods for load curves and bus load uncertainty and correction in Section 5.5.1 are used to select bus load states.

5.7.2 Fault Probability Models

The uncertainty of fault events is represented using five models.

5.7.2.1 Probability of Fault Occurrence. The probability of fault occurrence is represented using the same model as described in Section 5.6.3.2.

5.7.2.2 Probability of Fault Location. A fault can occur either at a bus or on a line. If it occurs at a bus, its location is this bus. If it occurs on a line, the fault location can be modeled using a discrete probability distribution, which can be derived from historical data. A line is divided into M_L segments, and the probability of a fault occurring in the ith segment is calculated by

$$P_i = \frac{f_i}{\sum_{i=1}^{M_L} f_i} \tag{5.62}$$

where f_i is the number of the faults occurring in the ith segment in historical fault records.

The number of segments may be varied for different lines and depends on data available. For instance, a line can be divided into the following three segments with respect to its master end:

- Close-end (first 20% of the line)
- Midline (middle 60% of the line)
- Far-end (last 20% of the line)

5.7.2.3 Probability of Fault Type. Fault types can be classified into the following categories:

- Single-phase-to-ground
- Double-phase-to-ground
- Three-phase
- Phase-to-phase

Similarly, a discrete probability distribution for the fault type can be obtained from historical fault data. The formula has the same form as Equation (5.62) with the following definition changes: P_i is the probability of the ith fault type, f_i is the number of the ith fault type, and $M_L = 4$.

5.7.2.4 *Probability of Unsuccessful Automatic Reclosure.* Most high-voltage overhead lines are equipped with automatic reclosure devices. A "false" fault will not create any severe consequence if the automatic reclosure is successful. The second fault clearing is needed only when the automatic reclosure fails. Conceptually, the probability of unsuccessful automatic reclosure may be obtained from historical data. However, many data collection systems do not record the information on the relative success of reclosure but do provide a description of the cause of faults. Generally, there is a strong correlation between the cause of faults and the probability of successful automatic reclosure. For instance, a typical statistic analysis indicates that the automatic reclosure is successful for over 90% of the faults caused by lightning but for only about 50% of the faults due to other causes. The probability of unsuccessful automatic reclosure can be estimated using the conditional probability concept:

$$P_{ru} = P(L)P(U|L) + P(O)P(U|O) \tag{5.63}$$

where P_{ru} is the probability of unsuccessful automatic reclosure; $P(L)$ and $P(O)$ are the probabilities of faults, respectively, due to lightning and other causes, which can be easily obtained from historical records; and $P(U|L)$ and $P(U|O)$ are the conditional probabilities of unsuccessful reclosure given that the fault is caused by lightning and other causes, respectively, which can be estimated through a statistic analysis. The equation can be extended to distinguish more fault causes if there are sufficient data records.

5.7.2.5 *Probability of Fault Clearing Time.* The process of clearing a fault is composed of three tasks: fault detection, relay operation, and breaker operation. The fault detection can be assumed to be completed instantly. The times of relay and breaker operations are both random. There are two modeling approaches for the fault clearing time:

- The relay and breaker operation times are modeled using two separate probability distributions, and the probability distribution of the total fault clearing time is obtained by a convolution.
- The total fault clearing time is assumed to directly follow a probability distribution.

Generally, a normal distribution assumption is used for the relay or breaker operation time, or the total fault clearing time. The estimates of the mean and standard deviation for the normal distribution can be determined from historical data. A system loses transient stability only when the fault clearing time is longer than the critical clearing time.

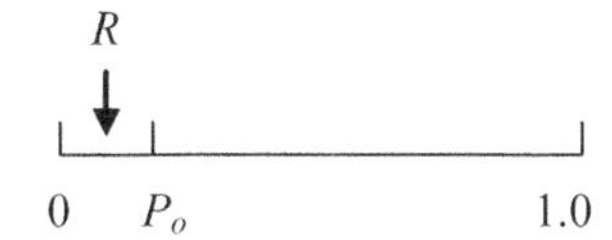

Figure 5.7. Sampling fault occurrence.

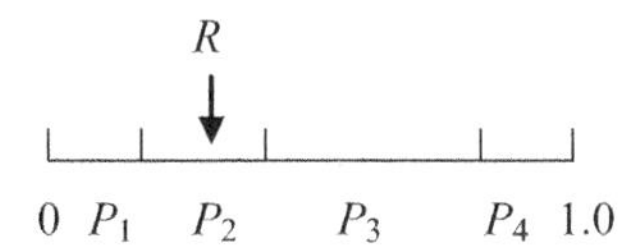

Figure 5.8. Sampling fault type.

5.7.3 Selection of Fault Events

Selection of fault events depends on the five fault probability models described above. Monte Carlo simulations of fault events can be methodologically classified into the following three categories:

1. *Sampling with Two Possibilities.* This category includes determination of a fault occurrence or a reclosure failure. Consider the fault occurrence as an example. A uniformly distributed random number R in the interval [0,1] is created for each bus or line to be considered. If $R < P_o$, the fault occurs; otherwise the fault does not occur. This is shown in Figure 5.7.
2. *Sampling with Multiple Possibilities.* This category includes determination of a fault location or a fault type. Consider the fault type as an example. The probability values of the four fault types are successively placed in the interval [0,1], as shown in Figure 5.8. A uniformly distributed random number R in the interval [0,1] is created. The location of R denotes which fault type is randomly selected in sampling.
3. *Sampling with a Normally Distributed Random Variable.* The fault clearing time sampling belongs to this category, which includes the following two steps:
   - Create a standard normally distributed random number X using the method described in Section A.5.4.2 of Appendix A.
   - The random fault clearing time is calculated by

$$\tau_c = X\sigma + \mu \tag{5.64}$$

where μ and σ are, respectively, the mean and standard deviation of the fault clearing time.

5.7.4 Transient Stability Simulation

For a sampled fault event, the transient stability time-domain simulation is performed to determine whether it causes system instability. The number of transient stability simulations is very large in the probabilistic transient stability assessment since prefault system states and fault events are randomly selected. The simulation methods have been briefly described in Section 4.8. In fact, commercial programs for transient stability can be used. Considerable techniques have been developed to speed up the simulation process, such as the variable step algorithm for differential equations, early termination criteria, and second kick method.

It is important to appreciate that a disturbance sequence must be specified in the transient stability simulation. The disturbance sequence is varied for different fault events and is usually defined in operation criteria of a control center. Various special protection systems (SPSs) are installed in many utility systems. These include a variety of remedial action schemes (RASs) such as the generation rejection, switching of VAR equipment, line tripping, and load shedding. Any remedial actions should be also defined in the disturbance sequence.

5.7.5 Assessing Average Transient Instability Risk

The procedure of assessing transient instability risk using the Monte Carlo simulation includes the following steps:

1. Prefault system states are selected using the Monte Carlo simulation methods described in Section 5.7.1. This includes the determination of network topology, generation unit states, and bus loads.
2. Fault events are randomly selected using the Monte Carlo simulation methods described in Section 5.7.3. This includes the determination of occurrence, location, and types of faults; successfulness of automatic reclosure; and fault clearing time. Note that only those lines and generating units in the crucial set that are not in a planned outage status during selection of prefault system states are considered in selecting occurrence of faults. See Section 5.6.3.1 for the concept of the crucial set.
3. A disturbance sequence in transient stability simulation is specified for each fault. If a special protection system (SPS) is installed for a fault, the remedial actions defined in the SPS are included in the disturbance sequence.
4. A time-domain transient stability simulation is performed for each fault event.
5. A consequence analysis is conducted for each fault event. This is the evaluation of damage costs due to transient instability or remedial actions and is associated with the following aspects:
   - Consequence of system instability
   - Load shedding
   - Generation rejections
   - Equipment damages due to transient overvoltage
   - Transfer limit reduction that may lead to losses of revenue
   - Penalties for violation on criteria or agreements

6. The following two transient instability risk indices are calculated:

$$\mathrm{PTI} = \frac{m}{M} \tag{5.65}$$

$$\mathrm{ECDTIR} = \frac{\sum_{i=1}^{m} R_{fi} + \sum_{i=1}^{M-m-n} R_{si}}{M} \tag{5.66}$$

where PTI is the probability of transient instability, ECDTIR is the expected cost damage due to transient instability risk, m is the number of faulted system states that are unstable, n is the number of faulted system states that are stable without any remedial action, M is the total number of sampled system states, R_{fi} is the cost damage estimate for the ith faulted system state that loses transient stability, and R_{si} is the cost damage estimate for the ith faulted system state that becomes stable because of a remedial action of special protection system. The cost damage R_{si} may be caused by a remedial action such as load shedding or generation rejection, a possible equipment problem, economic loss due to transfer limit reduction, or penalty due to criterion violation. The PTI index represents the average likelihood of the system losing transient stability in various fault events. The ECDTIR index represents the average cost damage caused by various fault events. Either item in Equation (5.66) can be separately calculated to obtain a subindex of ECDTIR, which corresponds to one of the two different cost damages.
7. Steps 1–6 are repeated until a variance coefficient of the PTI or ECDTIR index is smaller than a specified threshold.

The two indices can be evaluated with and without applying a special protection system (SPS). The difference between the two cases represents the effect of the SPS. An ideal situation is one in which special protection systems installed in the system are so complete and flawless that no transient instability occurs in sampled system faults. In this case, the first term in Equation (5.66) becomes zero and the ECDTIR index becomes the expected cost to avoid transient instability.

It should be pointed out that the failure probability of SPS has not been included in the assessment process described above. A special protection system includes various components (sensors, communication channel, computer software/hardware, and acting devices) and can fail with a probability. The effect of a SPS failure can be easily incorporated in the assessment process by sampling its failure probability. The sampling method is similar to that used for a fault occurrence or reclosure failure, as shown in Figure 5.7. Note that the failure probabilities of the special protection systems for different contingencies must be separately evaluated in advance.

The transient instability risk indices can be used as system dynamic performance indicators in transmission planning. For example, an acceptable PTI index target may be established. If the PTI index for a future year in planning is higher than the target, this indicates that the transient stability performance of the system needs to be improved by either reinforcing the system or adding a new SPS. The difference in the PTI or

ECDTIR index before and after a reinforcement or SPS project represents the improvement of system transient stability due to the project.

5.8 CONCLUSIONS

This chapter discussed the probabilistic reliability evaluation in transmission planning. Reliability of transmission systems includes adequacy and security. Adequacy reflects the static ability of the system to deliver power and energy from sources to load buses and is associated with network connectivity and power flow constraints. Security relates the ability of the system in response to dynamic disturbances and is associated with voltage and transient stability.

Reliability indices are the outcome of quantified reliability evaluation and represent the system risk level. The basic definitions of adequacy and security indices are given. It should be appreciated that the definitions provide only the general concepts of the indices. The reliability indices have concrete expressions when different evaluation techniques are used.

Reliability worth assessment provides a reliability measure in a monetary value. This is extremely important and useful in the economic analysis of planning projects as it places system reliability on a scale comparable with those of other economic components. The basic idea is to assess the unreliability costs using the unit interruption cost times the EENS index. The four approaches for estimating the unit interruption cost have been described, in which the customer damage function approach is the most poplar one. It should be noted that this is not a unique idea. For example, it has been pointed out in Section 5.7.5 that the unreliability costs can include those due to equipment damage, transfer limit reduction, and penalty.

Substation adequacy evaluation, composite generation–transmission adequacy evaluation, and probabilistic voltage and transient stability assessments are illustrated in detail. There are two fundamental reliability evaluation techniques: Monte Carlo simulation and state enumeration. The Monte Carlo methods have been used in composite system adequacy evaluation and probabilistic voltage and transient stability assessments, whereas the state enumeration method has been used in substation reliability evaluation. Undoubtedly, the Monte Carlo methods can also be applied to substation reliability evaluation, and the state enumeration methods can also be utilized in composite system evaluation. More materials can be found in References 6 and 10.

The common feature in probabilistic adequacy and security assessments is evaluation of both probabilities and consequences, although there are varied characteristics in different cases. In the substation adequacy evaluation, the component failure models are relatively complicated, particularly when breaker stuck conditions are considered. However, system analysis for the substation adequacy evaluation is straightforward since it is associated only with network connectivity identification. In composite system adequacy evaluation, the component failure models can be simplified, but the consequence assessment requires system contingency analyses and the optimization model for load curtailments. In probabilistic voltage and transient stability assessments, the determination of system states includes selections of precontingency states and contingencies or faults. The selection techniques of precontingency states are similar to those used in the composite system adequacy evaluation, whereas the selection of contingencies or faults depends on their probability models. A technique of combining the optimization model with the reduced Jacobian matrix method has been developed to recognize power flow solvability and to identify voltage instability. This technique can avoid the calculations of a sequence of power flow solutions for gradually approaching the collapse point, which are imperative in the continuation power flow method. Special protection systems (SPSs) or remedial-action schemes (RASs) are often used for mitigating or eliminating system instability. Including or excluding SPS in the probabilistic transient stability assessment will have very different effects on the instability risk indices. It should be also emphasized that a SPS is fault-specific and can fail itself.

Reliability evaluation is one of the most crucial steps in probabilistic transmission planning. More developments associated with reliability evaluation will be discussed further in the subsequent chapters for actual planning applications.

6

ECONOMIC ANALYSIS METHODS

6.1 INTRODUCTION

Economic analysis and reliability evaluation are two essential aspects in transmission planning. Probabilistic reliability evaluation has been discussed in Chapter 5. A number of references in engineering economics are available [81–84]. The intent of this chapter is to review the useful concepts of engineering economics, to extend them to applications in transmission planning, and particularly to incorporate unreliability costs in overall economic analysis models. This incorporation is a distinguishing feature for probabilistic transmission planning.

As mentioned in Chapter 2, there are three cost components in a transmission planning project: capital investment, operation cost, and unreliability cost. The composition of capital investment and operation cost, and the basic nature of unreliability cost in the economic analysis are outlined in Section 6.2. One of the most important concepts in engineering economics is the time value of money. Various costs are comparable only when they are valued at the same point of time. This concept is applicable for all the three cost components. The time value of money and the present value method are illustrated in Section 6.3. Depreciation is another important concept associated with economic analysis of capital investment and is discussed in Section 6.4. Economic

Probabilistic Transmission System Planning, by Wenyuan Li

assessment of investment projects is the fundamental analysis in selection and justification of portfolios or alternatives of a project and is addressed in Section 6.5. Retirement of equipment has significant impacts on system reliability, and therefore a decision on equipment replacement often requires system planning studies. Economic analysis of equipment replacement is discussed in Section 6.6. The uncertainties of parameters in economic analysis are the origin of financial risks. The probabilistic analysis method used to deal with uncertain factors in the economic analysis is explained in Section 6.7.

6.2 COST COMPONENTS OF PROJECTS

6.2.1 Capital Investment Cost

The capital investment of a project includes the following subcomponents:

1. *Direct Capital Costs.* These costs include
   - Equipment costs.
   - Costs for transportation, installation, and commissioning of equipment.
   - Land and right-of-way costs.
   - Removal costs of existing facilities.
   - Outsource costs for design and other services.
   - Contingency costs (unexpected costs).
   - Salvage values—these are negative costs if previous facilities have the residual values that can be utilized in a new project.
2. *Corporate Overhead.* This cost can be expressed as a percentage of total direct capital cost. It represents an allocation of additional corporate costs among capital projects, which are not included in the total direct capital cost.
3. *Financial Cost.* These costs consist of
   - *Interest during Construction (IDC).* The IDC reflects the cost of borrowing to finance the project for the period during which expenditures are made on the project before it is in service. This cost is also expressed in a percentage of the total direct capital cost. If the period during construction is included in the cash flow for present value calculation, it should be excluded from the estimate here to avoid a double count.
   - *Purchase and Service Taxes (PST).* The PST is the government's taxes on purchase of equipment and services, which is a percentage of the purchase price.

6.2.2 Operation Cost

The operation cost is estimated on an annual basis and includes the following subcomponents:

1. *Operation, Maintenance, and Administration (OMA) Cost.* The annual operating cost includes incremental system operation expenditures due to the project. A project may increase or reduce system operation expenditures. If increased,

it is a cost; if reduced, it is a negative cost or benefit. The *annual maintenance cost* refers to the yearly expenditure that will be made over the lifecycle of equipment, including repairs, overhauls, replacements of parts, and decommissioning. The annual administration cost is general administrative expenses.

2. *Taxes.* Annual taxes may include property and income taxes. Calculation of property taxes is often based on a percentage of the book value (a capital cost after depreciation) or a market value. *Income taxes* refer to those taxes charged to the project, such as sales tax. It should be noted that the taxes vary for different countries or provinces depending on their tax policies.
3. *Network Losses.* Some alternatives of a transmission project may either reduce or increase network losses. In the case of reduction, reduced losses can be treated as negative costs or benefits. The reduction or increase in network losses includes two implications: (1) it represents the reduction or increase of generation capacity (MW) requirement; (2) it also represents the reduction or increase of energy consumption (MWh). Utilities provide two unit values for network losses: dollars per MW and dollars per MWh. Accordingly, both elements should be included in calculating the cost due to loss increase or the benefit due to loss reduction. It should be emphasized that the allocation of the cost or benefit associated with network losses between a utility and its customers is a complex problem under the deregulated power market environment. This depends on the market model. The loss-related cost or benefit should be allocated or assigned with caution.

6.2.3 Unreliability Cost

The unreliability cost represents the damage cost caused by system component outages (or faults). This cost is the result of probabilistic reliability worth assessment, and its basic concept has been discussed in Chapter 5. It should be emphasized that the unreliability cost due to equipment is an incremental cost and its calculation method is varied in different cases. For example, the unreliability cost due to an equipment addition is the difference in the total system unreliability cost between the two cases with and without the addition, whereas that due to an equipment replacement is the difference in the total system unreliability cost between the two cases with old and new equipment. Note that an equipment addition or replacement generally leads to reduction of system unreliability cost, and therefore the unreliability cost due to equipment addition or replacement is usually a negative cost (benefit).

6.3 TIME VALUE OF MONEY AND PRESENT VALUE METHOD

More details of these topics can be found in general textbooks of engineering economics [81].

6.3.1 Discount Rate

The value of money varies over time. This concept is called the *time value of money*. If one borrows money from a bank, one pays interests to the bank. If one deposits money to a bank, one receives interests from the bank. The interests are generally increased with time. An interest rate is used to calculate the interests. Inflation also affects the time value of money. The nominal interest rate includes two components: (1) the *real interest rate*, which reflects the value of "waiting"—an investor should be compensated because his/her funds cannot be used by him/herself while they are used by a borrower; and (2) the *inflation rate*, which reflects the fact that inflation creates devaluation of money. One investment is compensated using an inflation rate in order to keep its original value.

In engineering economics, *discounting* refers to a calculation to convert a cost in the future to the present value. The rate used in discounting can be the nominal interest rate or the real interest rate. This rate is called the *discount rate*. If the real interest rate is used, the effect of inflation is excluded in discounting, and we call it the real discount rate. If the nominal interest rate is used, the effect of inflation is included, and we call it the *nominal discount rate*. There exists the following relationship:

$$(1 + r_{\text{rkf}}) = (1 + r_{\text{int}})(1 + r_{\text{inf}}) \tag{6.1}$$

where r_{rkf}, r_{int}, and r_{inf} are the risk-free nominal interest rate, real interest rate, and inflation rate, respectively.

From Equation (6.1), the risk-free nominal interest rate can be calculated by

$$r_{\text{rkf}} = r_{\text{int}} + r_{\text{inf}} + r_{\text{int}} \cdot r_{\text{inf}} \tag{6.2}$$

If both the real interest rate and the inflation rate are low, the product term in Equation (6.2) becomes negligible and therefore

$$r_{\text{rkf}} \approx r_{\text{int}} + r_{\text{inf}} \tag{6.3}$$

Although Equation (6.3) is often used, it should be remembered that the product term cannot be neglected when the real interest rate and/or inflation rate are high.

There always exists a financial risk for investments. Therefore a risk premium rate should be added to the risk-free nominal interest rate to obtain the risky nominal interest rate r_{rk}:

$$r_{\text{rk}} = r_{\text{rkf}} + r_{\text{prm}} \tag{6.4}$$

The real interest rate r_{int} is also a risk-free rate. Any bank has a risk in its financial activities. Similarly, the risky real interest rate can be obtained by adding a risk premium rate. It should be noted that the risk premium rates for the risky real interest rate and risky nominal interest rate may have different values.

6.3.2 Conversion between Present and Future Values

As mentioned earlier, money today has a different value from the same amount one year ago or one year later. In economic analysis for transmission planning projects, the discount rate r is used to calculate the present value of any cost component. The discount rate r can be a nominal or real interest rate depending on whether the effect of inflation is considered. It can also be a risky rate or risk-free rate depending on whether a project is risk-related or risk-free. Obviously, there is the following relationship between the prevent value PV_0 and the value one year later FV_1:

$$FV_1 = (1+r)\cdot PV_0 \tag{6.5}$$

$$PV_0 = (1+r)^{-1}\cdot FV_1 \tag{6.6}$$

The relationships in Equations (6.5) and (6.6) can be extended to the relationship between the present value and the value of n years later FV_n:

$$FV_n = (1+r)^n \cdot PV_0 \tag{6.7}$$

$$PV_0 = (1+r)^{-n} \cdot FV_n \tag{6.8}$$

The $(1+r)^n$ is called the *end value factor* for a single present value and is often denoted by (F|P, r, n). It is used to calculate the end value FV_n in year n from the present value PV_0 for the given discount rate r and year n. The $(1+r)^{-n}$ is called the *present value factor* for a single future value and often denoted by (P|F, r, n). It is used to calculate the present value PV_0 in the current year from the end value FV_n for the given discount rate r and year n.

6.3.3 Cash Flow and Its Present Value

The operation and unreliability costs of a transmission project can be evaluated year by year. In many cases, investment capital costs are incurred only in the first year or a few years at the beginning of a project. In other cases, the investment costs can be incurred in different years (multistage investment projects). Each investment can be distributed to annual equivalent costs over the years in its useful life. The annual costs in a period form a cash flow. Figure 6.1 indicates the three cases of cash flows. Figure 6.1a shows a cash flow with equal annual values; Figure 6.1b, a cash flow with increasing annual values; and Figure 6.1c, a general cash flow pattern in which the annual values can fluctuate up and down or even become negative. For example, when net benefits (gross benefits minus costs) are used, an annual value in the net benefit cash flow may be either positive or negative. A positive value represents a net benefit, whereas a negative value represents a net cost. In economic analysis for transmission planning, two approaches can be applied. In the first one, the cash flows of costs and gross benefits are created separately first and then a benefit/cost analysis is performed using the present values of the two cash flows. In the second one, the annual net benefit, which equals to the annual benefit minus the annual cost, is calculated first to form a cash flow, and then the present value of the net benefit cash flow is calculated.

Each annual value A_k ($k = 1,\ldots,n$) on a cash flow can be discounted to its present value using Equation (6.8), and the total present value of the cash flow over n years is expressed by

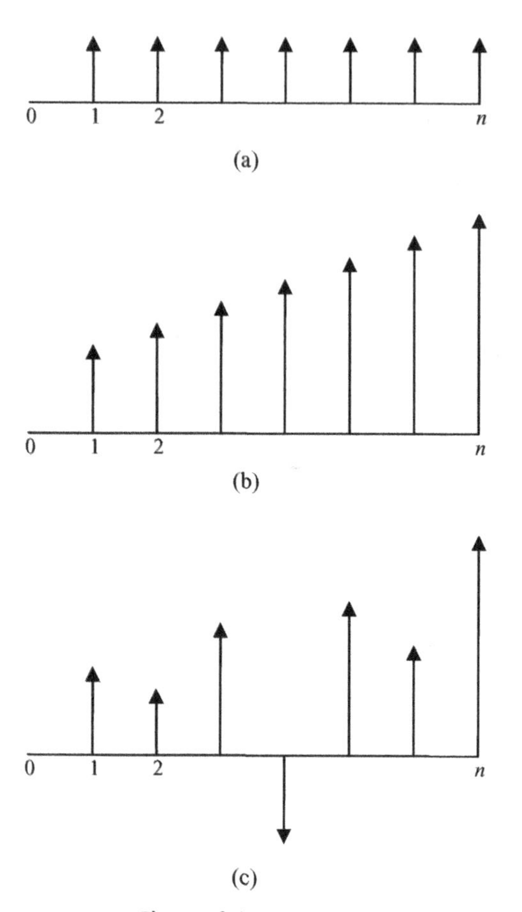

Figure 6.1. Cash flows.

$$\mathrm{PV}_0 = \sum_{k=1}^{n} A_k \cdot (1+r)^{-k} = \sum_{k=1}^{n} A_k \cdot (\mathrm{P}|\mathrm{F}, r, k) \tag{6.9}$$

Similarly, each annual value on a cash flow can also be converted to the future value in year n, and the total future value in year n is calculated by substituting Equation (6.9) into Equation (6.7):

$$\mathrm{FV}_n = \sum_{k=1}^{n} A_k \cdot (1+r)^{n-k} = \sum_{k=1}^{n} A_k \cdot (\mathrm{F}|\mathrm{P}, r, n-k) \tag{6.10}$$

6.3.4 Formulas for a Cash Flow with Equal Annual Values

The cash flow with equal annual values as shown in Figure 6.1a is often used to represent equivalent annual investment costs for a capital project. In this case, the simplified relationship between the present, future, and annual values can be obtained.

6.3.4.1 *Present Value Factor.* Assuming that each year has the equal (or equivalent) annual value A, Equation (6.9) becomes

$$\mathrm{PV}_0 = A \cdot \sum_{k=1}^{n} (1+r)^{-k} \tag{6.11}$$

Multiplying $(1 + r)$ on both sides of Equation (6.11) yields

$$\mathrm{PV}_0(1+r) = A \cdot \sum_{k=1}^{n} (1+r)^{1-k} \tag{6.12}$$

Subtracting Equation (6.11) from Equation (6.12), we have

$$\mathrm{PV}_0 \cdot r = A \cdot \left[1 - \frac{1}{(1+r)^n}\right] \tag{6.13}$$

Equivalently

$$\mathrm{PV}_0 = A \cdot \left[\frac{(1+r)^n - 1}{r \cdot (1+r)^n}\right] \tag{6.14}$$

The $[(1+r)^n - 1]/[r \cdot (1+r)^n]$ is called the *present value factor* for an equal annual value cash flow and is often denoted by (P|A,r,n). It is used to calculate the present value PV_0 in the current year from the equal annual values in the cash flow for the given discount rate r and the length of cash flow (n years).

6.3.4.2 *End Value Factor.* Substituting Equation (6.14) into Equation (6.7) yields

$$\mathrm{FV}_n = A \cdot \left[\frac{(1+r)^n - 1}{r}\right] \tag{6.15}$$

The $[(1+r)^n - 1]/r$ is called the *end value factor* for an equal annual value cash flow and often denoted by (F|A,r,n). It is used to calculate the end value FV_n in year n from the equal annual values in the cash flow for the given discount rate r and the length of cash flow (n years).

6.3.4.3 *Capital Return Factor.* Equation (6.14) can be expressed in its inverse form:

$$A = \mathrm{PV}_0 \cdot \left[\frac{r \cdot (1+r)^n}{(1+r)^n - 1}\right] \tag{6.16}$$

The $[r \cdot (1+r)^n]/[(1+r)^n - 1]$ is called the *capital return factor* (CRF) and is often denoted by (A|P,r,n). It is used to calculate the annuity (equal or equivalent annual value) A from the present value for the given discount rate r and the length of cash flow (n years).

When n is sufficiently large, the annuity approaches to perpetuity and the CRF approaches to the discount rate r so that

$$A = \mathrm{PV}_0 \cdot r \tag{6.17}$$

This approximation is often acceptable for a quick estimation of the equivalent annual cost in transmission planning since the life of transmission equipment is sufficiently long (30–60 years).

The financial significance of the CRF is that if an investment (PV_0) has a useful life of n years, the average annual return should not be lower than A. Otherwise, the primary and interests of the investment cannot be recovered within n years. The CRF has the following two applications in transmission planning:

- The equivalent annual investment cost for a capital investment project can be calculated using Equation (6.16).
- Even for a cash flow with unequal annual costs (such as the annual operation or unreliability cost of a capital project), once the present value of the cash flow is obtained using Equation (6.9), an equivalent equal annual cost can be estimated.

6.3.4.4 Sinking Fund Factor. Similarly, Equation (6.15) can be expressed in its inverse form:

$$A = \mathrm{FV}_n \cdot \left[\frac{r}{(1+r)^n - 1} \right] \tag{6.18}$$

The $r/[(1+r)^n - 1]$ is called the *sinking fund factor* (SFF) and is often denoted by (A|F,r,n). It is used to calculate the annuity (equal or equivalent annual value) A from the end value FV_n in year n for the given discount rate r and the length of cash flow (n years).

The general financial significance of the SFF is that if a fund (FV_n) is needed in the future year n, the average annual amount collected should not be lower than A. Otherwise, the fund cannot be achieved by the end of year n.

6.3.4.5 Relationships between the Factors. The six factors for dealing with the time value of money were presented above and are summarized as follows:

The present value factor for a single future value: $(\mathrm{P}|\mathrm{F}, r, n) = (1+r)^{-n}$

The end value factor for a single present value: $(\mathrm{F}|\mathrm{P}, r, n) = (1+r)^{n}$

The present value factor for an annuity cash flow: $(\mathrm{P}|\mathrm{A},r,n)=\left[\frac{(1+r)^n-1}{r\cdot(1+r)^n}\right]$

The capital return factor for an annuity cash flow: $(\mathrm{A}|\mathrm{P},r,n)=\left[\frac{r\cdot(1+r)^n}{(1+r)^n-1}\right]$

The end value factor for an annuity cash flow: $(\mathrm{F}|\mathrm{A},r,n)=\left[\frac{(1+r)^n-1}{r}\right]$

The sinking fund factor for an annuity cash flow: $(\mathrm{A}|\mathrm{F},r,n)=\left[\frac{r}{(1+r)^n-1}\right]$

The factor symbols imply that the quantity in front of the sign "|" in a symbol is the product of the quantity after the sign "|" and the factor. For example, (P|F,r,n) indicates that the present value (P) is the product of the future value (F) and (P|F,r,n), (A|P,r,n) indicates that the equal annual value (A) is the product of the present value (P) and (A|P,r,n), and so on. The r and n in the symbol indicate that the factor is a function of r and n.

It is apparent that there are product relationships between the factors. A factor equals to the product of the other two. For example

$$(\mathrm{A}|\mathrm{P},r,n)=(\mathrm{A}|\mathrm{F},r,n)\cdot(\mathrm{F}|\mathrm{P},r,n) \tag{6.19}$$

$$(\mathrm{P}|\mathrm{F},r,n)=(\mathrm{P}|\mathrm{A},r,n)\cdot(\mathrm{A}|\mathrm{F},r,n) \tag{6.20}$$

Besides, there is the following relationship between the CRF and SFF:

$$(\mathrm{A}|\mathrm{P},r,n)=(\mathrm{A}|\mathrm{F},r,n)+r \tag{6.21}$$

6.4 DEPRECIATION

6.4.1 Concept of Depreciation

Any equipment has its lifetime. It ages and wears out with time. The term *depreciation of equipment* refers to the financial calculation through which the initial value of equipment is gradually transferred to expenses by annual decreases in its book value. In engineering economics, it is called *accounting depreciation.* Depreciation is necessary because it reflects the wearing process of equipment; it is required for the purpose of taxation and it is a means to collect the fund for equipment maintenance and replacement since the depreciated values become the cost of products.

The following three input data are required in the depreciation calculation:

- Basic cost of an asset
- Salvage value
- Depreciable life

The *basic cost of an asset* is the total cost incurred over its lifecycle, including the initial investment cost of the asset and maintenance cost over its life. The *salvage value* is an estimated value at the time of disposal. It is the value recovered through sale or reuse. If an asset does not have any value at disposal, the salvage value can be specified to be zero. It is not easy to accurately estimate the salvage value. Besides, adjustments can be made in the process of depreciation if necessary.

There are the following four concepts of equipment life [84,85]:

- *Physical Lifetime.* A piece of equipment starts to operate from its brand new condition to a status in which it can no longer be used in the normal operating state and has to be retired. Maintenance activities can prolong its physical lifetime.
- *Technical Lifetime.* A piece of equipment may have to be replaced because of a technical progress, although it may still be used physically. For example, a new technology is developed for a type of equipment, and manufacturers no longer produce spare parts. This may result in such a situation that utilities cannot obtain necessary parts or the parts become too expensive for maintenance. Mechanical protection relays and mercury arc converters in HVDC systems are examples of the technical lifetime.
- *Depreciable Lifetime.* This refers to the timespan specified for depreciation. By the end of the depreciable life, the financial book value of equipment is depreciated to be a small salvage value. The depreciable life is sometimes called the *useful life*. A government specifies the depreciable life in terms of different categories of properties for the purpose of taxation. Utilities specify the depreciable life of various equipments on the basis of their financial model.
- *Economic Lifetime.* By the end of the economic life, using a piece of equipment is no longer economical, although it may be still used physically and technically. This concept will be discussed further in Section 6.6.

Several depreciation methods are discussed in the following subsections. It should be pointed out that maintenance and repairs can have effects on the basic cost and depreciable life. For instance, a maintenance activity may increase the basic cost of equipment and extend its depreciable life. In this case, an appropriate adjustment may be needed at some point of time in the depreciation calculation.

6.4.2 Straight-Line Method

The straight-line method [84] depreciates the value of an asset using an equal fraction of the basic cost in each year during the depreciation, which can be expressed as follows:

$$D_k = \frac{I - S}{m} \qquad (k = 1, \ldots, m) \tag{6.22}$$

where D_k is the depreciated value for year k; I and S are the basic investment cost and salvage value, respectively; and m is the depreciable life (in years).

Obviously, the depreciated value is the same for each year during the depreciable life. The depreciation rate is defined by

$$\alpha = \frac{1}{m} \tag{6.23}$$

In some literature, the depreciation rate in the straight-line method is defined as D_k/I, which may not be accurate since the total value to be depreciated is $I - S$ but not I.

The book value at the end of the kth year is

$$B_k = I - \sum_{j=1}^{k} D_j \qquad (k = 1, \ldots, m) \tag{6.24}$$

6.4.3 Accelerating Methods

In accelerating methods [84], the book value of equipment is depreciated more in early years than in later years. The accelerating methods are often used for the taxation purposes.

6.4.3.1 Declining Balance Method. In this method, the depreciated value in year k is calculated by

$$D_k = \alpha \cdot I(1 - \alpha)^{k-1} \qquad (k = 1, \ldots, m) \tag{6.25}$$

where α is the depreciation rate, I is the basic investment cost, and m is the depreciable life. Note that the concept of the depreciation rate in Equation (6.25) is different from that in the straight-line method.

The total depreciation (TD) up to year k can be calculated by

$$\begin{aligned} \text{TD}_k &= \alpha \cdot I \sum_{j=1}^{k} (1 - \alpha)^{j-1} \\ &= I \cdot [1 - (1 - \alpha)^k] \end{aligned} \tag{6.26}$$

The book value at the end of the kth year is

$$B_k = I - \text{TD}_k = I(1 - \alpha)^k \qquad (k = 1, \ldots, m) \tag{6.27}$$

There are two approaches to determining the depreciation rate:

1. At the end of depreciable life (year m), the book value should be equal to the salvage value. From Equation (6.27), substituting the salvage value S for B_k and m for k yields

$$S = I(1-\alpha)^m \tag{6.28}$$

The depreciation rate can be calculated by

$$\alpha = 1 - \left(\frac{S}{I}\right)^{1/m} \tag{6.29}$$

The deficiency of this approach is that it cannot handle the situation in which the salvage value is zero.

2. The depreciation rate is prespecified. For example, it can be specified as

$$\alpha = \beta \cdot \frac{1}{m} \tag{6.30}$$

where β is a given factor, such as 1.5 or 2.0. This second approach can handle any situation in which the salvage value is or is not zero. In using the approach, however, the book value at the end of depreciable life (year m) may be larger or smaller than the salvage value. If $B_m > S$, an adjustment is conducted by switching to the straight-line method starting from the year in which the depreciated value obtained using the straight-line method is just larger than that obtained using the second approach. If $B_m < S$, the book value is readjusted by stopping the depreciation in some year. This means that if the book value is lower than S in year k, the depreciated value in the kth year is adjusted so that $B_k = S$ and there is no more depreciation after year k.

6.4.3.2 *Total Year Number Method.* The *total year number* (TYN) is defined as follows:

$$\text{TYN} = \sum_{j=1}^{m} j = \frac{m(m+1)}{2} \tag{6.31}$$

where m is the depreciable life.

The depreciated value in year k is calculated by

$$D_k = \frac{m-k+1}{\text{TYN}}(I-S) \qquad (k = 1,\ldots,m) \tag{6.32}$$

The depreciation rate in year k is

$$\alpha_k = \frac{m-k+1}{\text{TYN}} \tag{6.33}$$

Apparently, the depreciation rate is decreased every year. It is the remaining depreciable life divided by the total year number defined in Equation (6.31).

The book value in each year can be calculated using Equation (6.24).

6.4.4 Annuity Method

In the annuity method, the depreciated value in each year is equal to the equivalent annual cost minus the interest of the book value at the end of the previous year. The time value of cost is considered in this method.

The equivalent annual value can be calculated by the present value times the capital return factor, as shown in Equation (6.16). The salvage value is the value at the end of depreciable life (year m) and should be converted to its present value using Equation (6.8). Therefore the equivalent annual value is

$$A = \left[I - S\frac{1}{(1+r)^m}\right] \cdot \left[\frac{r\cdot(1+r)^m}{(1+r)^m - 1}\right] \tag{6.34}$$

where I and S are the basic initial investment cost and salvage value, respectively. Note that r should be the risk-free discount rate. In other words, it should be the r_{rkf} or r_{int} described in Section 6.3.1.

The depreciated value in each year k is calculated by

$$\begin{aligned} D_1 &= A - r\cdot I \\ D_k &= A - r\cdot\left[I - \sum_{j=1}^{k-1} D_j\right] \qquad (k = 2,\ldots,m) \end{aligned} \tag{6.35}$$

This method results in a decelerating depreciation process. The equipment value is depreciated less in early years than in later years.

6.4.5 Numerical Example of Depreciation

Which depreciation method should be used is a financial decision and depends on the purpose and the business model of a company. Table 6.1 presents a numerical example of depreciation using the four different methods. This example is given for illustrative purposes only. The input data are as follows:

- Basic investment cost: $6000
- Salvage value: $100

TABLE 6.1. Depreciated and Balance Values (in $) Using the Four Depreciation Methods

	Straight-Line Method		Total Year Number Method		Declining Balance Method		Annuity Method	
Year	Depreciated Value	Balance	Depreciated Value	Balance	Depreciated Value	Balance	Depreciated Value	Balance
1	590.00	5410.00	1072.73	4927.27	2015.85	3984.15	407.27	5592.73
2	590.00	4820.00	965.45	3961.82	1338.57	2645.58	439.86	5152.87
3	590.00	4230.00	858.18	3103.64	888.85	1756.73	475.04	4677.83
4	590.00	3640.00	750.91	2352.73	590.22	1166.52	513.05	4164.78
5	590.00	3050.00	643.64	1709.09	391.92	774.60	554.09	3610.69
6	590.00	2460.00	536.36	1172.73	260.24	514.35	598.42	3012.27
7	590.00	1870.00	429.09	743.64	172.81	341.54	646.29	2365.97
8	590.00	1280.00	321.82	421.82	114.75	226.79	698.00	1667.98
9	590.00	690.00	214.55	207.27	76.20	150.60	753.84	914.14
10	590.00	100.00	107.27	100.00	50.60	100.00	814.14	100.00

- Depreciable life: 10 years
- Interest rate: 0.08

It can be seen that all the four methods exactly reached the salvage value after 10 years' depreciation. The straight-line method provides the equal depreciated value in each year, the total year number and declining balance methods depreciate more in early years than in later years, and the annuity method depreciates less in early years than in later years.

6.5 ECONOMIC ASSESSMENT OF INVESTMENT PROJECTS

The purpose of conducting an economic assessment of investment projects is to provide an overall economic comparison between alternatives of a project or a ranking list for different projects. This information is the basis for decisionmakers to select the best alternative or project and justify or not justify an investment. There are three assessment methods:

- Total cost method
- Benefit/cost analysis method
- Internal rate of return method

The basic concept of economic assessment has been briefly summarized in Section 2.3.2. This section provides more details. Note that all the methods focus on the relative economic comparison. If a project has a target for the improvement in system reliability or operation efficiency, the target should be used as an additional condition in selecting a project or an alternative of a project.

6.5.1 Total Cost Method

The cash flows of three cost components (investment, operation, and unreliability costs) for each alternative are created. The present value of total cost (PVTC) is calculated by

$$\mathrm{PVTC} = \sum_{k=0}^{n} (I_k + O_k + R_k) \cdot (\mathrm{P}|\mathrm{F}, r, k) \tag{6.36}$$

where I_k, O_k, and R_k are the annual investment, operation, and unreliability costs in year k, respectively; r is the discount rate; and n is the length of cash flows (the number of years considered). Note that (P|F,r,k) denotes the present value factor for a single future value. Its definition and the definition of (A|P,r,n) used in Equation (6.37) have been given in Section 6.3.4.5. Also, note that the costs at the year zero are included in Equation (6.36).

Once the present value is obtained, it can be converted to the equivalent annual value (EAV) by

$$\text{EAV} = \text{PVTC} \cdot (\text{A}|\text{P}, r, n) \tag{6.37}$$

The alternatives can be compared using either PVTC or EAV. It is important to appreciate the following three points:

1. In Equation (6.36), I_k is the annual investment cost required by a reinforcement alternative. There are two approaches for the annual operation cost O_k and the annual unreliability cost R_k. In the first one, O_k and R_k are systemwide costs, meaning that these are the cost estimates for the whole system after an alternative of a project is implemented. In the second one, O_k (or R_k) is the incremental cost caused by a reinforcement alternative, that is, the difference between the system operation (or unreliability) costs with and without the alternative. The first approach requires less calculation effort. Although the total cost obtained using the first approach does not provide any clue about the cost of a project alternative, the total cost is meaningful for comparison purposes as the original system cost without any alternative is automatically canceled out in the comparison. The second approach requires more calculations. This is because the operation or unreliability cost of a reinforcement alternative (such as addition of a line or transformer) is embedded in the system cost. The system analysis for calculating the incremental operation or unreliability cost due to a reinforcement alternative must be performed twice, once with the alternative and then without the alternative. The total cost obtained using the second approach provides information on pure cost of an alternative.
2. The investment cost of an alternative is always positive, whereas the operation or unreliability cost caused by an alternative may be either positive or negative. The operation cost includes different subcomponents as described in Section 6.1, and some subcomponents may carry a negative cost. For example, an alternative may reduce network losses, which can be translated into a benefit or a negative cost. If this benefit is larger than other operation costs, the net operation cost becomes a negative value. In general, a transmission reinforcement alternative improves system reliability, but a system unreliability index cannot be reduced to be zero. Therefore, even if the system unreliability cost is reduced by an alternative, it is still positive when the first approach is used. However, the incremental unreliability cost is normally negative when the second approach is used since the incremental reliability improvement due to an alternative is a benefit. In some cases, a nonreinforcement alternative in transmission planning may lead to deterioration of system reliability. For example, when an independent power producer (IPP) is connected to the system, it may either improve or deteriorate system reliability depending on its location and connection manner.

3. The total cost method can be applied only to the ranking or comparison among alternatives of a project for resolving the same system problem. It cannot be applied to rank or compare projects for resolving different system problems.

6.5.2 Benefit/Cost Analysis

6.5.2.1 Net Benefit Present Value Method. The *net benefit* is the difference between gross benefits and costs. The *net benefit present value* (NBPV) is calculated by

$$\text{NBPV} = \sum_{k=0}^{n} (B_k - C_k - I_k) \cdot (\text{P}|\text{F}, r, k) \tag{6.38}$$

where B_k, C_k, and I_k are the annual benefit, positive operation cost, and investment cost of an alternative in year k, respectively; r is the discount rate; and n is the length of cash flows (the number of years considered). (P|F,r,k) is the present value factor for a single future value.

The annual investment is always the positive cost. The effect of an alternative on operation includes both additional costs and benefits. These two portions are evaluated separately and contribute to C_k and B_k, respectively. Considering the fact that a reinforcement alternative always improves system reliability, the effect of such an alternative on system reliability is assumed to be a part of B_k. The benefit due to the reliability improvement is the reduction in the unreliability cost caused by the alternative. However, if a nonreinforcement alternative causes deterioration in system reliability, this leads to a negative benefit component in B_k. Obviously, each component in C_k or B_k should be an incremental cost or benefit due to the alternative.

An alternative with a negative NBPV cannot be financially justified. An alternative with a larger NBPV is generally better if there is no constraint on the capital investment. However, it should be appreciated that the net benefit present value represents an absolute economic profit but does not reflect a relative economic return. The net benefit present value method can be applied to the comparison among alternatives of a project but may not be appropriate for ranking projects for resolving different system problems.

6.5.2.2 Benefit/Cost Ratio Method. The benefit/cost ratio (BCR) is calculated by

$$\text{BCR} = \frac{\sum_{k=0}^{n} (B_k - C_k) \cdot (\text{P}|\text{F}, r, k)}{\sum_{k=0}^{n} I_k \cdot (\text{P}|\text{F}, r, k)} \tag{6.39}$$

where all the quantities and symbols are the same as defined in Equation (6.38).

The benefit cost ratio (BCR) represents the relative economic return, and the BCR method can be applied to the comparison among alternatives of a project or ranking among projects for resolving different system problems. The larger the BCR is, the better the project or alternative is. An alternative or project with a BCR <1.0 cannot be financially justified. Utilities usually set a threshold value of BCR higher than 1.0 for justification of projects or alternatives. For example, BCR >1.5 or 2 is a frequently used threshold.

6.5.3 Internal Rate of Return Method

In general, an investment project has a positive net benefit present value (NBPV). It can be proved from Equation (6.38) that the NBPV decreases as the discount rate increases. This indicates that if a larger discount rate is used, it will be more difficult to justify a project or alternative. Selecting an appropriate discount rate is an important financial decision for utilities. Utilities select the value of discount rate on the basis of multiple factors, including their business model.

The internal rate of return (IRR) is defined as the break-even discount rate leading to NBPV = 0. In other words, the IRR is the solution r^* of the following equation

$$\text{NBPV} = \sum_{k=0}^{n} (B_k - C_k - I_k) \cdot (\text{P}|\text{F}, r^*, k) = 0 \tag{6.40}$$

where all the quantities and symbols are the same as defined in Equation (6.38), except that r^* is the unknown IRR.

A bisection algorithm can be used to find IRR. Two initial r values are selected such that one leads to a positive NBPV and another to a negative NBPV. The average of these two r values is used to calculate a new NBPV. The original r that causes NBPV to have a sign opposite that of the new NBPV is kept and the other is relinquished. The new r and the retained original r are used in the next iteration until a solution is found.

Mathematically, Equation (6.40) can have multiple solutions if the annual amounts in a net benefit cash flow change their signs (from positive to negative or vice versa) more than once. This is a demerit of the IRR method. However, the majority of actual projects have a net benefit cash flow in which the annual amounts do not change the sign or change it only once, resulting in a unique real number solution of IRR.

If the IRR for a project or alternative is smaller than the minimum discount rate that has been selected by a utility for its financial analysis, the project or alternative is not financially justifiable since this implies that the utility's discount rate will result in a negative NBPV.

When two projects or alternatives are compared, the concept of incremental internal rate of return (IIRR) is introduced. The IIRR is defined as the discount rate that makes the net benefit present values of the two projects or alternatives equal, that is

$$\sum_{k=0}^{n} (B_{1k} - C_{1k} - I_{1k}) \cdot (\text{P}|\text{F}, r', k) - \sum_{k=0}^{n} (B_{2k} - C_{2k} - I_{2k}) \cdot (\text{P}|\text{F}, r', k) = 0 \tag{6.41}$$

where all the quantities and symbols are similar to those defined in Equation (6.38), the subscript 1 or 2 represents the first or second project or alternative, and r' is the IIRR that satisfies Equation (6.41).

Equation (6.41) can be rewritten as follows:

$$\sum_{k=0}^{n}(B_{1k}-B_{2k})\cdot(\mathrm{P}|\mathrm{F},r',k)=\sum_{k=0}^{n}(C_{1k}+I_{1k})-(C_{2k}+I_{2k})\cdot(\mathrm{P}|\mathrm{F},r',k) \qquad (6.41\text{a})$$

Equation (6.41a) indicates that the present value of incremental benefit between the two projects or alternatives is equal to the present value of incremental cost between the two at the IIRR.

If a project or alternative requires a lower cost but creates a higher return (benefit) than the other does, this project or alternative is certainly better. However, a conflict situation may occur: a project or alternative may require a higher cost with a higher return. In this case, the IIRR can be used as a criterion as follows. If r' is larger than the minimum discount rate selected by the utility for its financial analysis, the project or alternative with a higher investment and a higher return is better. Otherwise, if r' is smaller than the utility's discount rate, the project or alternative with a lower investment and a lower return is better.

When multiple projects or alternatives are compared, the following procedure is used:

1. The internal rates of return for all alternatives or projects are calculated. The alternatives or projects with the IRR smaller than the utility's discount rate are not justifiable and are excluded from further consideration.
2. The remaining alternatives or projects are listed in order of increasing investment cost.
3. The first two alternatives or projects are compared using their IIRR, and the better one is determined.
4. The alternative or project with the better IIRR selected in step 3 is compared with the third alternative or project. The process proceeds until all alternatives or projects have been compared.

6.5.4 Length of Cash Flows

The length of cash flows needs to be predetermined in using any of the methods discussed above. It is a challenge issue to determine the number of years to be considered [i.e., n in Equations (6.36)–(6.41)]. In engineering economics, it has been suggested that n should be the useful (depreciable) life of a project or alternative. Unfortunately, there are difficulties in using the useful life as the length of cash flows for transmission planning projects. This is not just because different alternatives or projects may have different useful lives. A more crucial reason is because the useful life of transmission equipment in a project or alternative is generally very long, such as 40–50 years for a transformer and much longer for an overhead line. It is impossible to evaluate operation and unreliability costs of the system over such a long time period since there is no system information (load forecast and other system conditions) available beyond the planning period, which may range from only several to 20 years. Usually, other future projects will be added to the system after a project is implemented but the future projects are unknown at the current time when a project is planned. This results in difficulty or inability to evaluate the effect of a project or alternative on operation and unreliability costs beyond the planning period.

It is suggested that the planning period be used for the length (the number of years) of cash flows for transmission projects. However, the effect of initial total investment cost spreads over the useful life and does not stop at the end of the planning period. When the initial total investment cost is converted into equivalent annual investment costs on a cash flow using the capital return factor, the timespan of this cash flow is still its useful life. Two approaches can be used to deal with this inconsistency. In the first one, the equivalent annual investment costs beyond the planning period are considered by introducing an equivalent salvage value at the end of the planning period. The demerit of doing so is the fact that the effects of a project or alternative on system operation and reliability beyond the planning period are ignored, but the effect of its investment beyond the planning period is still included. This implies that the possible positive effects of a project or alternative on system operation and reliability is underestimated. In the second approach, the effect of investment cost beyond the planning period is also ignored. This approach is normally acceptable in actual applications, particularly for a long planning period since the effect of a salvage value is small. In general, the second approach is better, although an adjustment may be needed for a short planning period in some cases.

6.6 ECONOMIC ASSESSMENT OF EQUIPMENT REPLACEMENT

The majority of system planning projects are associated with equipment addition. However, an economic analysis is also performed on the projects for replacement of existing equipment. Two frequent issues in replacement planning are (1) whether an old or aged facility should be replaced and (2) if so, when it should be replaced. The purpose of economic assessment for equipment replacement is to address these two issues.

6.6.1 Replacement Delay Analysis

The current year is used as a reference point. If a replacement is performed in the current year, the net investment cost is the capital cost of new equipment minus the salvage value of old equipment. If the replacement is delayed by n^* year, there are two impacts. On one hand, the operation and unreliability costs incurred by continuously using the old equipment are higher than those incurred by using the new equipment. We have to pay the increased operation and unreliability costs. On the other hand, we save the interests of the investment cost of new equipment for n^* years. Note that the capital investment is still required n^* later and we cannot save the whole investment by delaying the replacement. The salvage value of old equipment will be decreased and the difference in the salvage value between the current year and year n^* should be taken away from the saved interests. If the increased costs are higher than the saved interests, the replacement should not be delayed further. This idea can be mathematically expressed by

$$\sum_{k=1}^{n^*}(OE_k - OR_k) + (RE_k - RR_k) > \sum_{k=1}^{n^*} i \cdot I \cdot (1+i)^{k-1} - (S_0 - S_{n^*}) \tag{6.42}$$

where OE_k and OR_k are the operation costs required by existing (old) and replacement (new) equipment in year k, respectively; RE_k and RR_k are the unreliability costs incurred by using existing and replacement equipments in year k, respectively; I is the initial total investment cost of replacement equipment; S_0 and S_{n^*} are the salvage values of existing equipment in the current year and year n^*, respectively; i is the risk-free annual real interest rate (i.e., r_{int} discussed in Section 6.3.1); and n^* is the number of years to be determined for replacement delay. The saved interests are calculated by considering a compound rate approach.

The annual operation cost includes the subcomponents that are different in value for existing and replacement equipment (such as maintenance and repair costs), whereas the subcomponents that have the same value for existing and replacement equipment (such as network losses) can be excluded from the annual operation cost. The *annual unreliability cost* refers to the incremental damage cost due to random failures of existing or replacement equipment. Obviously, the existing equipment will cause a higher unreliability cost than will the new equipment because an old facility generally has greater unavailability. A replacement occurs at the stage near the end of life of existing equipment, and the salvage value is always small. In most cases, the term $(S_0 - S_{n^*})$ is negligible.

The minimum n^* that satisfies inequality (6.42) is the number of years for which the replacement should be delayed. The procedure to find the minimum n^* is simple. Let $n^* = 1$ first. If the inequality in Equation (6.42) is not satisfied, let $n^* = 2$ and recheck it, and so on. The process proceeds until the inequality in Equation (6.42) is just satisfied.

6.6.2 Estimating Economic Life

The equivalent annual investment cost (EAIC) over m^* years, which is often called the *annual capital recovery cost*, can be calculated by

$$\text{EAIC} = I \cdot (\text{A}|\text{P}, r, m^*) - S_{m^*} \cdot (\text{A}|\text{F}, r, m^*) \tag{6.43}$$

where I is the initial total investment, S_{m^*} is the salvage value at the end of m^* years, r is the discount rate, and m^* is the economic life (in years) that needs to be found.

The equivalent annual operation and unreliability cost (EAOUC) incurred by a piece of equipment over m^* years can be calculated by

$$\text{EAOUC} = \left(\sum_{k=0}^{m^*} (O_k + R_k) \cdot (\text{P}|\text{F}, r, k) \right) \cdot (\text{A}|\text{P}, r, m^*) \tag{6.44}$$

where O_k and R_k are the annual operation and unreliability costs incurred by the equipment in year k. Note that the R_k should be an incremental cost, which is the difference in the system unreliability cost between the two cases with considering and not considering the failure of the equipment. If the network loss due to the equipment is a part of O_k, this part should also be an incremental network loss cost due only to the equipment, which could be either positive or negative.

The total equivalent annual cost (TEAC) is the sum of EAIC and EAOUC. The salvage value is always a small amount at the end of economic life and is often negligible compared to the investment cost. The EAIC is a decreasing function of the parameter m^*. In other words, the longer a piece of equipment is used, the smaller the EAIC becomes. On the other hand, the EAOUC is an increasing function of the parameter m^*. This is because both the operation and unreliability costs increase as equipment ages because of increased OMA costs and unavailability of equipment in an aging status. Therefore, the TEAC is a convex function of m^* with a unique minimum point. According to the definition given in Section 6.4.1, the economic life is the length of time in years by the end of which the total equivalent annual cost (TEAC) reaches the minimum point. In other words, the economic life is the m^* that minimizes the TEAC.

The procedure to find the economic life m^* is straightforward. The TEAC is calculated by letting $m^* = 1,2,3,\ldots$, and so on. The minimum TEAC can be observed in the series of TEAC values. The year corresponding to the minimum TEAC is the economic life.

Conceptually, the economic life can be regarded the year in which a replacement should take place. However, this criterion alone should be used with caution if the economic life has been calculated at an early stage of equipment in service. The estimated operation and unreliability costs for many future years will be very inaccurate. The discount rate is also uncertain and may vary over the life. Nevertheless, the estimated economic life is useful information even if it is not accurate. It can be reestimated as time advances. The economic life can be combined with the replacement delay analysis in actual applications. The economic life of equipment is estimated first, and then the replacement delay analysis method is applied when the service year of equipment approaches the end of the economic life.

6.7 UNCERTAINTY ANALYSIS IN ECONOMIC ASSESSMENT

There are uncertainties for the input data in the economic analysis, including the discount rate, useful life, salvage value, and estimates of investment, operation, and unreliability costs. The uncertainties of the input data create the uncertainties of the outputs, including the depreciated and book values, present and annual values, and outputs used for comparisons between alternatives or projects. Two methods can be used to deal with the uncertainties: sensitivity analysis and probabilistic techniques.

6.7.1 Sensitivity Analysis

Sensitivity analysis is straightforward. This is based on the concept of incremental change—if a variation of an input data is given, what would be the variation in an output?

For example, if we want to obtain the sensitivity of the present value of total cost (PVTC) to the discount rate in Equation (6.36), we can specify different variations of discount rate and calculate a series of values of PVTC. The same idea can apply to the sensitivity between any input and any output.

The advantage of sensitivity analysis is simplicity. However, it has two demerits: (1) it provides the sensitivity of an output to only one input variable at a time—any change in output caused by simultaneous variations of two or more input variables will create unclear information; and (2) it cannot provide a single mean value of output considering uncertainties of all input data.

6.7.2 Probabilistic Analysis

The probabilistic analysis method in the economic assessment includes the following steps:

- A discrete probability distribution representing the uncertainty of an input data is obtained either by the estimation from engineering judgment (subjective probability) or by statistical records (objective probability).
- Multiple values of an output are calculated for all possible values of input data defined in the discrete probability distribution using the relevant formulas or calculation processes in the economic assessment.
- The mean value and standard deviation of the output are calculated.
- A financial risk assessment for planning projects or alternatives is performed.

The net benefit present value (NBPV) in Equation (6.38) is used as an example. It is assumed that the discount rate r has the following discrete probability distribution

$$p(r = r_i) = p_i \qquad (i = 1, \ldots, N_r) \tag{6.45}$$

where r is a random variable of discount rate. The discrete probability distribution indicates that r has N_r possible values and each one has a probability of p_i. This probability distribution can be estimated from the historical records of interest and inflation rates that were used in the financial analysis of a utility.

The N_r values of NBPV for a planning alternative are calculated using Equation (6.38) from N_r values of r. The mean and standard deviation of NBPV for the alternative is calculated by

$$E(\text{NBPV}) = \sum_{i=1}^{N_r} \text{NBPV}_i(r_i) \cdot p_i \tag{6.46}$$

$$\text{Std(NBPV)} = \sqrt{\sum_{i=1}^{N_r} [\text{NBPV}_i - E(\text{NBPV})]^2 \cdot p_i} \tag{6.47}$$

Let us look at how to consider the uncertainties of both useful life and discount rate for a planning alternative. The equivalent annual investment cost for the alternative can be calculated using the initial total investment cost and the capital return factor by

$$I_k = I \cdot \left[\frac{r \cdot (1+r)^m}{(1+r)^m - 1} \right] \tag{6.48}$$

where I and I_k are the initial total investment and equivalent annual investment costs, respectively; r is the discount rate; and m is the useful life of the alternative.

It is assumed that the useful life has N_s possible estimates m_j ($j = 1,\ldots,N_s$) and each estimated value has an equal possibility (i.e., the subjective probability for each value is $1/N_s$). With N_r values of r and N_s values of m, the $N_r \times N_s$ values of I_k can be obtained from Equation (6.48) first and then the $N_r \times N_s$ values of NBPV for the alternative are calculated using Equation (6.38). The mean and standard deviation of NBPV for the alternative is calculated by

$$E(\text{NBPV}) = \frac{1}{N_s} \sum_{i=1}^{N_r} \sum_{j=1}^{N_s} \text{NBPV}_{ij}(r_i, m_j) \cdot p_i \tag{6.49}$$

where NBPV_{ij} is the value of NBPV for r_i and m_j. Its standard deviation is estimated by

$$\text{Std(NBPV)} = \sqrt{\frac{1}{N_s} \sum_{i=1}^{N_r} \sum_{j=1}^{N_s} [\text{NBPV}_{ij} - E(\text{NBPV})]^2 \cdot p_i} \tag{6.50}$$

The method described above can be easily extended to a case where the uncertainties of more than two input data need to be considered.

In comparing different alternatives, the same discrete probability distribution of discount rate should be used. However, the discrete probability distributions of other input data (such as the useful life) can be different. With the results from the probabilistic analysis, both the mean value and standard deviation of NBPV should be considered in the comparison between alternatives. Obviously, if two alternatives have very close standard deviations of NBPV, the one with a larger mean value of NBPV is better as it creates more net benefits. If two alternatives have the very close mean values of NBPV, the one with a smaller standard deviation is better because it carries a lower financial risk. It is possible that one alternative has a larger mean value of NBPV with a larger standard deviation but another has a smaller mean value of NBPV with a smaller standard deviation. In this case, an engineering judgment is needed. This is associated with a compromise between benefit and financial risk.

The procedure for the probabilistic analysis method for other economic assessments is similar to that for the NBPV described above.

6.8 CONCLUSIONS

This chapter discussed the economic analysis methods in transmission planning. The investment and operation costs are the basic components in general engineering economics, whereas the unreliability cost is a special component that needs to be added in the economic assessment for transmission system planning. The incorporation of this component requires modified concepts in the economic analysis. It is important to appreciate that the unreliability cost and some subcomponents of operation cost due to an alternative or a piece of equipment should be calculated using an incremental approach.

The time value of money, which originates from the interest and inflation, is the basis of engineering economics. The discount rate in discounting calculations can be a nominal or real interest rate depending on whether an inflation rate is included. It can be either a risk-free rate or risky rate depending on whether the financial risk is considered in a project. The six factors for conversion between the present, annual, and future values are all based on the time value of money. Depreciation is another important concept in the economic analysis. The investment cost, depreciable (useful) life, and salvage value are the three input data needed to calculate the annual depreciated and book values of equipment. Different depreciation methods provide varied (equal, accelerating, or decelerating) annual depreciation rates. Which method should be used depends on financial considerations and business models of utilities.

The economic assessment of investment projects is at the core of economic analysis in transmission planning. The purpose is to provide the decisionmaking information in determining the final alternative of a planning project or ranking different projects. The total cost method and the net benefit present value method can be used for the comparison between alternatives of a project but are not appropriate for ranking projects intended to resolve different system problems because they provide the information only on absolute economic profit. The benefit/cost ratio method and the internal rate of return method can be used in the justification of a project or alternative, comparison between alternatives of a project, and ranking among different projects as they provide the information of relative economic return.

The best time of replacement can be decided using the economic assessment of equipment replacement. The replacement delay analysis method is used when a piece of equipment approaches the end-of-life stage at which more accurate data become available. Although the economic life of equipment can be estimated even at an early stage of equipment service, it is not accurate because of inaccuracy of data for future years. The combined use of the economic life estimation and the replacement delay analysis method provides a better result.

There are always uncertainties of input data in the economic assessment. Sensitivity and probabilistic analysis methods can be used to deal with the uncertainties. The probabilistic techniques create the mean value and standard deviation of economic indices for a planning project or alternative. Both the mean and standard deviation provide essential information in decisionmaking. The mean value is used to judge the economic profit of a project or alternative, whereas the standard deviation is an indicator of its financial risk level.

7

DATA IN PROBABILISTIC TRANSMISSION PLANNING

7.1 INTRODUCTION

Preparing valid data is an essential step in probabilistic transmission planning. The quality of planning decision depends not only on the methods and computing tools but also on the quality of data. In other words, the data are as important as are methods and tools. It can be seen from the discussions in the previous chapters that the load forecast, power system analyses (including power flow, contingency analysis, optimal power flow, voltage stability, and transient stability), quantified reliability evaluation, and economic assessment are basic tasks in probabilistic transmission planning. Each task has specific data requirements.

In general, there is no concern about the data of existing system equipment, as these data were previously prepared and validated. However, a planning project is usually associated with equipment additions and thus requires the preparation of new data. In addition to the equipment data, system operating limits and load data are of importance in system analysis. More data are required for the reliability evaluation and economic analysis. Other data include the information on generation sources and interconnection. All the data need to be reviewed and updated regularly or when any modification becomes necessary. In fact, many data such as those regarding bus loads and outage information change over the years.

Probabilistic Transmission System Planning, by Wenyuan Li

Load forecast data consists of historical load records and impacting parameters, and have been discussed together with load forecast methods in Chapter 3. The basic data requirements for power system analysis are illustrated in Section 7.2. The data for reliability evaluation is based on a huge amount of outage records and related statistical analysis. The reliability data, which can be collected for both equipment and delivery points in a transmission system, are discussed in Section 7.3. Data for economic analysis have been described in Chapter 6 and are briefly summarized together with other data in Section 7.4.

7.2 DATA FOR POWER SYSTEM ANALYSIS

The data in power system analysis cover a wide range. This section focuses on equipment parameters, equipment ratings, system operation limits, and load coincidence factors.

7.2.1 Equipment Parameters

For most equipment, the parameters provided by manufacturers can be directly used with the following exceptions:

- The parameters of overhead lines need to be calculated because they depend on actual arrangements of lines and environment factors (such as temperature).
- Although manufacturers provide the parameters of each type of cable, a circuit may contain multiple mixed line and cable sections.
- The parameters of transformers need to be calculated using the information provided by manufacturers or obtained from field tests.

7.2.1.1 Parameters of Overhead Line. There are four basic parameters for an overhead line: series resistance (R_L), series inductive reactance (X_L), shunt capacitive susceptance (B_L), and shunt conductance (G_L).

1. *Series Resistance.* The resistance per unit length is calculated by

$$R_L = \frac{\rho}{S} \tag{7.1}$$

where R_L is the resistance (Ω/km); ρ is the resistivity ($\Omega \cdot \text{mm}^2$/km); and S is the effective cross-sectional area (mm^2). The resistivity at 20°C is 17.2 $\Omega \cdot \text{mm}^2$/km for copper and 28.3 $\Omega \cdot \text{mm}^2$/km for aluminum. A coefficient should be introduced to reflect stranding and skin effects. As an approximate estimate, the stranding and skin effects can increase the resistivity by 8–12%. In fact, the resistance values for various types of lines are provided in manufacturers' specifications. However, the resistance values provided by manufacturers are the ones estimated at a reference temperature (usually 20°C). The resistance at an actual conductor temperature can be calculated by

$$R_{Lt} = R_{L\text{ref}}[1 + \alpha_{\text{ref}}(t - t_{\text{ref}})] \tag{7.2}$$

where t and t_{ref} are the conductor temperature and reference temperature (in °C), respectively; R_{Lt} and $R_{L\text{ref}}$ are the resistances at the temperature t and reference temperature, respectively; and α_{ref} is the thermal coefficient of resistance. The value 20°C is often used as the reference temperature. As an approximate estimate, the thermal coefficient at 20°C is 0.0038 for copper and 0.0036 for aluminum.

2. *Series Reactance.* The reactance per unit length is calculated by

$$X_L = k_x \cdot \log \frac{D_m}{r_{\text{eq}}} \tag{7.3}$$

where X_L is the reactance (Ω/km), D_m is the geometric mean distance between phase conductor bundles (m), r_{eq} is the equivalent geometric mean radius of phase conductor bundle (m), and k_x is a coefficient associated with system frequency. For 60 Hz, $k_x = 0.1737$ and for 50 Hz, $k_x = 0.1448$. D_m is calculated by

$$D_m = (D_{ab} D_{bc} D_{ca})^{1/3} \tag{7.4}$$

where D_{ab}, D_{bc}, and D_{ca} are the distances (m) between the three centers of three-phase bundles of a line, respectively. For a single cylindrical conductor, r_{eq} is calculated by

$$r_{\text{eq}} = (e^{-\mu/4}) \cdot r \approx 0.779 \cdot r \tag{7.5}$$

where r is the effective radius (m) of the single cylindrical conductor and μ is the relative permeability of the conductor, which approximately equals 1.0 for copper and aluminum. For a symmetrically bundled conductor, r_{eq} is calculated by

$$r_{\text{eq}} = [r_{\text{eq}s} \cdot n \cdot (r_b)^{n-1}]^{1/n} \tag{7.6}$$

where $r_{\text{eq}s}$ is the equivalent radius of each subconductor, which can be estimated using Equation (7.5); r_b is the radius of the bundle of conductors; and n is the number of subconductors per phase.

3. *Shunt Susceptance.* The susceptance per unit length for a single cylindrical conductor is calculated by

$$B_L = \frac{k_b}{\log(D_m / r)} \tag{7.7}$$

where B_L is the susceptance (μs/km), D_m is the same as defined in Equation (7.3), r is the effective radius of a single cylindrical conductor, and k_b is a coefficient associated with system frequency. For 60 Hz, $k_b = 9.107$ and for 50 Hz,

TABLE 7.1. Typical Overhead Transmission Line Parameters

Voltage Level (kV)	R_L (Ω/km)	X_L (Ω /km)	B_L (μs/km)
60	0.244	0.425	3.870
138	0.140	0.441	3.719
230	0.076	0.480	3.304
360	0.040	0.382	4.274
500	0.024	0.330	4.898

$k_b = 7.590$. For a symmetrically bundled conductor, r in Equation (7.7) is replaced by the following equivalent geometric mean radius r_{eq}

$$r_{eq} = [r \cdot n \cdot (r_b)^{n-1}]^{1/n} \tag{7.8}$$

where r is the effective radius of each single subconductor, r_b is the radius of the bundle of conductors, and n is the number of subconductors per phase.

4. *Shunt Conductance.* The line conductance is caused by corona that will occur only when the voltage level exceeds a critical value. The line conductance is a very small value and is often ignored in the calculations for planning purposes.

The typical overhead transmission-line parameters at 60 Hz (for X_L and B_L) and 20°C (for R_L) are shown in Table 7.1. Note that the values in Table 7.1 are based on sampled actual lines and provide only a general idea about the order of magnitude.

7.2.1.2 Parameters of Cable. The formulas for cable parameters are similar to those for overhead lines. However, the values of cable parameters are different from those of overhead lines because of structures, arrangements, and insulation materials:

- Conductors of cable are much closer from each other.
- Conductors of cable have varied cross-sectional shapes depending on cable types.
- Conductors of cable have metallic or polymeric shields, sheaths, sheets, or pipes.
- Insulation materials between conductors of cable are not air.

Accurate calculations of cable parameters are more complex than those of overhead-line parameters. Fortunately, not only do manufacturers provide the parameter values for various types of cable in their specifications; commercial programs for calculating cable parameters are also available.

The typical transmission cable parameters at 60 Hz (for X_C and B_C) and at 20°C (for R_C) are shown in Table 7.2. Note that the values in Table 7.2 are based on sampled actual cables and provide only a rough idea about the order of magnitude. The sampled cables for 60, 138, and 230 kV are underground cables, and those for 500 kV are submarine cables.

TABLE 7.2. Typical Transmission Cable Parameters

Voltage Level (kV)	R_C (Ω /km)	X_C (Ω /km)	B_C (μs/km)
60	0.055	0.238	151.96
138	0.064	0.228	122.41
230	0.019	0.220	131.12
500[a]	0.026	0.066	108.17

[a]The parameters for 500-kV submarine cables.

7.2.1.3 Parameters of Transformer. The manufacturers of transformers provide the rated voltage (V_N), capacity rating (S_N), no-load loss (P_c), load loss (P_{cu}), impedance in percentage ($Z_{\%}$) with regard to the base impedance, excitation current in percentage ($I_{\%}$) with regard to the base current, and other data. The base impedance or base current is calculated from the rated voltage and capacity. These data can be used to estimate the parameters of transformers in the power system analysis. Parameters of two- and three-winding transformers are calculated as follows:

1. *Parameters of Two-Winding Transformers.* The *resistance* of a two-winding transformer is estimated by

$$R_T = \frac{P_{cu} \cdot (V_N)^2}{1000 \cdot (S_N)^2} \tag{7.9}$$

where R_T is the resistance (Ω) of the transformer; P_{cu} is the load loss (kW) at the rated current; and V_N and S_N are the rated voltage (kV) and capacity rating (MVA), respectively. The *reactance* of a two-winding transformer is estimated by

$$X_T \approx Z_T = \frac{Z_{\%} \cdot (V_N)^2}{100 \cdot S_N} \tag{7.10}$$

where X_T and Z_T are the reactance (Ω) and impedance (Ω) of transformer, respectively; $Z_{\%}$ is the impedance in percentage with regard to the base impedance; and V_N and S_N are the same as defined in Equation (7.9). Note that the reactance is assumed to be approximately equal to the impedance in this equation because R_T is always a small value compared to X_T. Otherwise, the reactance can be calculated by $X_T = \sqrt{(Z_T)^2 - (R_T)^2}$. The *susceptance* of a two-winding transformer is estimated by

$$B_T \approx \frac{I_{\%} \cdot S_N}{100 \cdot (V_N)^2} \tag{7.11}$$

where B_T is the susceptance (1/Ω), $I_{\%}$ is the excitation current in percentage with regard to the base current, and V_N and S_N are the same as defined in Equation

(7.9). It is assumed in Equation (7.11) that the excitation current is approximately equal to the current flowing through the susceptance branch in the equivalent circuit of transformer since the current flowing through the conductance branch is extremely small. The *conductance* of a two-winding transformer is estimated by

$$G_T \approx \frac{P_c}{1000 \cdot (V_N)^2} \tag{7.12}$$

where G_T is the conductance (1/Ω), P_c is the no-load loss (kW), and V_N is the same as defined in Equation (7.9). It is assumed in Equation (7.12) that the core loss is approximately equal to the no-load loss.

2. *Parameters of Three-Winding Transformers.* The equivalent circuit of a three-winding transformer is shown in Figure 7.1. Manufacturers provide the load losses and impedances in percentage with regard to the base impedance for each pair of windings. The estimation of *resistance* for each winding includes the following two steps:

   - The load losses for each winding are calculated first by

$$\left.\begin{aligned} P_{cu1} &= 0.5 \cdot [P_{cu(1-2)} + P_{cu(1-3)} - P_{cu(2-3)}] \\ P_{cu2} &= 0.5 \cdot [P_{cu(1-2)} + P_{cu(2-3)} - P_{cu(1-3)}] \\ P_{cu3} &= 0.5 \cdot [P_{cu(2-3)} + P_{cu(1-3)} - P_{cu(1-2)}] \end{aligned}\right\} \tag{7.13}$$

   where P_{cu1}, P_{cu2}, and P_{cu3} are the load losses (kW) at the rated current for the three windings respectively; $P_{cu(1-2)}$, $P_{cu(1-3)}$, and $P_{cu(2-3)}$ are the load losses (kW) at the rated current for each pair of windings.

   - The resistance for each winding is estimated by

$$R_{Ti} = \frac{P_{cui} \cdot (V_N)^2}{1000 \cdot (S_N)^2} \qquad (i = 1,2,3) \tag{7.14}$$

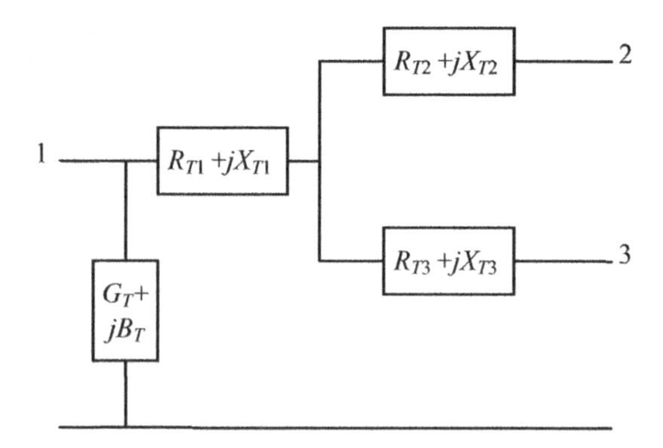

Figure 7.1. Equivalent circuit of three-winding transformer.

where R_{Ti} is the resistance (Ω) of the ith winding; V_N and S_N are the rated voltage and capacity rating of three-winding transformer, respectively. The estimation of *reactance* for each winding includes the following two steps:

- The impedances in percentage with regard to the base impedance for each winding are calculated first by

$$\left.\begin{aligned} Z_{\%T1} &= 0.5\cdot[Z_{\%(1-2)} + Z_{\%(1-3)} - Z_{\%(2-3)}] \\ Z_{\%T2} &= 0.5\cdot[Z_{\%(1-2)} + Z_{\%(2-3)} - Z_{\%(1-3)}] \\ Z_{\%T3} &= 0.5\cdot[Z_{\%(2-3)} + Z_{\%(1-3)} - Z_{\%(1-2)}] \end{aligned}\right\} \tag{7.15}$$

where $Z_{\%T1}$, $Z_{\%T2}$, and $Z_{\%T3}$ are the impedances in percentage with regard to the base impedance for three windings, respectively; $Z_{\%(1-2)}$, $Z_{\%(1-3)}$ and $Z_{\%(2-3)}$ are the impedances in percentage for each pair of windings.

- The reactance (Ω) for each winding is estimated by

$$X_{Ti} \approx Z_{Ti} = \frac{Z_{\%Ti}\cdot(V_N)^2}{100\cdot S_N} \qquad (i = 1,2,3) \tag{7.16}$$

- The estimation method of susceptance and conductance for three-winding transformers is the same as that for two-winding transformers.

7.2.1.4 Parameters of Synchronous Generator. The parameters of synchronous generators can be found in the specification of manufacturers. The ranges of the parameters are shown in Table 7.3 [55,86,87]. Note that all the reactance or resistance values are given in per unit with regard to the rated base values of generators. The parameters of each generator may be quite different. The values in Table 7.3 provide an idea of the order of magnitude which follows that in Equations (7.17)–(7.19):

$$X_d \geq X_q > X_q' \geq X_d' > X_q'' \geq X_2 \geq X_d'' > X_a > X_0 \tag{7.17}$$

$$T_{do}' > T_d' > T_{do}'' > T_d'' \tag{7.18}$$

$$T_{qo}' > T_q' > T_{qo}'' > T_q'' \tag{7.19}$$

7.2.1.5 Parameters of Other Equipment. In general, the parameters of other equipment can be found from the equipment nameplate or specification of manufacturers. The impedance of reactors can be calculated from field test data using the same formulas for transformers. The susceptance of shunt capacitors can be calculated using the rated voltage (kV) and capacity rating (MVAR).

7.2.2 Equipment Ratings

The equipment ratings are important data in transmission planning because a considerable number of planning projects deal with overloading issues of lines/cables,

TABLE 7.3. Range of Synchronous Generator Parameters

Parameter	Symbol	Unit	Hydraulic unit	Thermal Unit	Condenser
Stator leakage reactance	X_a	p.u.	0.1–0.2	0.1–0.2	0.10–0.16
Stator resistance	R_a	p.u.	0.002–0.02	0.0015–0.005	N/A
Synchronous reactance (*d* axis)	X_d	p.u.	0.6–1.5	1.0–2.3	1.6–2.4
Synchronous reactance (*q* axis)	X_q	p.u.	0.4–1.0	1.0–2.3	0.8–1.2
Transient reactance (*d* axis)	X'_d	p.u.	0.2–0.5	0.15–0.4	0.25–0.5
Transient reactance (*q* axis)	X'_q	p.u.	0.4–0.9	0.3–1.0	N/A
Sub-transient reactance (*d* axis)	X''_d	p.u.	0.14–0.35	0.10–0.25	0.15–0.30
Sub-transient reactance (*q* axis)	X''_q	p.u.	0.2–0.45	0.10–0.25	0.15–0.30
Negative sequence reactance	X_2	p.u.	0.15–0.45	0.10–0.25	0.15–0.30
Zero sequence reactance	X_0	p.u.	0.04–0.15	0.04–0.15	0.05–0.10
Transient time constant (*d* axis) (open-circuit)	T'_{do}	s	1.5–9.0	3.0–11.0	N/A
Transient time constant (*d* axis) (short-circuit)	T'_d	s	0.8–3.0	0.4–1.6	0.8–2.4
Subtransient time constant (*d* axis) (open-circuit)	T''_{do}	s	0.01–0.3	0.02–0.2	N/A
Subtransient time constant (*d* axis) (short-circuit)	T''_d	s	0.01–0.06	0.02–0.11	0.01–0.03
Transient time constant (*q* axis) (open-circuit)	T'_{qo}	s	N/A	0.5–2.5	N/A
Transient time constant (*q* axis) (short-circuit)	T'_q	s	N/A	0.15–0.5	N/A
Subtransient time constant (*q* axis) (open-circuit)	T''_{qo}	s	0.01–0.3	0.02–0.2	N/A
Subtransient time constant (*q* axis) (short-circuit)	T''_q	s	0.01–0.06	0.02–0.11	0.01–0.03

transformers, or other equipment. Any occurrence of overloading in a normal or a contingency system state implies that reinforcement in the system is needed. Equipment ratings can be found from nameplates or manufacturers' specifications with the following exceptions:

- The ratings of overhead lines and cables need to be calculated because they rely on environmental factors, permissible conductor temperature, and overhead-line clearance. The criteria may be varied at different utilities, particularly for clearance requirements.
- Although the ratings of transformers are provided by manufactures, transformers can be operated in an overloading state for a limited time without reduction in

life expectancy. The overloading capacity depends on internal (top-oil and hottest-spot) and ambient temperatures.

The commonly recognized industry standards (such as IEEE and IEC standards) are available for calculating the equipment ratings. Commercial programs using the standards are also available for estimating the ratings of lines and cables. This section provides an overview on capacity limits of overhead lines, cables, and transformers.

7.2.2.1 Current Carrying Capacity of Overhead Line. The heat created by a current flowing through the resistance of conductor and by solar radiation on the conductor is dissipated by convection into the surrounding air and radiation to surrounding objects. Therefore the current carrying capacity of an overhead line [88] can be approximately estimated by

$$I = \sqrt{\frac{(W_c + W_r - W_s)S_l}{R_l}} \tag{7.20}$$

where I is the current carrying capacity (A) of conductor; S_l is the conductor surface area per unit length (cm^2/m); R_l is the conductor resistance per unit length (Ω/m) at the permissible conductor temperature; W_c and W_r are the watts (W/cm^2) dissipated by convection and radiation, respectively; and W_s is the watts (W/cm^2) due to solar radiation on the conductor. The W_s value is normally very small and negligible; W_c can be approximately estimated by

$$W_c = \frac{0.00573\sqrt{pv}}{(T_{\text{avg}} + 273)^{0.123}\sqrt{d}}(T_c - T_a) \tag{7.21}$$

where p is the pressure in atmospheres ($p = 1.0$ for atmospheric pressure); v is the velocity of surrounding air (m/s); T_{avg} is the average of conductor temperature and air temperature (°C); d is the outside diameter of the conductor (cm); T_a is the air temperature (°C), which varies for different locations and seasons; and T_c is the permissible conductor temperature (°C), which is determined by the rule (such as the clearance requirement) of a utility.

The value of W_r can be approximately estimated by

$$W_r = 5.704 \cdot E \cdot \left[\left(\frac{T_c + 273}{1000}\right)^4 - \left(\frac{T_a + 273}{1000}\right)^4\right] \tag{7.22}$$

where E is the relative emissivity of the conductor surface, which is 0.5 for average oxidized copper.

These formulas provide an approximate estimation of the current carrying capacity of an overhead line. More accurate methods can be found in the IEEE standard book [89].

TABLE 7.4. Permissible Conductor Temperature (T_c) for Cables

Insulation Material	Permissible Normal Temperature (°C)	Permissible Emergency Temperature(°C)
Low-density polyethlene	75	90
Others	90	105

7.2.2.2 Current Carrying Capacity of Cable. This topic is discussed in detail in the EPRI report [90]. The heat flowing through a thermal resistance generates a temperature rise. In an underground cable, the conductor, each of the other cable layers, trench backfill, or native earth has a thermal resistance. Therefore there exists the following relationship:

$$T_c - T_a = \sum_i W_i \cdot (R_{\text{th}})_i \tag{7.23}$$

Here, W_i is the heat loss per unit length (W/m) caused by the ith layer (the conductor is one of the layers); $(R_{\text{th}})_i$ is the thermal resistance (°C·m/W)of the ith layer; T_a is the ambient temperature (°C), which is varied for different locations and seasons; and T_c is the permissible conductor temperature (°C), which depends on the insulation material in a cable and operational conditions as shown in Table 7.4.

Equation (7.23) can be rewritten as

$$T_c - T_a = W_o \cdot \sum_i Q_i \cdot (R_{\text{th}})_i \tag{7.24}$$

where Q_i is the ratio of the heat loss in the ith layer with regard to the heat loss caused only by the conductor. W_o is the ohmic losses caused by the conductor, which can be calculated by

$$W_o = I^2 R_l \tag{7.25}$$

where I is the current carrying capacity (A) of the conductor and R_l is the conductor resistance per unit length (Ω/m) at the permissible conductor temperature.

By combining Equations (7.24) and (7.25), we can estimate the current carrying capacity of cable as follows:

$$I = \sqrt{\frac{T_c - T_a}{R_l \cdot \Sigma_i Q_i \cdot (R_{\text{th}})_i}} \tag{7.26}$$

The Q_i and $(R_{\text{th}})_i$ depend on the type, structure, and material of each layer in a cable, and can be obtained from the information provided by manufacturers.

A more accurate method for calculating the current carrying capacity of underground cable can be found in the IEC 60287 standard book [91].

7.2.2.3 *Loading Capacity of Transformer.* The transformer rating can be found on the nameplate. However, it is important to appreciate that transformers have a relatively high ability to withstand overloading in both normal and emergency conditions.

The loading limit of transformer depends on the oil temperature. The oil temperature rise after the time t is given by

$$T_t = T_f(1 - e^{-t/\tau}) \tag{7.27}$$

where T_t is the oil temperature rise (°C) at time t; T_f is the final temperature rise (°C) at a load and at $t = \infty$; and τ is the thermal time constant of oil, which is equal to the thermal capacity of oil divided by the radiation constant.

It can be seen that the oil temperature is dynamic and depends on multiple factors, including the operation history and conditions. It is not easy to conduct an accurate estimation on the loading capacity of a transformer. For the purpose of transmission planning, the loading capacity of a transformer can be approximately estimated using the following approaches [88] if there is no manufacturer specification available:

1. *Loading Capacity in Normal Operation Condition.* The maximum loading limit of a transformer in the normal operation condition can be estimated by

$$S_{\max} = \{1 + \min[K, k(1.0 - \text{ACF})\} \cdot S_N \tag{7.28}$$

where S_N is the rated MVA; ACF is the average capacity factor (smaller than 1.0) of a transformer in the normal operation condition, which is defined as the average operating MVA during a 24-h period divided by the rated MVA (S_N); k is 0.4 for a transformer with a forced-air-cooled or forced-oil-cooled system, or 0.5 for a transformer with a self-cooled or water-cooled system; and K is the permissible maximum loading increase in a fraction of S_N, which is normally provided by manufacturers. If no manufacturer information is available, K can be approximately chosen as 0.2 for a transformer with a forced-air-cooled or forced-oil-cooled system and 0.25 for a transformer with a self-cooled or water-cooled system. A loading level within S_{max} in the normal operation condition will not lead to reduction in the life expectancy of transformers.

2. *Loading Capacity in Short-Time Overloading Condition without Life Reduction.* In some contingency conditions, short-time overloading, which occurs only once during any 24-h period, can be allowed without sacrifice in the life expectancy of transformer. Figure 7.2 presents the relatively conservative (secure) estimates of allowable short-time overloading percentage without life reduction. Some transformers may have a higher overloading capacity than the percent values given in the figure. This depends on the design margin provided by manufacturers and loading conditions prior to the overloading period. The "initial loading condition" (ILC) in the figure refers to the average loading percentage with regard to the rated MVA for 24 h prior to the overloading period. It should be noted that the approach based on the average capacity factor given

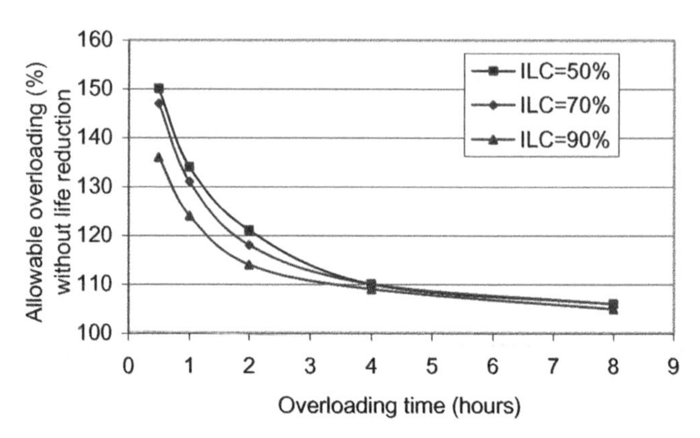

Figure 7.2. Allowable overloading without life reduction.

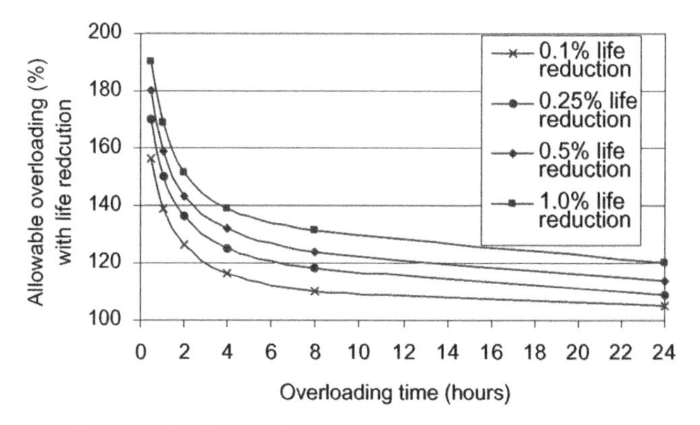

Figure 7.3. Allowable overloading with life reduction.

in Equation (7.28) and the approach based on short-time overloading shown in Figure 7.2 cannot be applied concurrently. It is necessary to choose either one.

3. *Loading Capacity in Short-Time Overloading Condition with Life Reduction.* In some contingency conditions, short-time overloading with reduction in the life expectancy of a transformer may be allowed. If the economic benefit obtained by delaying addition of a new transformer is greater than the economic loss caused by life reduction of an old transformer, accepting short-time overloading events due to unusual contingency conditions may be considered as an option in probabilistic transmission planning. Figure 7.3 presents relatively conservative (secure) estimates of short-time overloading percentage with life reduction. The percent values of life reduction in the figure correspond to each overloading event. Contingencies are unusual events. In probabilistic planning, the frequency and duration of contingency events are evaluated. Some transformers may have a higher overloading percentage than the values in the figure. This depends on the design margin of transformers.

7.2.3 System Operation Limits

The maximum capacities of equipment described in Section 7.2.2 are basically limited by permissible temperatures and are often called *thermal limits*. The system operation security limits include not only thermal limits but also other constraints such as voltage limits, allowable frequency fluctuations, and transient and voltage stability margins. Violation of the system operation limits caused by various reasons is the main driver for system reinforcement in transmission planning. The system operation limits reflect security requirements. Table 7.5 is an example of the system operation limits, which is based on NERC (North American Electric Reliability Corporation) and WECC (Western Electricity Coordinating Council) criteria [1,3]. These limits are mandatory system performance requirements in transmission planning and may be slightly different for individual utilities.

7.2.4 Bus Load Coincidence Factors

The load forecast and modeling methods have been discussed in Chapter 3. An issue that was not addressed is the concept of bus load coincidence factors. This is an important concept in power flow case preparation. The load forecast generally provides annual and seasonal peak values for the whole system, regions, and individual substations. Loads at substations never reach their peaks at the same point of time. This is called *noncoincidence*. In other words, we cannot and should not simply use forecasted peaks of individual substations as bus loads in a power flow case. A power flow study is often conducted at the system peak (for bulk system planning studies) or at a region peak (for regional network planning studies). It is necessary to calculate the bus loads corresponding to a system (or region) load level.

The concept of the bus load coincidence factor is illustrated in Figure 7.4. For simplicity, only two substation load curves are considered. The total load curve, which is the summation of the two substation load curves, can be viewed as the system (or region) load curve. Points *A*, *E*, and *D* are the peaks of the system and two substation loads respectively. Apparently, they do not occur at the same time. Points *C* and *B* represent the load points of the two substations corresponding to the system (or region) peak. The bus load coincidence factors (LCFs) for the two substations are defined as

$$\mathrm{LCF}_1 = \frac{L_C}{L_E} \tag{7.29}$$

$$\mathrm{LCF}_2 = \frac{L_B}{L_D} \tag{7.30}$$

where L_C and L_E are the loads at points *C* and *E* in load curve 1 for substation 1, and L_B and L_D are the loads at points *B* and *D* in load curve 2 for substation 2.

The bus load coincidence factors can be calculated using historical load curve data. In the calculations, attention should be paid to the following [92]:

TABLE 7.5. System Performance Requirements

Category[a]	System Limits						
	Equipment Rating Limits	Transient Voltage Dip	System Frequency	Voltage or Voltage Deviation	Voltage Stability Margin (%)	Transient Instability or Cascading Outages	Loss or Transfer of Firm Loads
A	Normal limits	—	Normal limits	Normal limits	10	Not allowed	Not allowed
B	Contingency limits	Not to exceed 25% at load buses or 30% at nonload buses; not to exceed 20% for more than 20 cycles at load buses	Not below 59.6 Hz for 6 cycles at load buses	Deviation not to exceed 5% at any bus	5	Not allowed	Not allowed
C	Contingency limits	Not to exceed 30% at any bus; not to exceed 20% for more than 40 cycles at load buses	Not below 59.0 Hz for 6 cycles at load buses	Deviation not to exceed 10% at any bus	2.5	Not allowed	Planned or controlled
D	Evaluate consequence and system risks						

[a]*Categories*: A—no contingency; B—loss of a single element; C—loss of multiple elements; D—extreme events: loss of multiple elements or cascading outages. Detailed descriptions of contingency categories A, B, C, and D can be found at the NERC or WECC Website [1,3].

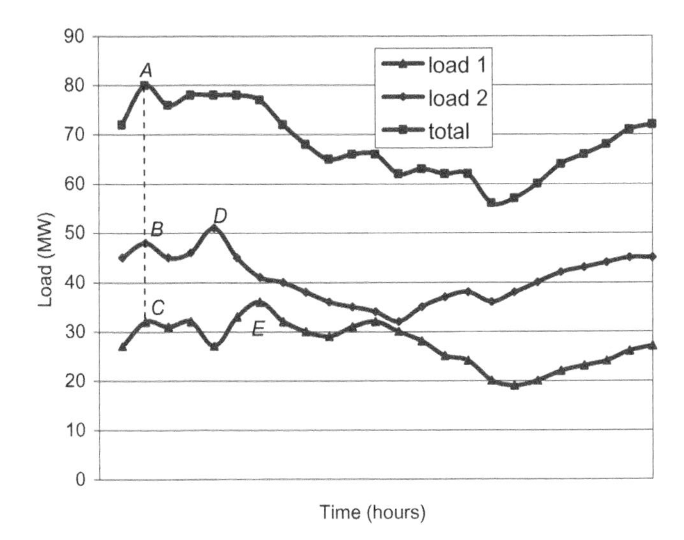

Figure 7.4. Concept of bus load coincidence factor.

- Invalid data on each substation load curve, which may be caused by wrong records, unreasonable spikes, random and unrepeatable shutdowns, and so on, should be filtered out first using a statistical method with relevant rules to capture the regular patterns of the load curve.
- The load value at each point (such as *A*, *B*, *C*, *D*, or *E*) should be replaced using an average of loads at several points around it to reduce the uncertainty of load record at a single point.
- The load curve data and bus load coincidence factors should be updated each year.
- The bus load coincidence factors can be calculated for different reference points on the system or region load curve (such as annual peak, winter or summer peak, winter or summer minimum) and for different load categories (such as industrial, residential, or commercial customers or combinations thereof).

Once the bus load coincidence factors (LCFs) are available, the bus loads for any reference point on the system or regional load curve can be calculated using the LCF values for that point and peak forecasts of substation loads in preparing different power flow cases.

7.3 RELIABILITY DATA IN PROBABILISTIC PLANNING

7.3.1 General Concepts of Reliability Data

As mentioned in Chapter 5, reliability indices are divided into two categories: indices for predicting future system performance and indices for reflecting historical performance. *Future indices* are evaluated using the quantified reliability assessment methods that have been presented in Chapter 5. The *reliability data* discussed in this section refer to the historical indices and statistical records required for calculating the historical indices.

The *historical performance* can be indicated at either equipment level or power delivery point level. The equipment indices reflect the average performance of individual pieces of equipment or an equipment group. These indices are important information in asset management and are used as the input data in the reliability evaluation for future systems. The delivery point indices reflect the average performance of individual delivery points, delivery point groups, or the whole system and thus include the effects of network configurations and consequences of load interruptions. The delivery point indices can also be used to establish the reliability criteria (i.e., target indices) in probabilistic transmission planning.

An equipment outage may or may not cause curtailments of loads. The outage information can be divided into two groups: one resulting in loss of load and the other not resulting in loss of load. The system delivery point indices are calculated using the outage data only in the first group, whereas the equipment indices are calculated using all equipment outage data regardless of consequences to customers.

The quality of data is a key in the data collection process. Outage statistics is a huge data pool in which bad or invalid records cannot be fully avoided. A data validation procedure is necessary to filter out invalid data. Another characteristic of reliability data is their dynamic feature. The volume of outage records increases with time, and therefore historical reliability indices change from year to year. In most cases, historical reliability indices are in the form of average values, although a discrete probability distribution of a reliability index can be calculated if there are sufficient raw data. It is important to appreciate that there is always some uncertainty regarding an average value. A standard deviation is often used to express the uncertainty of reliability data. In analysis of historical outage data, the standard deviation provides an idea about the degree of dispersion of the data and is useful information for a sensitivity analysis in reliability assessment. The interval estimation on the historical reliability indices can also be conducted using statistical records. As will be discussed in Chapter 8, information on interval estimation can be applied to building fuzzy models of equipment reliability indices to deal with the uncertainty of data in system reliability evaluation. The features of reliability data sharply distinguish these data from the data for power system analysis discussed in Section 7.2. It is necessary to build a specific reliability database for collection, storage, processing, calculation, and analysis of reliability data [93].

7.3.2 Equipment Outage Indices

Three basic indices are used for historical equipment performance: outage duration, outage frequency, and unavailability. The indices for forced outages and planned outages are calculated separately. It should be noted that there may be some confusion regarding definitions of the two terms *outage* and *failure*. They may or may not have different implications depending on specific cases. A *failure* may refer only to a real failure, whereas an *outage* may refer to out of-service condition caused by any reason. For example, a healthy component can be in an outage state by a switching action. A judgment needs to be made on which outage event should be accounted for and which one should be excluded from the calculation of indices.

Outages are often categorized into sustained and momentary outages. A commonly recognized definition is that a sustained outage is equal to or longer than one minute whereas a momentary outage is shorter than one minute. In general, momentary outages are associated only with automatic reclosure or temporary switching events. In most reliability databases, the outage frequency index is calculated separately for sustained and momentary outage events whereas momentary outages are ignored in calculating the outage duration index.

In an advanced reliability database such as the one of Canadian Electricity Association (CEA), outages are classified as equipment-related outages and terminal-related outages. An *equipment-related outage* is caused by failure of equipment itself, whereas a *terminal-related outage* is caused by failure of terminal devices of the equipment. *Terminal devices* are auxiliaries of equipment whose failures are not separately recorded and reported. The range of terminal devices is clearly defined for each type of equipment in the outage database. There are a few reasons for such a classification: (1) the calculation methods for line/cable-related and their terminal-related outage indices are different. (2) it is inappropriate to assign all outage events to equipment-related outages. For example, an outage caused by misoperation of personnel or protection devices can be assigned to terminal-related outages. Besides, as seen in Chapter 5, the reliability evaluations for transmission networks and substation configurations are performed separately. If necessary, the outages of lines/cables caused by substation equipment may be incorporated into the terminal-related outage indices for the lines/cables to approximately represent the effects of substation equipment failure in transmission network reliability evaluation.

7.3.2.1 Outage Duration (OD).

This index is also called the *average outage duration* or *mean time to repair* (recovery) in some literature and can be calculated in hours per occurrence (h/occ) by

$$\mathrm{OD} = \frac{\sum_{i=1}^{M} D_i}{M} \tag{7.31}$$

where D_i is the outage time (hours) for each outage event of equipment in a given transmission equipment group and M is the number of outage events in the timespan considered.

The equipment group is often classed by voltage. Equation (7.31) can also be used to calculate the OD index for individual equipment. This is a special case where an equipment group contains only one piece of equipment. This formula applies to either equipment-related or terminal-related outages, or both.

7.3.2.2 Outage Frequency (OF). This index is the average number of outages in one year. Conceptually, the outage frequency is different from the outage rate [6]. Unfortunately, these two terms have been mistakenly mixed up or used interchangeably in many documents and papers. The outage rate can be calculated from the outage frequency and outage duration or vice versa. Equation (C.11) in Appendix C clarifies the relationship between outage frequency and outage rate. From an engineering viewpoint, the outage frequency and outage rate are numerically very close because the outage duration is very short compared to the operating time in most practical engineering cases. Although the mixup is generally acceptable, caution should be taken when outage duration is long. In this situation, replacing the outage frequency by the outage rate or vice versa may lead to a relatively large error.

The calculations of outage frequency for lines/cables and other equipment are different as the concept of length applies for transmission lines and cables but not for other equipment.

The line/cable-related outage frequency for lines and cables in the number of occurrences per 100 km per year (occ./100 km/year) is calculated by

$$\mathrm{OF}_l = \frac{\sum_{i=1}^{K}\sum_{j=1}^{NY_i}(M_l)_{ij} \times 100}{\sum_{i=1}^{K} L_i \cdot NY_i} \tag{7.32}$$

where L_i is the length (in km) of the ith line (or cable) in a given line (or cable) group; K is the number of lines (or cables) in the group; $(M_l)_{ij}$ is the number of line/cable-related outages of the ith line (or cable) in the jth year; and NY_i is the number of service years for the ith line (or cable).

Equation (7.32) applies only to the line/cable-related outage frequency index. A line/cable may have multiple terminals (substations). The terminal-related outage frequency index for lines/cables, which is expressed in number of occurrences per terminal per year, is calculated by

$$\mathrm{OF}_t = \frac{\sum_{i=1}^{K}\sum_{j=1}^{NY_i}(M_t)_{ij}}{\sum_{i=1}^{K} NT_i \cdot NY_i} \tag{7.33}$$

where K and NY_i are the same as defined as in Equation (7.32), NT_i is the number of terminals of the ith line (or cable), and $(M_t)_{ij}$ is the number of terminal-related outages of the ith line (or cable) in the jth year.

It should be noted that very few outage databases available at present have a classification between equipment-related and terminal-related outages, due to the difficult requirements in raw data collection. In this case, a line outage caused by terminal (substation) facilities (such as misoperation of protection relays) may have been included as parts of statistics of line outages or just ignored. Even if included, however, it is not conceptually correct because terminal-related outages should not be weighted by line/cable lengths. This inaccuracy will result in some errors in the reliability data.

The outage frequency for other equipment (transformers, breakers, capacitors, reactors, etc.) is calculated by

$$\mathrm{OF}_e = \frac{\sum_{i=1}^{K} \sum_{j=1}^{NY_i} (M_e)_{ij}}{\sum_{i=1}^{K} NY_i} \tag{7.34}$$

where K is the number of equipment pieces in a given equipment group, $(M_e)_{ij}$ is the number of outages of the ith equipment in the jth year, and NY_i is the number of service years for the ith equipment. Equation (7.34) can be separately applied to either equipment-related or terminal-related outages as nonline/cable equipment generally has only one defined terminal. The equipment-related and terminal-related outage frequencies for nonline/cable equipment are expressed in occurrences/equipment/year and occurrences/terminal/year, respectively.

7.3.2.3 Unavailability (U). The forced outage unavailability is also called the *forced outage rate* (FOR). Note that FOR is a probability value that is not expressed by a unit of measure. The term FOR can create confusion as it may give a misimpression that it resembles an outage rate (numbers of outages in a unit time) rather than a probability. Unfortunately, the term FOR was chosen to represent the forced unavailable probability many years ago, and it is impossible to abandon this term now since it has been used for a long time. The most important thing is to recognize its significance.

The unavailability of any equipment is calculated by

$$U = \frac{\mathrm{OD} \times \mathrm{OF}}{8760} \tag{7.35}$$

Here, OF is OF_l for line/cable-related outages or OF_t for terminal-related outages of lines/cables or OF_e for other equipment. In the case of line/cable-related outages, because the OF_l obtained from Equation (7.32) is expressed in occurrences per 100 km per year, the U obtained using the OF_l is also in "per 100 km". If the unavailability of a specific line or cable is calculated from the average unavailability of a line or cable group, its unavailability is the unavailability per 100 km times its length (in km) and divided by 100. Similarly, because the OF_t for terminal-related outages is expressed in occurrences per terminal per year, the U obtained using the OF_t is also in "per terminal." It should be appreciated that if outage statistics of individual lines or cables are used to directly calculate the unavailability of each line or cable, it is unnecessary to use the concept of "per 100 km" or "per terminal."

It should be pointed out that Equation (7.35) can be applied to calculate the unavailability for either forced outages or planned outages depending on which outages have been used in calculating the OD and OF indices. It can be seen that only two in the three indices of OD, OF, and U are independent. As long as any two are obtained from statistics, the other one can be calculated.

7.3.2.4 Calculating Equipment Outage Indices. In general, a reliability database can create the equipment outage indices in two ways: the outage indices for individual equipment and the average outage indices for an equipment group in terms of voltage levels and/or different regions or the whole system. In reliability evaluation studies, individual equipment outage indices are preferred as long as they are based on sufficient statistics. If there are no or too few outage records available for some equipment, the average outage indices in an equipment group can be used. In the latter case, the line/cable-related outage frequency for each individual line or cable is calculated using the average line/cable-related outage frequency in occurrences/100 km/year times its length. The terminal-related outage frequency for each individual line or cable is calculated using the average terminal-related outage frequency in occurrences/terminal/year times the number of terminals. This step is not needed for nonline/cable equipment since there is no concept of length and a piece of equipment has only one terminal. The total equivalent outage indices of a line/cable or a piece of other equipment can be calculated by combining its equipment-related and terminal-related indices. This includes the following:

- In most data collection systems, outage frequencies rather than outage rates are provided as raw data. The outage rate is calculated by

$$\lambda = \frac{\mathrm{OF}}{1 - \mathrm{OF} \cdot r} \tag{7.36}$$

  where λ is the equipment-related (or terminal-related) outage rate in occ/year, OF is the equipment-related (or terminal-related) outage frequency in occ/year, and $r = \mathrm{OD}/8760$ is the equipment-related (or terminal-related) outage duration in years/repair.

- The total equivalent outage indices considering both equipment-related and terminal-related outages for a piece of equipment can be calculated by

$$\mathrm{OF}_{\mathrm{total}} = \mathrm{OF}_1(1 - \mathrm{OF}_2 r_2) + \mathrm{OF}_2(1 - \mathrm{OF}_1 r_1) \tag{7.37}$$

$$\lambda_{\mathrm{total}} = \lambda_1 + \lambda_2 \tag{7.38}$$

$$r_{\mathrm{total}} = \frac{\lambda_1 r_1 + \lambda_2 r_2 + \lambda_1 r_1 \lambda_2 r_2}{\lambda_1 + \lambda_2} \tag{7.39}$$

$$U_{\mathrm{total}} = \mathrm{OF}_1 r_1 + \mathrm{OF}_2 r_2 - \mathrm{OF}_1 \mathrm{OF}_2 r_1 r_2 \tag{7.40}$$

  where U, λ, r, and OF represent the unavailability, outage rate (occ/year), outage duration (years/occ), and outage frequency (occ/year), respectively. Subscripts 1 and 2 represent equipment-related and terminal-related indices, respectively.

If both the outage frequency and outage duration are small values, which is true in most cases, the outage rate is numerically very close to the outage frequency so that the outage frequency can be used to replace the outage rate and the following approximate formulas can be applied:

$$\lambda_{\mathrm{total}} \approx \mathrm{OF}_{\mathrm{total}} \approx \mathrm{OF}_1 + \mathrm{OF}_2 \tag{7.41}$$

$$r_{\text{total}} \approx \frac{\text{OF}_1 r_1 + \text{OF}_2 r_2}{\text{OF}_1 + \text{OF}_2} \tag{7.42}$$

$$U_{\text{total}} \approx \text{OF}_1 r_1 + \text{OF}_2 r_2 \tag{7.43}$$

It should be noted that when the outage duration is long (such as for a submarine cable), the approximate formulas may create relatively large errors and should be used with caution in such cases.

7.3.2.5 Examples of Equipment Outage Indices. Examples of the three outage indices [94] for various types of transmission equipment are shown in Tables 7.6–7.14. The indices are based on statistics collected by CEA from all major utilities across Canada and are the average values of the three indices for Canadian utilities. The timespan in calculating the three outage indices was a 5-year period from January 1, 2001 to December 31, 2005. Note that the indices based on historical outage records are varied with the start/end years and length of the period considered. The line/cable-related and terminal-related indices for lines or cables are shown in separate tables, whereas the total indices of combining equipment-related and terminal-related outages

TABLE 7.6. Transmission-Line Outage Indices for Line-Related Sustained Forced Outages (Canada)

Voltage Level (kV)	Outage Frequency (occ/100 km/year)	Average Outage Duration (h/occ)	Unavailability (per 100 km)
≤109	2.6151	11.0	0.003295
110–149	1.0089	8.5	0.000980
150–199	0.6836	78.7	0.006138
200–299	0.3396	35.2	0.001364
300–399	0.2026	15.4	0.000357
500–599	0.2199	14.2	0.000356
600–799	0.1965	43.1	0.000966

TABLE 7.7. Transmission-Line Outage Indices for Terminal-Related Sustained Forced Outages (Canada)

Voltage Level (kV)	Outage Frequency (occ/term/year)	Average Outage Duration (h/occ)	Unavailability (per terminal)
≤109	0.2735	28.7	0.000895
110–149	0.1421	23.5	0.000382
150–199	0.0491	20.2	0.000113
200–299	0.1648	17.3	0.000326
300–399	0.0925	62.5	0.000659
500–599	0.2012	18.9	0.000433
600–799	0.1607	23.3	0.000427

TABLE 7.8. Transmission-Line Outage Indices for Line-Related Momentary Forced Outages (Canada)

Voltage Level (kV)	Outage Frequency (occ/100 km/year)
≤109	2.4686
110–149	0.9923
150–199	0.6064
200–299	0.4378
300–399	0.1435
500–599	0.6736
600–799	0.1107

TABLE 7.9. Transmission-Cable Outage Indices for Cable-Related Sustained Forced Outages (Canada)

Voltage Level (kV)	Outage Frequency (occ/100 km/year)	Average Outage Duration (h/occ)	Unavailability (per 100 km)
≤109	1.4085	337.0	0.054184
110–149	1.6496	250.6	0.047188
200–299	1.6153	611.9	0.112831
300–399	12.9032	70.8	0.104213
500–599	1.0695	3.5	0.000427

TABLE 7.10. Transmission-Cable Outage Indices for Terminal-Related Sustained Forced Outages (Canada)

Voltage Level (kV)	Outage Frequency (occ/term/year)	Average Outage Duration (h/occ)	Unavailability (per terminal)
≤109	0.0056	2.0	0.000001
110–149	0.0361	304.3	0.001256
200–299	0.0588	2.1	0.000014
300–399	0.1275	49.5	0.000720
500–599	0.2500	0.8	0.000021

TABLE 7.11. Transformer Outage Indices for Sustained Forced Outages (Canada)

Voltage Level (kV)	Outage Frequency (occ/year)	Average Outage Duration (h/occ)	Unavailability
≤109	0.1073	210.2	0.002574
110–149	0.1439	188.6	0.003098
150–199	0.2349	341.1	0.009148
200–299	0.1284	167.1	0.002450
300–399	0.0825	163.4	0.001539
500–599	0.0841	153.4	0.001473
600–799	0.0424	279.4	0.001352

TABLE 7.12. Circuit Breaker Outage Indices for Sustained Forced Outages (Canada)

Voltage Level (kV)	Outage Frequency (occ/year)	Average Outage Duration (h/occ)	Unavailability
≤109	0.1297	494.9	0.007327
110–149	0.1015	160.1	0.001856
150–199	0.0502	213.9	0.001226
200–299	0.1530	132.5	0.002315
300–399	0.1200	211.2	0.002893
500–599	0.2658	100.1	0.003036
600–799	0.1794	138.3	0.002832

TABLE 7.13. Shunt Reactor Outage Indices for Sustained Forced Outages (Canada)

Voltage Level (kV)	Outage Frequency (occ/year)	Average Outage Duration (h/occ)	Unavailability
≤109	0.0705	334.9	0.002694
110–149	N/A	N/A	N/A
150–199	N/A	N/A	N/A
200–299	0.0067	2.0	0.000002
300–399	0.0619	382.4	0.002700
500–599	0.0344	332.0	0.001302
600–799	0.0317	1059.8	0.003835

TABLE 7.14. Shunt Capacitor Outage Indices for Sustained Forced Outages (Canada)

Voltage Level (kV)	Outage Frequency (occ/year)	Average Outage Duration (h/occ)	Unavailability
≤109	0.0887	1248.2	0.012633
110–149	0.2440	261.6	0.007287
150–199	N/A	N/A	N/A
200–299	0.4015	52.3	0.002398
300–399	0.0656	27.4	0.000205
600–799	N/A	N/A	N/A

are given for other equipment. More information on various equipment indices of utilities in Canada, the United States, and China can be found in References 94 and 95.

7.3.3 Delivery Point Indices

A *delivery point* is a point in the transmission network where electric power is transferred from the bulk electricity system to distribution systems or transmission customers. This point is usually defined at the low-voltage side of stepdown substations. There are two types of delivery points. In the first type, a delivery point is supplied from the bulk electricity system through one single circuit. In the second type, a delivery point is supplied from the bulk electricity system through multiple circuits. The delivery points are also classified in terms of their voltage levels.

Delivery point interruptions can be categorized in terms of two criteria: (1) forced or planned interruptions, or (2) momentary or sustained interruptions (using the same criteria for equipment outages, i.e., shorter or longer than one minute). The delivery point indices are therefore calculated separately for the different categories.

The sustained interruption time at delivery points can be accounted for by means of the two approaches: bulk electricity system interruption time and customer load interruption time. The bulk electricity system interruption duration may be longer than the customer load interruption duration, as the power supply for some customers may be recovered earlier through a load transfer in a distribution system or a self-generation facility of a transmission customer.

A reliability database collects the information of outage events of resulting in delivery point interruptions (loss of loads). The delivery point model, which is based on the relationship between delivery point interruptions and equipment outages, must be established. The duration, causes, and amounts of interruptions due to individual outage events can be analyzed and reported in the database.

7.3.3.1 Definitions of Delivery Point Indices.

There are five main delivery point indices [96]: T-SAIDI, T-SAIFI-MI, T-SAIFI-SI, SARI, and DPUI. These are system-based indices and depend on interruptions at delivery points. It should be noted that the SAIDI and SAIFI indices for transmission systems are different from the SAIDI and SAIFI that have been used for distribution systems, although they use the same terms. In order to avoid possible confusion, a prefix T has been added in the abbreviations of delivery point indices for transmission systems in this book:

1. *T-SAIDI (System Average Interruption Duration Index).* This is a measure of the average interruption duration that a delivery point experiences in a given period and is calculated in minutes per delivery point per year (min/DP/year) by

$$\text{T-SAIDI} = \frac{\sum_{i=1}^{K_D} \sum_{j=1}^{N_y} D_{ij} \cdot 60}{K_D N_y} \tag{7.44}$$

   where D_{ij} is the interruption time (in hours) at the ith delivery point in the jth year, K_D is the number of delivery points monitored in the system, and N_y is the number of years considered. Note that N_y is totally different from NY_i in Equations (7.32), (7.33) and (7.34).

2. *T-SAIFI-MI (System Average Interruption Frequency Index—Momentary Interruptions).* This is a measure of the average number of momentary interruptions

that a delivery point experiences in a given period and is calculated in interruptions per delivery point per year by

$$\text{T-SAIFI-MI} = \frac{\sum_{i=1}^{K_D} \sum_{j=1}^{N_y} (M_m)_{ij}}{K_D N_y} \tag{7.45}$$

where K_D and N_y are the same as defined in Equation (7.44); and $(M_m)_{ij}$ is the number of momentary interruptions at the ith delivery point in the jth year.

3. *T-SAIFI-SI (System Average Interruption Frequency Index—Sustained Interruptions).* This is a measure of the average number of sustained interruptions that a delivery point experiences in a given period and is calculated in interruptions per delivery point per year by

$$\text{T-SAIFI-SI} = \frac{\sum_{i=1}^{K_D} \sum_{j=1}^{N_y} (M_s)_{ij}}{K_D N_y} \tag{7.46}$$

where all the quantities are the same as defined in Equation (7.45) except that $(M_s)_{ij}$ is the number of sustained interruptions at the ith delivery point in the jth year. Note that $(M_m)_{ij}$ or $(M_s)_{ij}$ is essentially different from $(M_l)_{ij}$ in Equation (7.32), $(M_t)_{ij}$ in Equation (7.33), or $(M_e)_{ij}$ in Equation (7.34). $(M_m)_{ij}$ is associated with interruption (loss of load) events at a delivery point, whereas $(M_l)_{ij}$ or $(M_t)_{ij}$ or $(M_e)_{ij}$ is associated with equipment outage events that may or may not cause interruptions at delivery points.

4. *SARI (System Average Restoration Index).* This is a measure of the average duration of a delivery point interruption. In essence, it represents the average restoration time for a delivery point interruption and is calculated in minutes per interruption by

$$\text{SARI} = \frac{\sum_{i=1}^{K_D} \sum_{j=1}^{N_y} D_{ij} \cdot 60}{\sum_{i=1}^{K_D} \sum_{j=1}^{N_y} (M_s)_{ij}} \tag{7.47}$$

where all the quantities are the same as defined in Equations (7.44) and (7.46). Note that only sustained interruptions are considered in calculating the SARI.

5. *DPUI (Delivery Point Unreliability Index).* This is a measure of overall bulk electricity system performance in terms of a composite unreliability index in a given period and is calculated in system minutes per year by

$$\text{DPUI} = \frac{\sum_{i=1}^{K_D} \sum_{j=1}^{N_y} \sum_{k=1}^{(M_s)_{ij}} C_{ijk} D_{ijk} \cdot 60}{P_s N_y} \tag{7.48}$$

where K_D, N_y, and $(M_s)_{ij}$ are the same as defined in Equation (7.46); C_{ijk} and D_{ijk} are, respectively, the average load curtailment (MW) and interruption

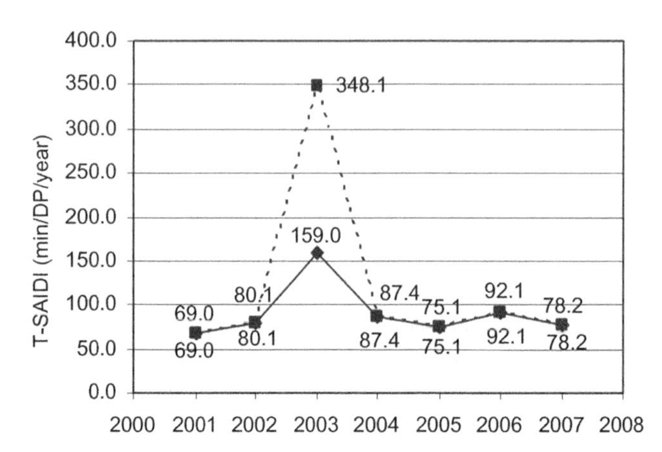

Figure 7.5. Annual trend of T-SAIDI index (Canada).

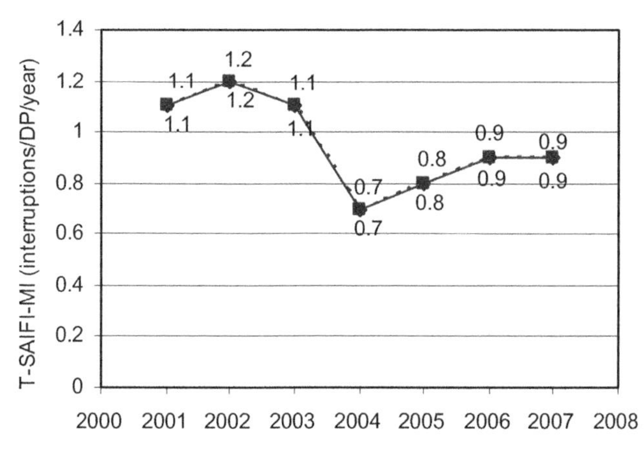

Figure 7.6. Annual trend of T-SAIFI-MI index (Canada).

duration (in hours) during the kth interruption event in the jth year at the ith delivery point; P_s is the average annual system peak load (MW) in N_y years considered.

Another method for the average annual DPUI is to calculate the DPUI for each year first and then take the average. This second method leads to a value slightly different from the DPUI obtained using Equation (7.48) if the annual system peak is different in each year. Obviously, the DPUI index is an equivalent duration (in minutes) for which, if the total system load at the time of annual system peak was interrupted, the energy not supplied due to such an interruption would be equal to the total unsupplied energy due to interruption events at all monitored delivery points in one year.

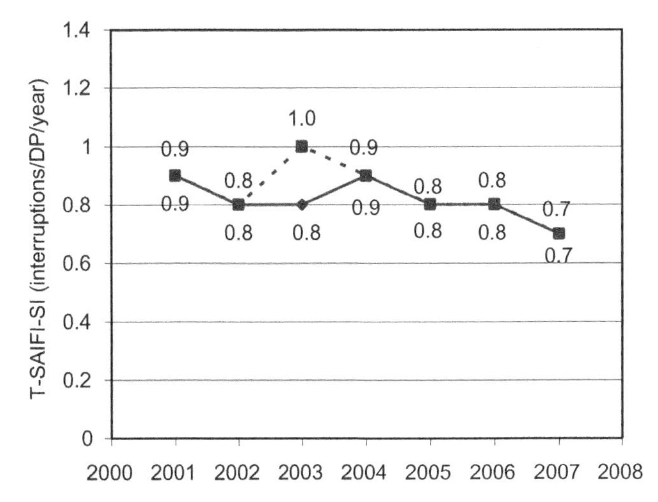

Figure 7.7. Annual trend of T-SAIFI-SI index (Canada).

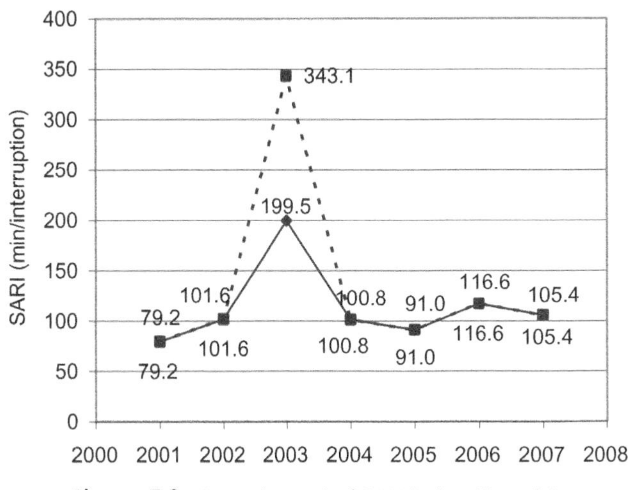

Figure 7.8. Annual trend of SARI index (Canada).

It is worth to pointing out that the period considered in calculating the five delivery point indices is often one year (i.e., $N_y = 1$), although the equations have been given in the expression that can accommodate data in multiple years.

7.3.3.2 Examples of Delivery Point Indices. Figures 7.5–7.9 show the annual trend of the five delivery point indices for major Canadian utilities from 2001 to 2007 [96]. The dashed lines in the figures show the delivery point indices including all interruption events, whereas the solid lines show those excluding the August 14, 2003 blackout event in the eastern part of North America.

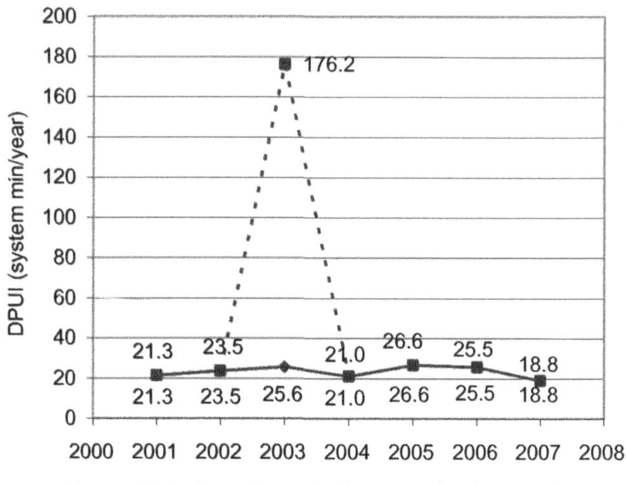

Figure 7.9. Annual trend of DPUI index (Canada).

7.4 OTHER DATA

Other data required in probabilistic transmission planning include the data of generation sources obtained from generation planning, information of interconnection of loads, independent power producers (IPPs) and wheeling customers (see Section 7.4.2, fourth paragraph, for definition), and economic data.

7.4.1 Data of Generation Sources

Transmission planning and generation planning have mutual dependence. The integrated planning is an iterative process. The initial results obtained in generation source planning are the input information for transmission planning. Conversely, the final decision in generation planning also relies on the results obtained in transmission network planning.

The data of generation sources required for transmission planning includes the types, locations, capacities (MW), monthly energy generated (MWh), reactive capacities (*P–Q* curves), and parameters of all new generators; the data of their auxiliary equipment (such as excitation systems); and their reliability data (outage frequency, outage duration, and unavailability). If a new generating plant is considered, the data of generating station configuration and relevant station facilities (transformers, breakers, etc.) is also required. In long-term planning, these data are often based on estimation and may not be accurate. The information of multiple possibilities for the types and locations of new generators and the range of capacities, energy profiles, and machine parameters are necessary data. If possible, subjective probability values for data uncertainty should be estimated. For example, if there are two or more possible types or locations of new generators, even a subjective probability for each type or each location will be useful information for probabilistic transmission planning. The reliability data can be based on the average values of existing same-type generators with a similar size. Additional data are required for nontraditional generators. For instance, historical hourly records of windspeeds at a site for windfarms should be collected and prepared.

7.4.2 Data for Interconnections

The generators discussed in Section 7.4.1 are those owned by utilities or generation companies. In addition to utility's generators, there are other types of customers for a transmission company under the power market environment: independent power producers (IPPs), new load connections, and wheeling customers.

On one hand, an IPP is a generator and conceptually can be treated in the same way as a utility generator. On the other hand, there are differences between IPPs and utility generators in transmission planning studies: (1) the type and location of an IPP generator are determined by its owner, and the IPP is not a part of integrated generation and transmission planning; and (2) there are special issues for IPP access to the transmission system. For instance, multiple IPP generators with different sizes and at different locations may request connection to the transmission system at the same time. In this case, a task of transmission planners is to investigate the feasibility and impacts of IPP interconnections. The feasibility is generally associated with technical assessment, whereas the impact studies include both technical and economic aspects. The purpose of investigation is to provide a short list for decisionmakers to finalize which IPP's connection is accepted and which one is rejected. This is often associated with a public hearing process. The data for an IPP generator are basically the same as those for a utility-owned generator. It is more difficult to obtain accurate data from an IPP.

A new load connection may or may not require an enhancement of the existing transmission system. It depends on the access location, size, and voltage level of the new load. Feasibility and impact studies are needed. The data required include the characteristics of the load, location of access, estimated MW and MVAR, load factor, information of connecting substation (if the customer has its own substation), and requirements in protections and harmonics.

The term "wheeling" customer refers to whoever requests point-to-point power transfer services. Its power is injected into the transmission system at one point and delivered to its own customer at another point. Similarly, a wheeling customer may or may not require an enhancement of the existing transmission system. The purpose of planning studies is to investigate the feasibility and impacts. The data required includes the wheeling size (MW), estimated loading pattern, transaction requirements, injection and delivered points, and voltage level. In many cases, a wheeling service is associated with the connection of a neighboring transmission network to the bulk system. In such cases, the relevant data of the neighboring system is needed.

7.4.3 Data for Economic Analysis

It can be seen from the economic analysis methods discussed in Chapter 6 that the data required for the economic analysis of a planning project include

- Subcomponents of capital investment
- Subcomponents of operation cost
- Unit interruption cost
- Unreliability cost, which can be estimated using the reliability worth evaluation techniques in Chapter 5
- Discount rate and its probability distribution, which can be estimated using its historical statistics and financial analysis
- Economic life, which can be estimated using the method in Section 6.6.2
- Depreciable life, which is specified by the financial department of a utility
- Planning timespan, which is specified in terms of the planning purpose

7.5 CONCLUSIONS

This chapter discussed the data required in probabilistic transmission planning. These data are always as important as the planning methods. If data quality is not ensured, the accuracy of system study, reliability evaluation, and economic analysis cannot be guaranteed, and this would result in a misleading decision in planning.

The data preparation is associated with efforts in many aspects and is sometimes tedious. However, this is an essential task for successful planning. The data for power system analysis and reliability evaluation have been discussed in specific detail, whereas other data are briefly summarized.

The equipment parameters are the basic data in power system analysis. The majority of the equipment parameters can be directly obtained from manufacturers' specifications. However, additional calculations are required to obtain the parameters of overhead lines, cables, and transformers. The equipment ratings and system operation limits are crucial criteria in determining whether a system reinforcement project is needed. In general, additional evaluations are required to acquire the maximum carrying capacities of overhead lines, cables, and transformers, whereas the system operation limits are the thresholds that are specified by utilities on the basis of system security requirements. Although several international standards (such as the IEEE and IEC standards) and commercial computing programs are available for calculating the maximum carrying capacities of equipment, it is still important for planners to understand the calculation methods used in the programs. The concept of bus load coincidence factors is presented. In the existing practice of many utilities, bus loads in power flow cases at a given system (or regional) load level are obtained using a proportional scaling approach that is apparently inaccurate. The use of bus load coincidence factors can help improve the accuracy of bus load models in power flow cases.

Reliability data are essential in the quantified reliability evaluation of transmission systems. The term *reliability data* refers to historical outage statistics and reliability indices based on the statistics. The historical reliability indices are categorized into equipment and delivery point indices. The equipment indices (outage frequency, outage duration, and unavailability) are the input data required for the reliability evaluation methods discussed in Chapter 5. Delivery point indices are indicators of historical system performance, which provide useful information for determination of investment strategies and probabilistic planning criteria. An application example will be demonstrated in Chapter 12. It should be appreciated that the historical delivery point indices discussed in this chapter are different from the reliability indices discussed in Chapter 5, which are used to measure future system reliability performance. The two most distinguishing characteristics of reliability data are the dynamic and uncertain features. Unlike the data for power system analysis, reliability data have to be updated from time to time since new outages result in changes in statistics and thus in both equipment and delivery point indices. The uncertainty of reliability data can be modeled using fuzzy variables, which will be discussed in Chapter 8.

With the methods and data presented in this and previous chapters, probabilistic transmission planning can be performed. Several actual applications will be demonstrated in the subsequent chapters.

8

FUZZY TECHNIQUES FOR DATA UNCERTAINTY

8.1 INTRODUCTION

As seen from Chapter 7, considerable data are required in probabilistic transmission planning. A successful planning decision depends on not only appropriate methods but also acceptable accuracy of data. There are uncertainties in various data, particularly for the data in load forecast and reliability assessment. Dealing with the uncertainties of data has been being a challenge in probabilistic transmission planning.

There are two types of uncertainty in power systems: randomness and fuzziness. A probabilistic model can be used for randomness but not for fuzziness. For instance, the two uncertainties exist in load forecast: the one that can be characterized by a probability model such as customer's regular consumption patterns, and the other one that does not follow any probability distribution such as an unexpected decision on the decrease or increase in production or relocation of industrial customers in the future. With existence of the second uncertainty factor, a fuzzy model is an appropriate method for representing the uncertainty in peak load forecast. For another instance, the outage data (outage frequency, outage duration, and unavailability) is usually modeled by a mean value in reliability evaluation. While this assumption is acceptable in many study cases, it may not be sufficient in some cases when the uncertainty of outage data

Probabilistic Transmission System Planning, by Wenyuan Li

becomes critical for results. Theoretically, the probability distribution of outage data is a better representation than only a mean value. However, it is not easy to obtain such a probability distribution because of limited statistical records for individual components. Also, the fuzziness, which cannot be represented by a probability distribution, does exist in raw outage data. It has been recognized for many years that the outage frequency or unavailability of a transmission overhead line is heavily related to weather conditions. Unfortunately, classification of weather conditions is fuzzy in nature because of vague language in weather descriptions such as normal or adverse weather, light or heavy rain or snow, and so on. When an outage is assigned to a normal or adverse weather condition in statistical data collection, it depends on a human being's fuzzy judgment. The outage frequencies or unavailability values of transmission equipment are also impacted by other environmental factors (such as animal activities) and operational conditions (such as loading levels). It is extremely difficult to precisely distinguish the effects of these factors and conditions on the outage data of individual components using a probability model since there is no or little correlation statistics available. Besides, other data such as line ratings or even line parameters are impacted by weather conditions as well. Many utilities may not have sufficient statistical records, but engineers generally can judge the range of data uncertainty. For all these cases, fuzzy models become a necessary complement to probabilistic models in order to copy with both types of uncertainty of input data in probabilistic transmission planning.

The purpose of this chapter is to address the uncertainties of load forecast and outage data using fuzzy models. Similar fuzzy modeling concepts can be also applied to the uncertainties of other data such as equipment parameters and ratings that are impacted by various environment factors. Although the discussion focuses on the application of combined fuzzy sets and probabilistic methods to transmission reliability assessment, the fuzzy modeling concept and techniques discussed here can be easily extended to other analyses in probabilistic transmission planning, such as fuzzy power flow, fuzzy contingency analysis, fuzzy optimal power flow, fuzzy network loss assessment, and so on.

The fuzzy models of system component outages and the mixed fuzzy and probabilistic models for loads are presented in Sections 8.2 and 8.3, respectively. A hybrid method of fuzzy sets and Monte Carlo simulation for system reliability assessment is proposed in Section 8.4. Two examples are given in Sections 8.5 and 8.6. The first one is an application to an actual utility system; the second one includes the impacts of fuzzy weather conditions.

8.2 FUZZY MODELS OF SYSTEM COMPONENT OUTAGES

The outage frequency and repair time (outage duration) are the two parameters of system component outages that can be directly collected in a data collection system. Other parameters such as unavailability can be calculated from these two parameters. The outage frequency and outage rate are numerically very close for transmission components, although they are conceptually different [6,11]. Therefore the difference between outage frequency and outage rate for transmission equipment is often ignored in practical engineering calculations of power system reliability evaluation. In this chapter, for simplicity, we will not differentiate them and will use the outage rate for the outage frequency. If differentiating them becomes necessary in some cases (such as a very long repair time for submarine cables), it is not difficult to calculate either one from the other [see Equation (C.11) in Appendix C for the relationship between the outage frequency and outage rate].

8.2.1 Basic Fuzzy Models

These models are discussed in detail in Reference 97.

8.2.1.1 Fuzzy Model for Repair Time. A sample mean of repair time can be easily calculated by a direct arithmetic average of repair times in different outage events

$$\bar{r} = \frac{1}{n}\sum_{i=1}^{n} r_i \tag{8.1}$$

where $\bar{r}$ is the point estimate of repair time (in hours), r_i is ith repair time, and n is the number of repairs in statistic data.

The confidence interval of the expected repair time can be estimated using a t-distribution or normal distribution criterion [98]. The estimation method is as follows. It is assumed that μ represents the real expected repair time and s is the sample standard deviation of repair time. If the t-distribution criterion is used, it can be affirmed that for a given significant level α, the random variable $(\bar{r}-\mu)\sqrt{n}/s$ is located between $-t_{\alpha/2}(n-1)$ and $t_{\alpha/2}(n-1)$ with the probability of $1-\alpha$, where $t_{\alpha/2}(n-1)$ is such a value that the integral of the t-distribution density function with $n-1$ degrees of freedom from $t_{\alpha/2}(n-1)$ to ∞ equals $\alpha/2$. Therefore we have

$$-t_{\alpha/2}(n-1) \le \frac{\bar{r}-\mu}{s/\sqrt{n}} \le t_{\alpha/2}(n-1) \tag{8.2}$$

Equation (8.2) can be equivalently expressed as

$$r' = \bar{r} - t_{\alpha/2}(n-1)\frac{s}{\sqrt{n}} \le \mu \le \bar{r} + t_{\alpha/2}(n-1)\frac{s}{\sqrt{n}} = r'' \tag{8.3}$$

Equation (8.3) indicates that the real expected repair time is located between the lower and upper bounds, which are determined by sampled repair times.

If the normal distribution criterion is used, the following lower and upper bounds for the real expected repair time can be obtained

$$r' = \bar{r} - z_{\alpha/2}\frac{s}{\sqrt{n}} \le \mu \le \bar{r} + z_{\alpha/2}\frac{s}{\sqrt{n}} = r'' \tag{8.4}$$

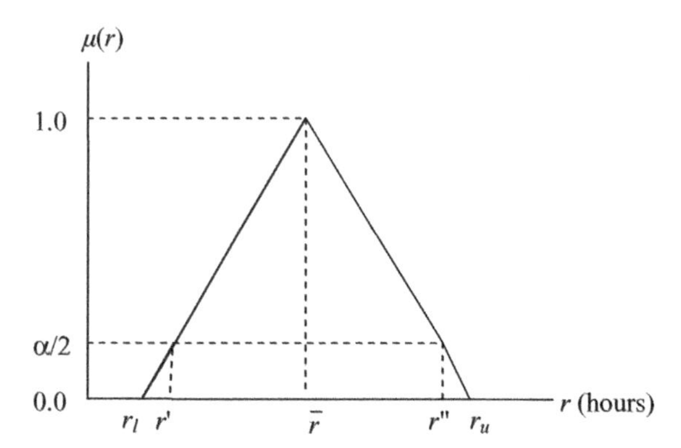

Figure 8.1. Membership function of repair time.

where $z_{\alpha/2}$ is such a value that the integral of the standard normal density function from $z_{\alpha/2}$ to ∞ equals $\alpha/2$.

The t-distribution and normal distribution criteria provide the two different interval estimates. In general, Equation (8.3) is used for a relatively small number of samples, whereas Equation (8.4) is used for a relatively large number of samples. In an actual application, both estimates can be made and a more conservative (wider) interval estimate can be obtained.

With the point and interval estimates, a triangle membership function of repair time r (in hours) can be easily created as shown Figure 8.1. The point estimate ($\bar{r}$ in the figure) corresponds to the membership function grade of 1.0. The significance level α is always a small fraction such as 0.05 (5%). Half of α corresponds to a small area at each of the two sides under the t-distribution or normal distribution density function curve. The significance level is conceptually similar to the fuzzy degree represented by a small membership function grade since both of them reflect a subjective confidence. Therefore it can be assumed that the lower and upper bounds obtained from Equation (8.3) or (8.4) correspond to the two points (r' and r'' in the figure) at the membership grade of $\alpha/2$ (such as 0.025) in the membership function. With the three points of (r', $\alpha/2$), ($\bar{r}$, 1.0), and (r'', $\alpha/2$), the two linear algebraic equations in the form of $y = a + bx$ can be built and the two endpoints (r_l, 0.0) and (r_u, 0.0) in the membership function can be calculated, as shown in Figure 8.1. Since $\alpha/2$ is such a small value (0.025), r_l and r' should be very close, and so are r_u and r''. Therefore, it is also acceptable to directly use r' and r'' as the two endpoints of the membership function. It can be seen that the membership function of repair time obtained from Equation (8.1) and (8.3) or (8.4) is a symmetric one. It should be noted that the two bounds of repair time can also be determined or modified according to the judgment of experienced maintenance engineers if necessary. For instance, historical repair times may not be good enough for estimating a repair time required in the future when resources and/or conditions for repairing work significantly change. In this case, the maintenance engineer's estimation on the range or an adjustment on the calculated range may be closer to reality.

8.2.1.2 Fuzzy Model for Outage Rate. The average outage rate of individual components cannot be obtained by a sample mean of outage records for each single component. The outage rate is estimated as average outages per year over a time period

$$\bar{\lambda} = \frac{n}{T} \tag{8.5}$$

where $\bar{\lambda}$ is the point estimate of outage rate (outages/year); n is the number of outages of a component in the exposed time T (in years), which refers to the total elapsed time minus the total outage time. In most cases, the total outage time is a very small portion and T can be approximated by the total elapsed time during which outage events are considered. This implies that the outage rate is approximated by the outage frequency.

The confidence interval of expected outage rate can be estimated using the following method. According to statistics theory, there exists the following relationship between χ^2 (chi square) distribution and Poisson distribution [99]:

$$\chi^2(2F) = 2\lambda T \tag{8.6}$$

Here, λ is the expected outage rate, T is the total time period considered, and F is the number of outages during T.

Equation (8.6) indicates that the quantity of 2 times the outages in the duration T follows the χ^2 distribution with $2F$ degrees of freedom. Therefore, for a given significance level α, it can be asserted that the outage rate λ falls into the following random confidence interval with the probability of $1-\alpha$:

$$\lambda' = \frac{\chi^2_{1-\alpha/2}(2F)}{2T} \le \lambda \le \frac{\chi^2_{\alpha/2}(2F)}{2T} = \lambda'' \tag{8.7}$$

Equation (8.7) gives the lower and upper bounds of the outage rate, which are determined using the outage data. A more conservative (wider) estimate for the upper bound can be expressed as

$$\lambda' = \frac{\chi^2_{1-\alpha/2}(2F)}{2T} \le \lambda \le \frac{\chi^2_{\alpha/2}(2F+2)}{2T} = \lambda'' \tag{8.8}$$

Similarly, the two endpoints (λ_l and λ_u) in the membership function of λ can be calculated using λ', λ'', $\bar{\lambda}$, and $\alpha/2$. The range represented by the two endpoints can also be adjusted by experienced engineers, particularly when the effects of weather, environment, or operation conditions on an outage rate are considered so that the outage rate range may be biased from the interval estimate from historical statistics.

A triangle membership function of outage rate can be easily created using the point and interval estimates of outage rate, as shown Figure 8.2. Note that the membership function of outage rate is not symmetric. Generally, the range between the point

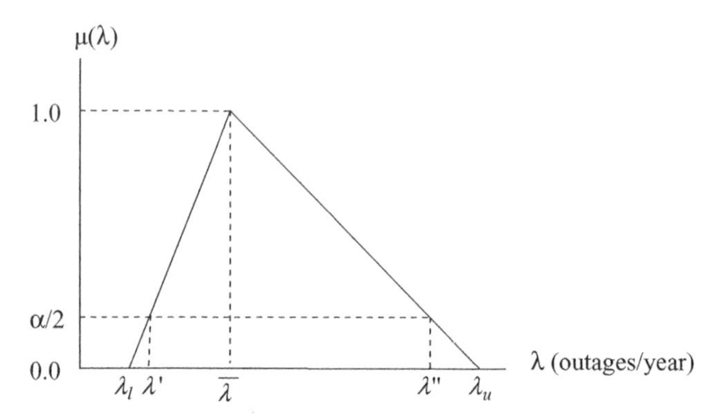

Figure 8.2. Membership function of outage rate.

estimate and its upper bound is larger than that between the point estimate and its lower bound.

8.2.1.3 Fuzzy Model for Unavailability. The unavailability U of system components can be calculated using the outage frequency f and repair time as

$$U = \frac{f \cdot r}{8760} \approx \frac{\lambda \cdot r}{8760} \tag{8.9}$$

where the units of outage frequency f (or outage rate λ) and repair time r are in outages/year and hours/outage, respectively. It has been assumed in Equation (8.9) that the outage frequency f approximately equals to the outage rate λ. This assumption is generally acceptable as the repair time r is very short compared to the operation time of a component. Otherwise, f needs to be calculated from λ using Equation (C.11) in Appendix C.

The membership function of unavailability can be obtained from those of λ and r in Figures 8.1 and 8.2. This is associated with the operation rules of intervals for a given membership grade (see Section B.2.2 in Appendix B). It should be noted that with the multiplication operation of the two triangle membership functions, the membership function of unavailability is no longer a strict triangle but one with some distortions leftward, as shown in Figure 8.3.

8.2.2 Weather-Related Fuzzy Models

These models are discussed in detail in References 100 and 101.

8.2.2.1 Exposure to One Weather Condition. A specific weather condition has a significant impact on outages and repairs of system components. Some transmission outage databases provide the information of weather conditions during outage events. The weather is generally divided into normal and adverse conditions. If a whole

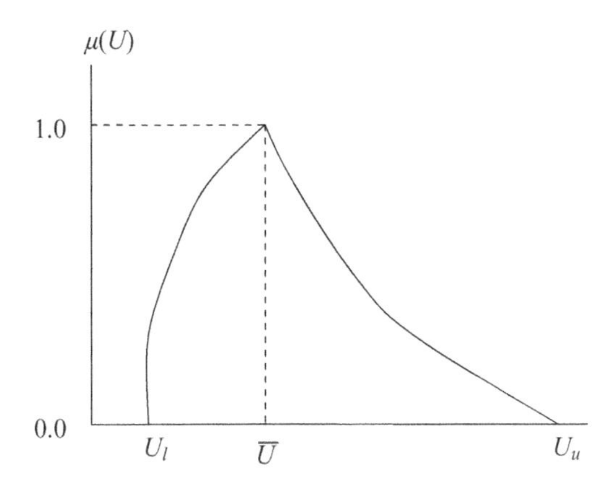

Figure 8.3. Membership function of unavailability.

overhead line is exposed to one weather condition, the basic fuzzy models described above can be directly used. Statistical outage samples are grouped in terms of weather conditions, and the membership functions of outage rate, repair time, and unavailability under different weather conditions can be estimated separately.

8.2.2.2 Exposure to Two Weather Conditions. A transmission line can be exposed to two weather conditions in real life. Assume that a transmission line traverses two weather forecast regions with a fraction R in region 1 and that region 1 is in the normal weather condition while region 2 is in the adverse weather condition. In terms of the series reliability relationship, the total equivalent outage rate λ_E and repair time r_E can be estimated by

$$\lambda_E = \varphi_1(\lambda_N, \lambda_A, R) = \lambda_N R + \lambda_A (1 - R) \tag{8.10}$$

$$r_E = \varphi_2(\lambda_N, r_N, \lambda_A, r_A, R) = \frac{\lambda_N R \cdot r_N + \lambda_A (1 - R) \cdot r_A}{\lambda_N R + \lambda_A (1 - R)} \tag{8.11}$$

where λ_N and λ_A are the outage rates in the normal and adverse weather conditions, respectively; r_N and r_A are the repair times in the normal and adverse weather conditions, respectively. These parameters can be estimated using the basic fuzzy models in Section 8.2.1. It should be appreciated that not only these four parameters but also R are fuzzy variables. This is due to the fact that it is impossible to have a crisp boundary between two weather conditions at exactly the same points as a political geographic division when the weather forecast is reported for each geographic region. In other words, the adverse weather may either extend into or withdraw from its neighboring region, and how much extension or withdrawal is a fuzzy range in nature.

The calculations of fuzzy numbers are essentially associated with finding the supremum and infimum of a fuzzy function. The following method is used to perform fuzzy operations for Equations (8.10) and (8.11). It can be seen by observing Equation

(8.10) that λ_E varies monotonically with each of λ_N, λ_A, and R, and that λ_N and λ_A must be the maximum for λ_E to reach its supremum and must be the minimum for λ_E to reach its infimum. For a given grade in the membership functions of λ_N, λ_A, and R, the infimum and supremum of λ_E can be found by

$$[\lambda_{E\min},\lambda_{E\max}] = [\min\{\varphi_1(\lambda_{N\min},\lambda_{A\min},R_{\min}),\varphi_1(\lambda_{N\min},\lambda_{A\min},R_{\max})\},\ \max\{\varphi_1(\lambda_{N\max},\lambda_{A\max},R_{\min}),\varphi_1(\lambda_{N\max},\lambda_{A\max},R_{\max})\}] \quad (8.12)$$

Equation (8.11) is equivalently rewritten as

$$r_E = \varphi_3(r_N,r_A,S) = \frac{r_N + S\cdot r_A}{1+S} \quad (8.13)$$

where

$$S = \varphi_4(\lambda_A,\lambda_N,R) = \frac{\lambda_A}{\lambda_N}\left(\frac{1}{R}-1\right) \quad (8.14)$$

Obviously, S varies monotonically with each of λ_N, λ_A, and R, and r_E varies monotonically with each of r_N, r_A, and S, which can be verified by calculating the first-order derivatives. The infimum and supremum of S can be found first by

$$[S_{\min},S_{\max}] = [\varphi_4(\lambda_{N\max},\lambda_{A\min},R_{\max}),\varphi_4(\lambda_{N\min},\lambda_{A\max},R_{\min})] \quad (8.15)$$

Then the infimum and supremum of r_E for the given grade in the membership functions of λ_N, λ_A and R can be found by

$$[r_{E\min},r_{E\max}] = [\min\{\varphi_3(r_{N\min},r_{A\min},S_{\min}),\varphi_3(\lambda_{N\min},\lambda_{A\min},S_{\max})\},\ \max\{\varphi_3(r_{N\max},r_{A\max},S_{\min}),\varphi_3(r_{N\max},r_{A\max},S_{\max})\}] \quad (8.16)$$

8.2.2.3 Exposure to Multiple Weather Conditions. The method in Section 8.2.2.2 can be extended to the case where a transmission line traverses more regions or is exposed to more weather conditions. Take three weather regions as an example. It is assumed that a transmission line traverses three regions, each of which is exposed to a different weather condition. The fractions of the line in regions 1 and 2 are denoted respectively by R_1 and R_2. The outage rates and repair times of the line in the three weather conditions are denoted respectively by $\lambda_1,\lambda_2,\lambda_3$ and r_1,r_2,r_3. The total equivalent outage rate and repair time can be estimated by

$$\lambda_E = \varphi_5(\lambda_1,\lambda_2,\lambda_3,R_1,R_2) = \lambda_1 R_1 + \lambda_2 R_2 + \lambda_3(1-R_1-R_2) \quad (8.17)$$

$$r_E = \varphi_6(r_1,r_2,r_3,S_1,S_2) = \frac{r_1\lambda_1 R_1 + r_2\lambda_2 R_2 + r_3\lambda_3(1-R_1-R_2)}{\lambda_1 R_1 + \lambda_2 R_2 + \lambda_3(1-R_1-R_2)} = \frac{r_1 + r_2 S_1 + r_3 S_2}{1+S_1+S_2} \quad (8.18)$$

where

$$S_1 = \varphi_7(\lambda_1, \lambda_2, R_1, R_2) = \frac{\lambda_2 R_2}{\lambda_1 R_1} \tag{8.19}$$

$$S_2 = \varphi_8(\lambda_1, \lambda_3, R_1, R_2) = \frac{\lambda_3}{\lambda_1}\left(\frac{1}{R_1} - \frac{R_2}{R_1} - 1\right) \tag{8.20}$$

Similarly, it can be confirmed that the functions φ_5, φ_6, φ_7, and φ_8 are all the monotonic functions of each of the respective variables in the defined ranges. For a given grade in the membership functions of λ_1, λ_2, λ_3, R_1, and R_2, the infimum and supremum of λ_E is found by

$$\begin{aligned}
[\lambda_{E\,\min}, \lambda_{E\,\max}] = [&\min\{\varphi_5(\lambda_{1\min}, \lambda_{2\min}, \lambda_{3\min}, R_{1\min}, R_{2\min}),\\
&\varphi_5(\lambda_{1\min}, \lambda_{2\min}, \lambda_{3\min}, R_{1\min}, R_{2\max}), \varphi_5(\lambda_{1\min}, \lambda_{2\min}, \lambda_{3\min}, R_{1\max}, R_{2\min}),\\
&\varphi_5(\lambda_{1\min}, \lambda_{2\min}, \lambda_{3\min}, R_{1\max}, R_{2\max})\},\\
&\max\{\varphi_5(\lambda_{1\max}, \lambda_{2\max}, \lambda_{3\max}, R_{1\min}, R_{2\min}),\\
&\varphi_5(\lambda_{1\max}, \lambda_{2\max}, \lambda_{3\max}, R_{1\min}, R_{2\max}),\\
&\varphi_5(\lambda_{1\max}, \lambda_{2\max}, \lambda_{3\max}, R_{1\max}, R_{2\min}),\\
&\varphi_5(\lambda_{1\max}, \lambda_{2\max}, \lambda_{3\max}, R_{1\max}, R_{2\max})\}]
\end{aligned} \tag{8.21}$$

and the infimums and supremums of S_1 and S_2 are found by

$$[S_{1\min}, S_{1\max}] = [\varphi_7(\lambda_{1\max}, \lambda_{2\min}, R_{1\max}, R_{2\min}), \varphi_7(\lambda_{1\min}, \lambda_{2\max}, R_{1\min}, R_{2\max})] \tag{8.22}$$

$$\begin{aligned}
[S_{2\min}, S_{2\max}] = [&\min\{\varphi_8(\lambda_{1\max}, \lambda_{3\min}, R_{1\min}, R_{2\max}), \varphi_8(\lambda_{1\max}, \lambda_{3\min}, R_{1\max}, R_{2\max})\},\\
&\max\{\varphi_8(\lambda_{1\min}, \lambda_{3\max}, R_{1\min}, R_{2\min}), \varphi_8(\lambda_{1\min}, \lambda_{3\max}, R_{1\max}, R_{2\min})\}]
\end{aligned} \tag{8.23}$$

With the ranges of r_1, r_2, r_3, S_1, and S_2 for the given grade in the membership functions, the infimum and supremum of r_E is found by

$$\begin{aligned}
[r_{E\,\min}, r_{E\,\max}] = [&\min\{\varphi_6(r_{1\min}, r_{2\min}, r_{3\min}, S_{1\min}, S_{2\min}), \varphi_6(r_{1\min}, r_{2\min}, r_{3\min}, S_{1\min}, S_{2\max}),\\
&\varphi_6(r_{1\min}, r_{2\min}, r_{3\min}, S_{1\max}, S_{2\min}), \varphi_6(r_{1\min}, r_{2\min}, r_{3\min}, S_{1\max}, S_{2\max})\},\\
&\max\{\varphi_6(r_{1\max}, r_{2\max}, r_{3\max}, S_{1\min}, S_{2\min}), \varphi_6(r_{1\max}, r_{2\max}, r_{3\max}, S_{1\min}, S_{2\max}),\\
&\varphi_6(r_{1\max}, r_{2\max}, r_{3\max}, S_{1\max}, S_{2\min}), \varphi_6(r_{1\max}, r_{2\max}, r_{3\max}, S_{1\max}, S_{2\max})\}]
\end{aligned} \tag{8.24}$$

A general operation rule for finding the infimum and supremum of the outage rate or repair time of a line that traverses multiple regions with different weather conditions can be summarized as follows. Assume that $Y = \varphi(X_1,\ldots,X_i,\ldots,X_m)$ with $X_1 = [x_{1\,\min}, x_{1\,\max}]$, …, $X_i = [x_{i\,\min}, x_{i\,\max}]$, …, $X_m = [x_{m\,\min}, x_{m\,\max}]$, where $X_1,\ldots,X_i,\ldots,X_m$ are the fuzzy outage rates or repair times in different weather conditions or the fuzzy division ratios between weather conditions, and that Y represents a fuzzy function expression for equivalent total outage rate or repair time of a line. For a membership function grade (indicated by μ), the two bounds of $Y(\mu)$ are calculated by

$$[Y(\mu)_{\min}, Y(\mu)_{\max}] = [\inf\{\varphi(X_1(\mu),\ldots,X_m(\mu))\}, \sup\{\varphi(X_1(\mu),\ldots,X_m(\mu))\}] \quad (8.25)$$

The symbols inf{ } and sup{ } indicate finding the infimum and supremum of $\varphi(X_1,\ldots, X_m)$ in all possible combinations when each $X_i(i = 1,\ldots,m)$ is assigned as its minimum value $x_{i\min}$ or maximum value $x_{i\max}$ at a given membership function grade of X_i $(i = 1,\ldots, m)$. As indicated above, this process needs only partial enumeration if appropriate intermediate variables similar to S_1 and S_2 are introduced.

It should be noted that the statistics in existing outage databases may have the information to support the division of only two weather conditions (normal and adverse). However, it is believed that weather conditions can be classified into more categories as more detailed weather information is collected and becomes available in the future.

8.3 MIXED FUZZY AND PROBABILISTIC MODELS FOR LOADS

8.3.1 Fuzzy Model for Peak Load

Utilities provide the most probable peak load forecast with its high and low bounds. This naturally corresponds to the fuzzy concept. An asymmetric triangle membership function of peak load can be created with the most probable peak (L_p) for the membership grade of 1.0 and the high (L_h) and low (L_l) bounds for the membership grade of zero (or a small value). This is shown in Figure 8.4. It should be pointed out that the two asymmetric bounds cannot be modeled using a normal distribution.

8.3.2 Probabilistic Model for Load Curve

This model is discussed in detail in Reference 102. A load curve reflects the load profile at different timepoints in a period such as one year. The load duration curve can be

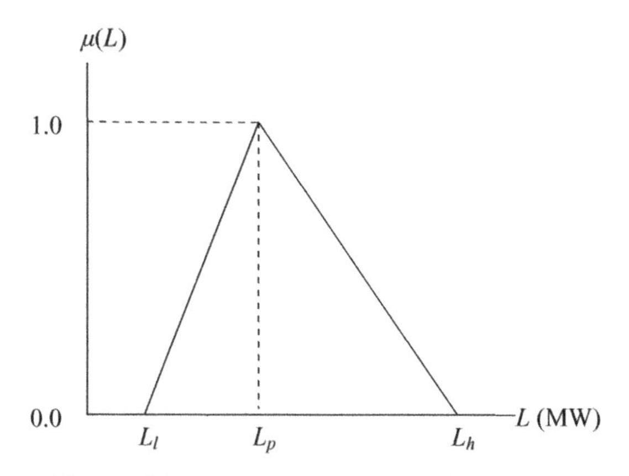

Figure 8.4. Membership function of peak load.

created using historical hourly load records, and then a discrete probability distribution for the load curve can be obtained. Assume that the load duration curve with *NP* hourly load points is divided into *NL* load levels. The usual approach is to specify the load levels in terms of per unit (i.e., a fraction) with respect to the peak load. The probability of each load level is calculated by

$$p_i = \frac{NS_i}{NP} \tag{8.26}$$

where NS_i is the number of load points located between the *i*th level and the next level.

The discrete probability distribution of loads for the load curve can be expressed as

$$p[\text{load} = L_i(\text{per unit})] = p_i \qquad (i = 1, \ldots, NL) \tag{8.27}$$

where L_i(per unit) represents the *i*th load level in a per unit with respect to the peak.

An improved approach is to apply the *K*-mean clustering technique described in Section 3.3.1, which can be used to calculate the load levels and their sample standard deviations. Each load level is the mean value of load points in a cluster with a sample standard deviation. The technique includes following steps:

1. Select initial cluster means M_i (in MW), where *i* denotes the *i*th cluster (i = 1,..., *NL*).
2. Calculate the distance D_{ki} from each hourly load point L_k (k = 1,...,*NP*) to each cluster mean M_i by

$$D_{ki} = |M_i - L_k| \tag{8.28}$$

3. Group load points by assigning them to the nearest cluster and calculate new cluster means by

$$M_i = \frac{1}{NS_i} \sum_{k=1}^{NS_i} L_k \qquad (i = 1, \ldots, NL) \tag{8.29}$$

   where NS_i is the number of load points in the *i*th cluster.
4. Repeat steps 2 and 3 until all cluster means remain unchanged between iterations.
5. Calculate the per unit values of load levels

$$L_i(\text{per unit}) = \frac{M_i}{L(\text{peak})} \qquad (i = 1, \ldots, NL) \tag{8.30}$$

   where *L*(peak) denotes the load peak (in MW) in the load duration curve.

6. Calculate the sample standard deviations of each load level:

$$\sigma_i = \sqrt{\frac{1}{NS_i - 1}\sum_{k=1}^{NS_i}(L_k - M_i)^2} \qquad (i = 1,\ldots,NL) \tag{8.31}$$

The uncertainty of loads around each level can be considered using σ_i. For example, if a Monte Carlo simulation method is used, a standard normal distribution random number Z_m is created using the approximate inverse transformation method (see Section A.5.4.2 in Appendix A). The sampled value of the system load at the ith level and in the mth sample is given by

$$L_{im} = Z_m\sigma_i + L_i(\text{MW}) \tag{8.32}$$

where L_i(MW) is the MW load at the ith level.

It can be seen that the load curve is represented by a composite probabilistic model in which a discrete probability distribution for multistep levels is combined with a normal distribution for the uncertainty around each level. When the pure probabilistic load model is used, L_i(MW) is merely M_i at the ith load level. When the probabilistic load curve model is combined with the fuzzy model of peak load forecast as shown in Figure 8.4, the L(peak) is a range for each membership function grade. In this case, L_i(MW) can be calculated from L_i(per unit) and the membership function of peak load [see Equation (8.34) below]. The implicit assumption is that the two bounds at each grade in the membership function for each load level proportionally vary as the mean value of the load level follows the load duration curve. This is a reasonable assumption. If necessary, however, independent membership functions at different load levels can be applied as long as there are sufficient data available to create them.

8.4 COMBINED PROBABILISTIC AND FUZZY TECHNIQUES

The fuzzy models of system peak load and component outages can be incorporated into the traditional methods for transmission system analyses, including power flow, contingency analysis, optimal power flow, and stability assessments. The application of the fuzzy models in transmission reliability evaluation is discussed in this section as an example, and two numerical examples are provided in Sections 8.5 and 8.6.

8.4.1 Probabilistic Representation for Region-Divided Weather States

A transmission system can spread over several regions in which a weather forecast is reported. Figure 8.5 shows a representative meteorological five-region division for a transmission system. Region-divided weather states for the transmission system can be recognized in light of weather forecast or meteorological records according to the

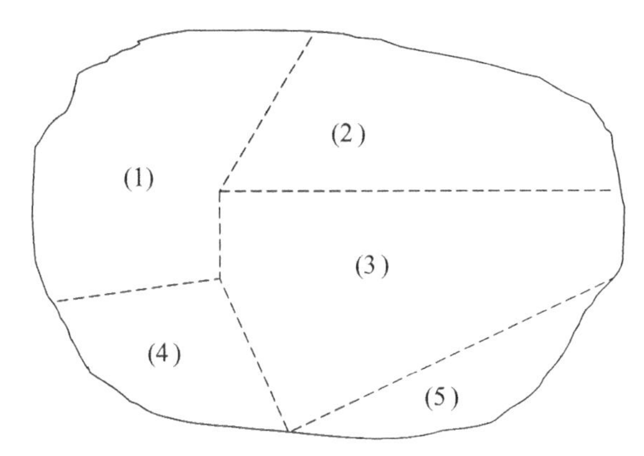

Figure 8.5. Meteorological region divisions.

following ways: a single region exposed to adverse weather, two adjoining regions existing in adverse weather, three adjoining regions encountering adverse weather, and so on. In the case of the five-region division, the maximum number of region-divided weather states for the transmission system is $C_5^0 + C_5^1 + C_5^2 + C_5^3 + C_5^4 + C_5^5 = 32$, where C_5^i represents the number of combination of i item from five items.

The probability for each region-divided weather state can be obtained from meteorological records. The number of regions basically depends on the available data rather than on the size of a transmission system. Assume that there are N region-divided weather states whose probabilities are not zero and that their probabilities are expressed by $P_1, P_2, \ldots, P_N$. In general, the probability of all the regions (the whole transmission system) exposed to the normal weather is much larger than the sum of the probabilities of all other weather states with at least one region exposed to the adverse weather. Either an enumeration method or a Monte Carlo simulation technique can be used to select the weather states that the transmission system experiences. In the Monte Carlo simulation, the values of $P_1, P_2, \ldots, P_N$ are set sequentially in the interval [0,1]. A uniformly distributed random number between [0,1] is created. If the random number falls in the segment corresponding to P_i, the transmission system is in the ith region-divided weather state for this sample. In the enumeration method, the weather states are simply considered one by one. The geographic range under the same weather condition is generally wide, and therefore a transmission system does not cover too many weather forecast regions. Even if the number of the regions is relatively large, many region-divided weather states will most likely not exist in real life. From this perspective, the enumeration method for weather state selection is more efficient. It should be emphasized that the weather states are probabilistic but the exposures of overhead transmission lines to the weather states are fuzzy, as explained and modeled in Section 8.2.2.

8.4.2 Hybrid Reliability Assessment Method

8.4.2.1 Evaluating Membership Functions of Reliability Indices. It can be seen from previous sections that the input data (system component outage parameters and system loads) are random fuzzy variables, and therefore reliability indices are also random fuzzy variables. The membership function of a fuzzy variable indicates the fuzzy ranges of the variable (minimum and maximum bounds) at different membership grades. In the power system reliability evaluation, there exists a monotonic relation between the system unreliability and the unavailability of each system component or the system load. In other words, the system reliability indices such as EENS (expected energy not supplied) or PLC (probability of load curtailment) monotonically vary with the unavailability of any component or system load. The reliability indices obtained when all the input data reach their two bounds for a membership grade will be the minimum and maximum bounds of reliability indices for that membership grade. Therefore, a traditional reliability evaluation method can be used to assess the values of reliability indices corresponding to a given membership grade, and the membership functions of reliability indices can be built up by considering sufficient membership grades. The proposed method is summarized as follows:

1. The outage parameters of all system components are prepared. This includes the following three categories:
   a. For the components that are not represented using a fuzzy model, the traditional model (one single average value for outage rate or repair time) is used. The unavailability values of all components in this category are calculated directly using Equation (8.9). Each of them is a single value and does not change in the following iterative process.
   b. For the components that are represented using a fuzzy model but are not affected by weather conditions, the basic fuzzy models in Section 8.2.1 are used. Their membership functions of outage rates and repair times are created. The membership functions of unavailability for the components in this category are calculated using Equation (8.9) and the fuzzy number operation rules.
   c. For the components (overhead lines) that are affected by weather conditions, the weather-related fuzzy models in Section 8.2.2 are used. Their membership functions of outage rates and repair times in a single weather condition (normal or adverse condition) are created.
2. The membership function of the system peak load and the discrete probability distribution model of a system load curve are created using the models given in Sections 8.3.1 and 8.3.2, respectively.
3. A region-divided weather state for the transmission system is selected using the method described in Section 8.4.1. For the selected weather state, the membership functions of outage rates and repair times for the components (overhead lines) in the third category are updated. If a line is exposed to a single weather condition, the membership functions of the outage parameters corresponding to that weather condition are used. If a line traverses multiple weather conditions, the membership functions of the outage parameters are calculated using the models in Section 8.2.2.2 or 8.2.2.3.

4. The membership functions of unavailability for the components in the third category, which vary with weather states selected in step 3, are calculated using Equation (8.9) and the fuzzy number operation rules.
5. For a given membership grade in the fuzzy load model of system peak created in step 2, the two inverse function values $\mu^{-1}(\mu(L))$ of the membership function $\mu(L)$ for the peak load are calculated.
6. For the same membership grade in the fuzzy models of component unavailability, the two inverse function values of the membership function $\mu_j(U)$ for unavailability of each component are calculated by

$$U_j = \mu_j^{-1}(\mu_j(U)) \tag{8.33}$$

   where U_j represents the unavailability of the jth component at the current membership function grade. There are two unavailability values for each grade, with one corresponding to the lower bound and the other to the upper bound. For simplicity, Equation (8.33) has not been expressed in a form to differentiate between the two.
7. For the load level L_i(in per unit) in the discrete probability distribution for the load curve created in step 2, the two MW values for the lower and upper bounds of the load level are calculated by

$$L_i(\text{MW}) = L_i(\text{per unit}) \cdot \mu^{-1}(\mu(L)) \tag{8.34}$$

   Similarly, Equation (8.34) has not been expressed in a form to differentiate between the two values.
8. A probabilistic reliability evaluation is performed for the load level obtained in step 7 using the traditional Monte Carlo simulation method, which is similar to the one described in Section 5.5.6. This includes the following:
   a. Select system contingency states by sampling states (up or down state) of all components using their unavailability values obtained in step 1 (for category 1) or step 6 (for categories 2 and 3).
   b. Sample system load states to incorporate the uncertainty of the load level using Equation (8.32).
   c. Perform contingency analyses to check limit violations (such as line ratings, bus voltage limits, etc.)
   d. If there exists any limit violation, solve an optimal power flow model given in Section 5.5.5 to minimize the total load curtailment while meeting all constraints and limits.
   e. Calculate average reliability indices for all system states randomly selected in the Monte Carlo simulation process.
9. The reliability indices obtained in step 8 correspond to load level L_i (MW) with the probability of p_i. Steps 7 and 8 are repeated to consider all the load

levels in the discrete probability distribution model. The total reliability indices are calculated by

$$\mathrm{RI}_l = \sum_{i=1}^{NL} \mathrm{RI}(L_i) \cdot p_i \tag{8.35}$$

where RI_l represents the total reliability index for a membership grade l; $\mathrm{RI}(L_i)$ is the reliability index corresponding to the load L_i(MW); NL is the number of load levels. Note that RI_l has two values for each membership grade and they are calculated separately.

10. Steps 5–9 are repeated for multiple membership grades of the membership functions $\mu(L)$ and $\mu_j(U)$ to obtain the membership functions of reliability indices corresponding to a selected weather state.
11. Steps 3–10 are repeated to obtain the membership functions of reliability indices for all region-divided weather states considered. The weather-related membership functions for a reliability index are weighted by the probabilities of corresponding weather states, and then the weighted ones are summed to create the total membership function of the reliability index.

8.4.2.2 Defuzzification of Membership Functions. There are two main defuzzification techniques in fuzzy theory: the centroid of gravity and composite maximum techniques. For the membership functions of reliability indices obtained using the method described above, the centroid of gravity technique is appropriate approach. This technique is similar to the weighted mean concept in probability theory, as it finds the balance point by calculating the weighted mean of a fuzzy membership function, which can be formulated as

$$\mathrm{RI(mean)} = \frac{\sum_{l \in V} \mathrm{RI}_l \cdot \mu(RI_l)}{\sum_{l \in V} \mu(\mathrm{RI}_l)} \tag{8.36}$$

where RI(mean) is the mean of reliability index, RI_l is the lth-domain point value in the membership function of reliability index, $\mu(\mathrm{RI}_l)$ is the membership function grade for that domain point, and V is the set of domain points to be considered. Obviously, RI(mean) is the point representing the gravity center of the fuzzy membership function for the reliability index.

8.5 EXAMPLE 1: CASE STUDY NOT CONSIDERING WEATHER EFFECTS

8.5.1 Case Description

The reliability of a regional system at BC Hydro (British Columbia Hydro and Power Authority) in Canada was evaluated using the method described in Section 8.4, except for the weather-related models. Weather effects will be illustrated using the second

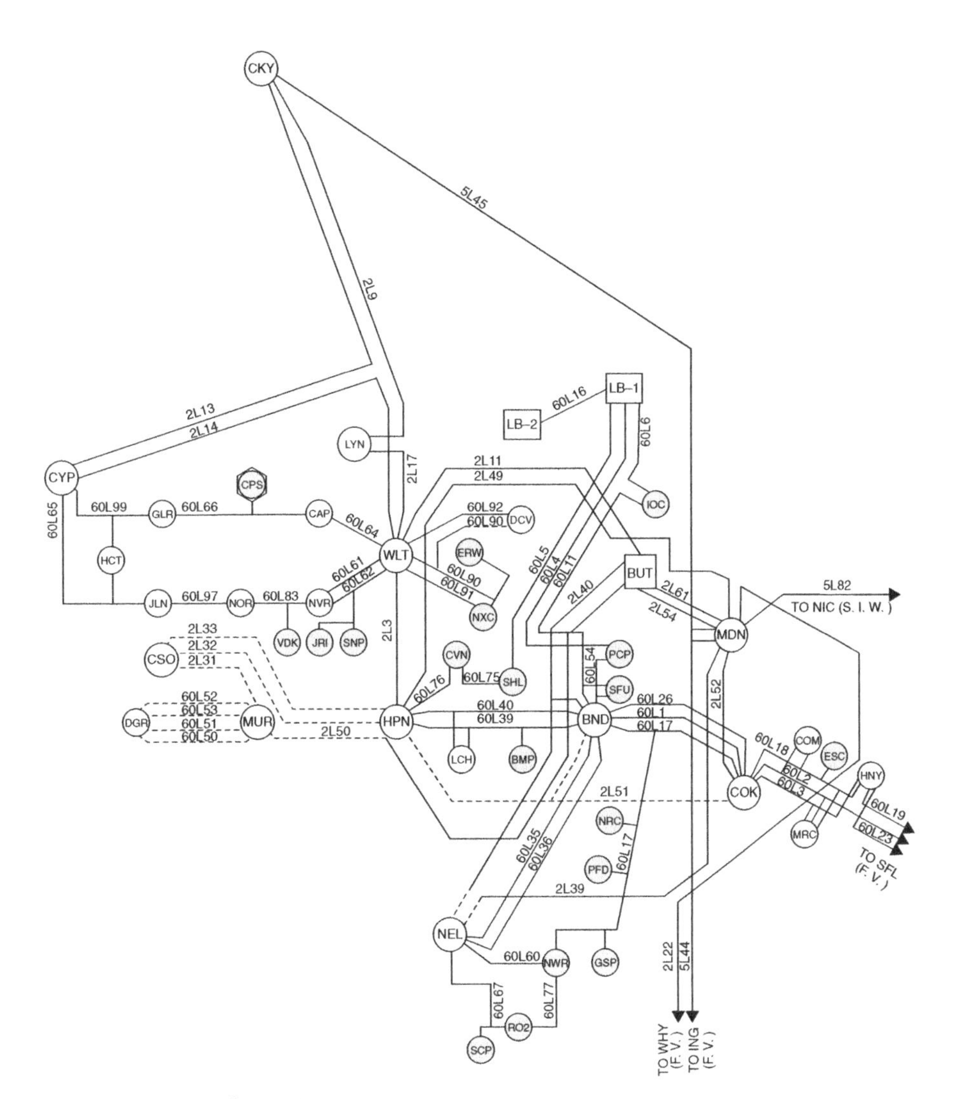

Figure 8.6. Single-line diagram of a regional system.

example in Section 8.6. The single-line diagram of the system is shown in Figure 8.6. The system has 104 buses, 167 branches, and 8 local generating units with a total generation capacity of 4580 MW (including injected powers from main generation sources outside the region). The membership function of the peak load was built using the most probable peak load forecast with its lower and upper bounds as discussed in Section 8.3.1. A discrete probability distribution model for the load curve was created using historical records and the clustering technique in Section 8.3.2. The membership functions of outage rates and repair times for system components were created using historical statistics and the basic fuzzy models in Section 8.2.1.

The following three cases were studied:

1. The fuzzy and probabilistic models for system component outage parameters were considered, whereas the load curve was represented only using the probability distribution model without the fuzzy model of peak load.
2. The mixed fuzzy and probabilistic models for loads were considered, whereas the outage parameters of system components were modeled using a traditional probabilistic approach (fixed mean values for outage rates and repair times).
3. The fuzzy and probabilistic models for both the loads and system component outage parameters were included.

8.5.2 Membership Functions of Reliability Indices

The membership functions of the following two reliability indices were evaluated using the method presented in Subsection 8.4.2 without considering weather effects:

- EENS—expected energy not supplied (MWh/year)
- PLC—probability of load curtailments

The membership functions of the two indices for the three cases were obtained and are plotted in Figures 8.7–8.12. The lower and upper bound values of EENS and PLC indices corresponding to five membership grades are also given in Tables 8.1 and 8.2, respectively. A step of 0.25 for the grade was used in the results. The step can be selected to be either larger or smaller depending on the requirement of accuracy in calculating membership functions. The total mean values of EENS and PLC indices considering both randomness (probability models) and fuzziness (fuzzy models) with more steps of grades are presented in Table 8.3. The mean values of the indices obtained using only the traditional probability approach without any fuzzy model, which should correspond to the points at the membership grade of 1.0 in the membership functions in this example, are also included in Table 8.3 for comparison. The variance coefficient of EENS was used as the convergence criterion in the Monte Carlo simulation and was set at 0.05.

The input data are random fuzzy variables, and so are the reliability indices. In other words, both the randomness characterized by probabilistic features and the fuzziness characterized by fuzzy sets are included in the indices. The differences among the three cases are contributed by the fuzziness modeling for data uncertainties since the probabilistic models are the same in all three cases. The following observations can be made:

- The fuzzy input data (fuzzy loads and fuzzy outage parameters) create fuzzy reliability indices with different membership functions. In the example given, the load fuzziness has a larger impact on the fuzzy ranges of reliability indices than does the fuzziness of outage parameters of system components. The two fuzzy uncertainties together result in the largest fuzzy ranges of reliability indices.

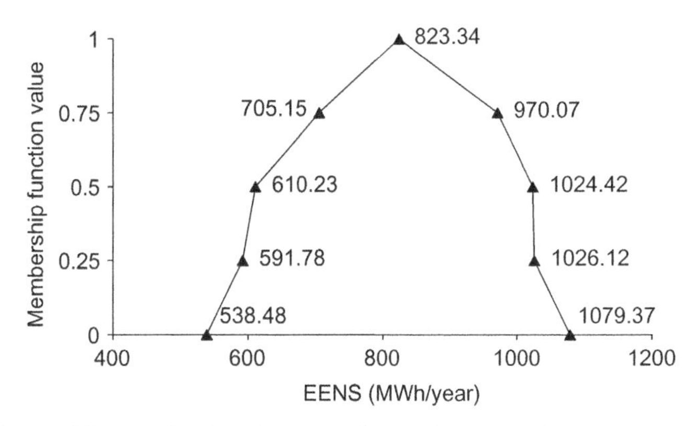

Figure 8.7. Membership function of EENS for case 1 (in Section 8.5.1).

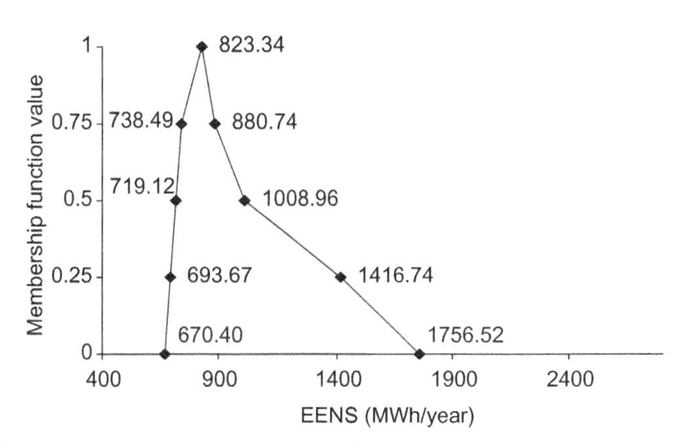

Figure 8.8. Membership function of EENS for case 2 (in Section 8.5.1).

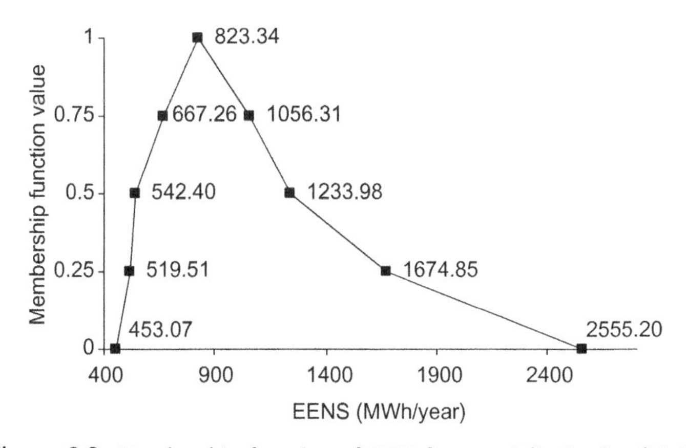

Figure 8.9. Membership function of EENS for case 3 (in Section 8.5.1).

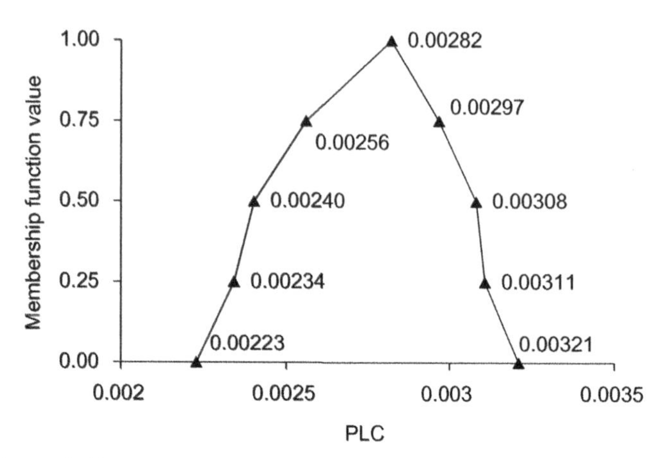

Figure 8.10. Membership function of PLC for case 1 (in Section 8.5.1).

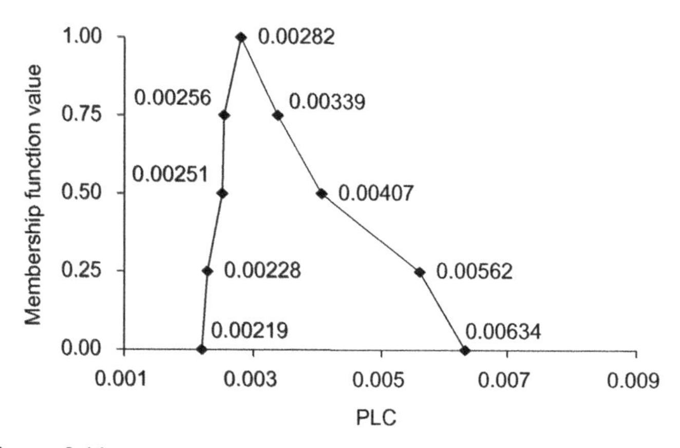

Figure 8.11. Membership function of PLC for case 2 (in Section 8.5.1).

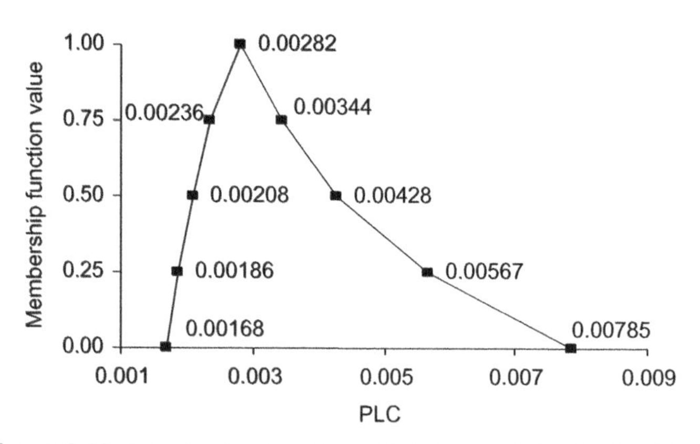

Figure 8.12. Membership function of PLC for case 3 (in Section 8.5.1).

TABLE 8.1. Membership Functions of EENS (MWh/year)

	Case 1		Case 2		Case 3	
Grade	Lower	Upper	Lower	Upper	Lower	Upper
1	823.34	823.34	823.34	823.34	823.34	823.34
0.75	705.15	970.07	738.49	880.74	667.26	1056.31
0.5	610.23	1024.42	719.12	1008.91	542.40	1233.98
0.25	591.78	1039.06	693.67	1416.74	519.51	1674.85
0	538.48	1079.37	670.40	1756.52	453.07	2555.20

TABLE 8.2. Membership Functions of PLC

	Case 1		Case 2		Case 3	
Grade	Lower	Upper	Lower	Upper	Lower	Upper
1	0.00282	0.00282	0.00282	0.00282	0.00282	0.00282
0.75	0.00256	0.00297	0.00256	0.00339	0.00236	0.00344
0.5	0.00240	0.00308	0.00251	0.00407	0.00208	0.00428
0.25	0.00234	0.00311	0.00228	0.00562	0.00186	0.00567
0	0.00223	0.00321	0.00219	0.00634	0.00168	0.00785

TABLE 8.3. Mean Values of Two Reliability Indices

Case or Condition	EENS (MWh/year)	PLC
Case 1	825.18	0.00277
Case 2	861.55	0.00315
Case 3	895.30	0.00308
No fuzziness	823.34	0.00282

- Although the membership function of peak load is symmetric, the corresponding membership functions of reliability indices are very asymmetric with much larger upper bounds. This indicates that the fuzziness of the reliability indices is much more sensitive to the high side of the peak load fuzzy uncertainty compared to the low side. This should be a general phenomenon.
- When both the fuzzy uncertainties of loads and system component outage parameters are considered, their combined impacts on the reliability indices are not a simple linear superposition.
- It is interesting to note that the mean of each reliability index after defuzzification is not very far away from the point value at the membership grade of 1.0, although the fuzzy range of the index is quite large. In this example, the mean of the EENS index for all three cases is larger than the point value at the membership grade of 1.0, whereas the mean of the PLC index for case 2 or 3 is larger than the point value at the membership grade of 1.0 but that for case 1 is smaller. This depends on the shape of the membership function and the number of grade steps.

- Another observation is that for the PLC index, the nearness degree between the mean and the point value at the membership grade of 1.0 is not necessarily proportional to the fuzzy range of the index. In this example, the PLC index has the larger fuzzy range in case 3 than in case 2, but the mean of the PLC index is closer to the point value at the membership grade of 1.0 in case 3 than in case 2. This also depends on the shape of the membership function and the number of grade steps.

8.6 EXAMPLE 2: CASE STUDY CONSIDERING WEATHER EFFECTS

8.6.1 Case Description

A reliability test system called RBTS (Roy Billinton Test System) was designed [103] and has been used to test a new model or method in composite system reliability assessment for years. The RBTS is used to demonstrate an application of the proposed method when weather effects on data uncertainties are incorporated. The single-line diagram of the RBTS is shown in Figure 8.13, and its basic data can be found in Reference 103.

In addition to the basic data, additional data are needed to incorporate the fuzzy weather models. It is assumed that the area covered by the whole transmission system is divided into three weather forecast regions, which is also shown in Figure 8.13 [divided by dashed lines and indicated by the numbers (1), (2), and (3)]. Each region can be in either normal or adverse weather condition. The region-divided weather state

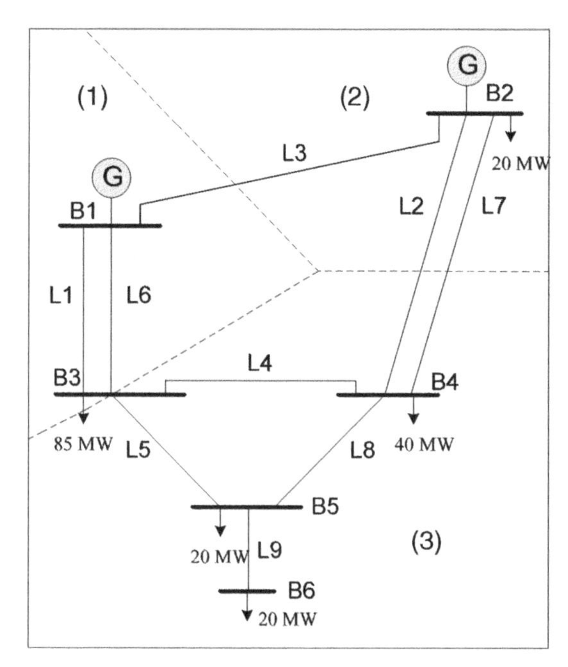

Figure 8.13. Single-line diagram and weather forecast regions of RBTS.

TABLE 8.4. Weather State Probability

Weather State	Probability
W0	0.84
W1	0.02
W2	0.02
W3	0.02
W12	0.03
W13	0.03
W23	0.03
W123	0.01

TABLE 8.5. Fuzzy Models of Outage Rates and Repair Times in Different Weather Conditions

Line	Normal Weather Condition	Adverse Weather Condition
	Outage Rate (outages/year)	
L1, L6	[0.7253, 0.9890, 1.3187]	[4.8889, 6.6667, 8.8889]
L4, L5, L8, L9	[0.3956, 0.6593, 1.1868]	[2.6667, 4.4444, 8.0000]
L3	[1.3736, 2.1978, 3.2967]	[13.8889, 22.2222, 33.3333]
L2, L7	[1.6484, 2.7473, 4.1209]	[16.6667, 27.7778, 41.6667]
	Repair Time (hours/outage)	
All lines	[7.5, 10, 12.5]	[10, 15, 20]

probabilities are given in Table 8.4. The symbol W0 indicates the whole system (all the three regions) under normal weather, and W123 represents the whole system in adverse weather. W1 refers only to region 1 in adverse weather, and W12 indicates that regions 1 and 2 are both in adverse weather but region 3 is in normal weather. Other symbols have similar meanings. Line 3 has 30% in region 1 and 70% in region 2 (i.e., $R = 0.3$), whereas lines 2 and 7 have 40% in region 3 and 60% in region 2 (i.e., $R = 0.4$). Lines 1 and 6 are fully located in region 1, whereas lines 4, 5, 8, and 9 are fully in region 3. The triangular fuzzy models for normal and adverse weather-related outage data of all the lines are given in Table 8.5. The midpoint values in the fuzzy models of outage rates are obtained by using the outage rates in the basic data as the average rates for the two weather conditions and assuming that outages occurring in the adverse weather condition account for 40% for lines 1, 4, 5, 6, 8, and 9, and 50% for lines 2, 3, and 7. These fuzzy data are consistent with the original data of the RBTS in the sense that the total average outage rate of each line calculated using the midpoints in their fuzzy models for normal and adverse weather conditions and probabilities of the line exposed to normal and adverse weather conditions, is the same as the original outage rate of that line given in Reference 103. It has been assumed that the midpoint value in the fuzzy model of repair time for normal weather condition is the same as the original repair time (10h) in Reference 103, whereas the midpoint in the fuzzy model of repair time for adverse weather condition is increased by 50% (15h), which reflects more difficulties in repair activities during adverse weather. A rectangular fuzzy model is assumed for the division ratio (i.e., R) between the two weather conditions. The fuzzy model of R is [0.2, 0.2, 0.4, 0.4] for line 3 and [0.3, 0.3, 0.5, 0.5] for lines 2 and 7. Compared to the crisp boundary between the regions mentioned above, this signifies that a weather condition does not strictly terminate at the geographic boundary between two weather forecast regions but has an equal possibility (10%) to either cross over (extend) or withdraw from the boundary. In order to focus on the weather effects on component outage parameters, the fuzziness of the peak load is not considered. The probability model of an annual load curve, which is based on a typical load curve shape from a utility, is still considered.

8.6.2 Membership Functions of Reliability Indices

The membership functions of the following three indices were evaluated by using the method described in Section 8.4 with the focus on weather effects:

- EENS—expected energy not supplied (MWh/year)
- PLC—probability of load curtailment(s)
- ENLC—expected number of load curtailments (occurrences per year)

The membership functions of the three indices for the eight different weather states were calculated and are presented in Tables 8.6–8.13. The membership functions of the

TABLE 8.6. Membership Functions of Reliability Indices in Weather State W0

	EENS (MWh/year)		PLC		ENLC (occ/year)	
Grade	Lower	Upper	Lower	Upper	Lower	Upper
1	131.8	131.8	0.00119	0.00119	1.05809	1.05809
0.75	111.6	149.7	0.00101	0.00135	0.93129	1.15187
0.5	93.7	158.8	0.00085	0.00143	0.81059	1.17054
0.25	80.2	210.5	0.00073	0.00182	0.72289	1.46146
0	60.1	219.5	0.00055	0.00198	0.54526	1.53449

TABLE 8.7. Membership Functions of Reliability Indices in Weather State W1

	EENS (MWh/year)		PLC		ENLC (occ/year)	
Grade	Lower	Upper	Lower	Upper	Lower	Upper
1	133.5	133.5	0.00123	0.00123	1.14050	1.14476
0.75	113.1	153.2	0.00104	0.00141	0.99627	1.27463
0.5	95.1	162.7	0.00088	0.00150	0.86028	1.30501
0.25	81.2	205.6	0.00075	0.00189	0.76151	1.60736
0	60.8	224.3	0.00057	0.00206	0.57678	1.70045

TABLE 8.8. Membership Functions of Reliability Indices in Weather State W2

	EENS (MWh/year)		PLC		ENLC (occ/year)	
Grade	Lower	Upper	Lower	Upper	Lower	Upper
1	132.8	136.1	0.00121	0.00123	1.18440	1.25183
0.75	112.4	154.5	0.00102	0.00140	1.01078	1.40802
0.5	93.8	170.0	0.00085	0.00152	0.87299	1.50432
0.25	80.3	220.0	0.00073	0.00196	0.76700	1.90490
0	60.1	250.7	0.00055	0.00221	0.58337	2.10465

TABLE 8.9. Membership Functions of Reliability Indices in Weather State W3

	EENS (MWh/year)		PLC		ENLC (occ/year)	
Grade	Lower	Upper	Lower	Upper	Lower	Upper
1	909.5	909.5	0.00807	0.00808	6.00357	6.18933
0.75	763.8	1147.3	0.00677	0.01018	5.35231	7.41023
0.5	611.5	1445.5	0.00543	0.01280	4.57435	9.00516
0.25	499.4	1750.8	0.00443	0.01546	3.99167	10.58921
0	380.6	2147.9	0.00341	0.01891	3.31165	12.76528

TABLE 8.10. Membership Functions of Reliability Indices in Weather State W12

	EENS (MWh/year)		PLC		ENLC (occ/year)	
Grade	Lower	Upper	Lower	Upper	Lower	Upper
1	144.0	151.6	0.00132	0.00137	1.38998	1.49633
0.75	115.1	183.6	0.00109	0.00164	1.12892	1.78766
0.5	96.8	200.9	0.00092	0.00179	0.97577	1.89475
0.25	82.5	267.1	0.00078	0.00233	0.84188	2.42368
0	61.1	298.7	0.00058	0.00262	0.63431	2.70341

TABLE 8.11. Membership Functions of Reliability Indices in Weather State W13

	EENS (MWh/year)		PLC		ENLC (occ/year)	
Grade	Low	Upper	Lower	Upper	Lower	Upper
1	912.1	920.7	0.00814	0.00819	6.32382	6.66289
0.75	766.0	1163.1	0.00683	0.01036	5.58725	8.07407
0.5	613.4	1472.7	0.00548	0.01306	4.73212	9.95216
0.25	500.5	1783.0	0.00446	0.01575	4.07963	11.70180
0	381.5	2190.3	0.00344	0.01928	3.37778	14.17432

TABLE 8.12. Membership Functions of Reliability Indices in Weather State W23

	EENS (MWh/year)		PLC		ENLC (occ/year)	
Grade	Lower	Upper	Lower	Upper	Lower	Upper
1	912.8	915.8	0.00811	0.00813	6.83599	6.90025
0.75	766.6	1161.3	0.00680	0.01028	5.99875	8.53997
0.5	613.4	1467.2	0.00545	0.01296	5.05401	10.67055
0.25	499.5	1781.4	0.00444	0.01568	4.32847	12.71015
0	380.7	2199.0	0.00341	0.01924	3.53526	15.40769

TABLE 8.13. Membership Functions of Reliability Indices in Weather State W123

	EENS (MWh/year)		PLC		ENLC (occ/year)	
Grade	Lower	Upper	Lower	Upper	Lower	Upper
1	939.9	939.9	0.00832	0.00832	7.44439	7.44439
0.75	787.4	1193.6	0.00697	0.01056	6.44321	9.28763
0.5	620.8	1516.3	0.00555	0.01334	5.30333	11.65119
0.25	502.2	1844.8	0.00450	0.01615	4.49659	13.89544
0	382.0	2284.7	0.00345	0.01985	3.67106	16.99034

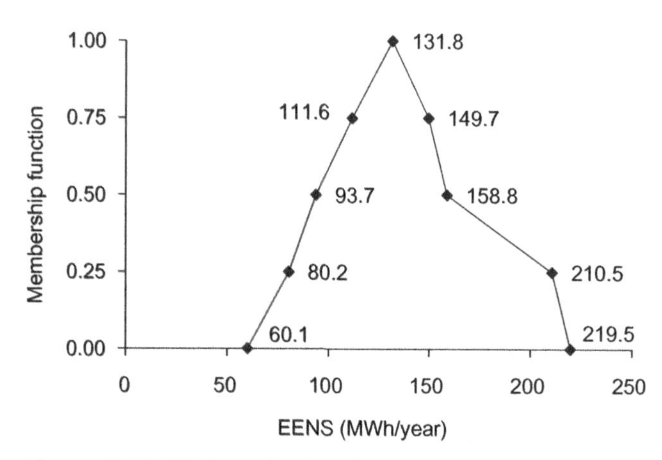

Figure 8.14. EENS membership function in weather state W0.

three indices for the whole system exposed to the normal (W0 state) or adverse (W123 state) weather condition are also plotted in Figures 8.14–8.19. The probabilistic average membership functions of the three indices considering all eight weather states are given in Table 8.14 and plotted in Figures 8.20–8.22. The following observations can be made:

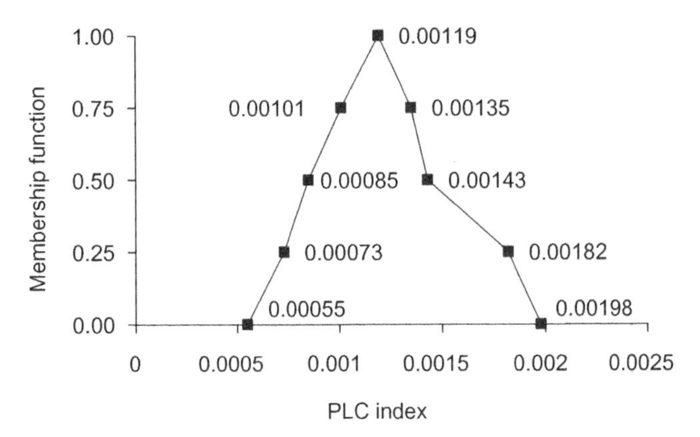

Figure 8.15. PLC membership function in weather state W0.

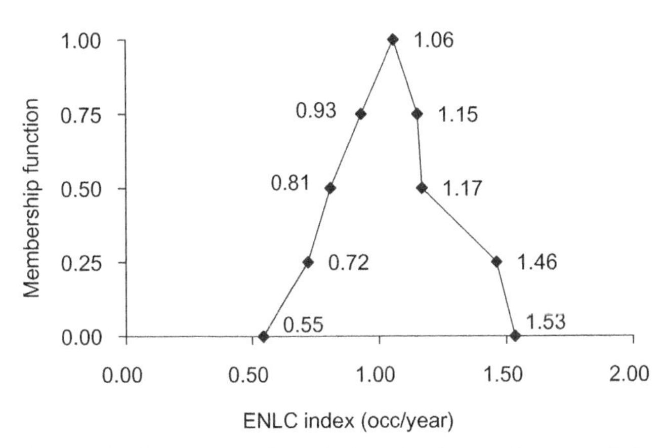

Figure 8.16. ENLC membership function in weather state W0.

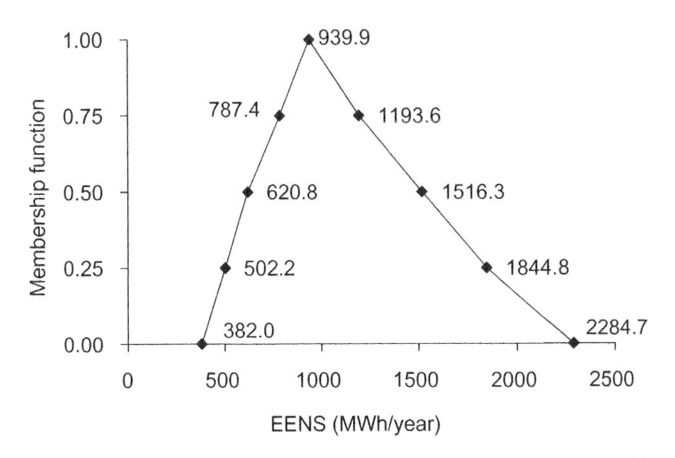

Figure 8.17. EENS membership function in weather state W123.

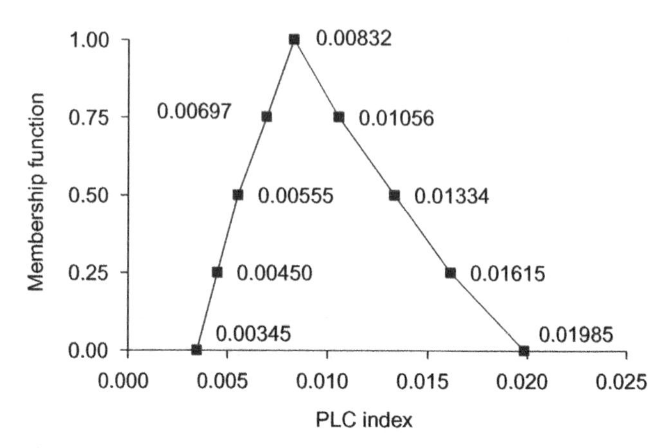

Figure 8.18. PLC membership function in weather state W123.

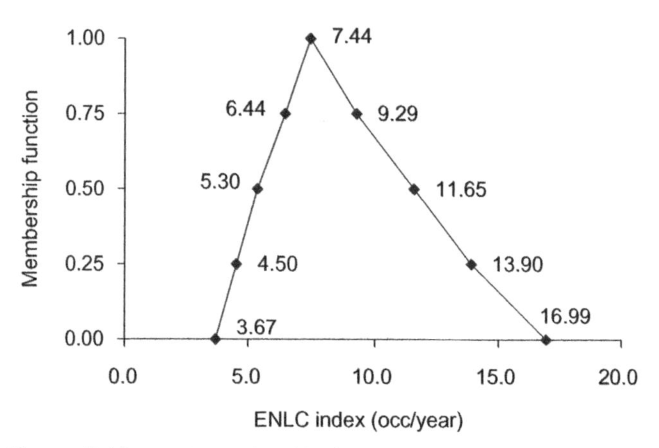

Figure 8.19. ENLC membership function in weather state W123.

TABLE 8.14. Membership Functions of Reliability Indices Considering All the Weather States

	EENS (MWh/year)		PLC		ENLC (occ/year)	
Grade	Lower	Upper	Lower	Upper	Lower	Upper
1	202.7	203.3	0.00182	0.00182	1.56630	1.58674
0.75	170.8	242.0	0.00154	0.00217	1.37535	1.81436
0.5	140.6	278.3	0.00127	0.00249	1.18294	2.01158
0.25	118.1	353.7	0.00107	0.00309	1.04010	2.45368
0	89.0	400.3	0.00081	0.00356	0.81059	2.75885

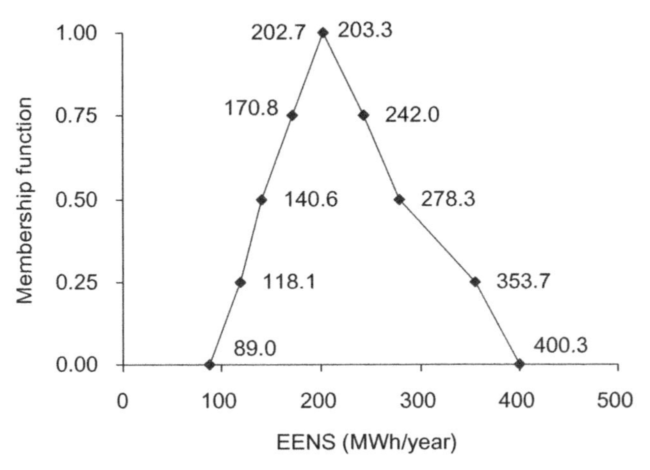

Figure 8.20. EENS membership function considering all the weather states.

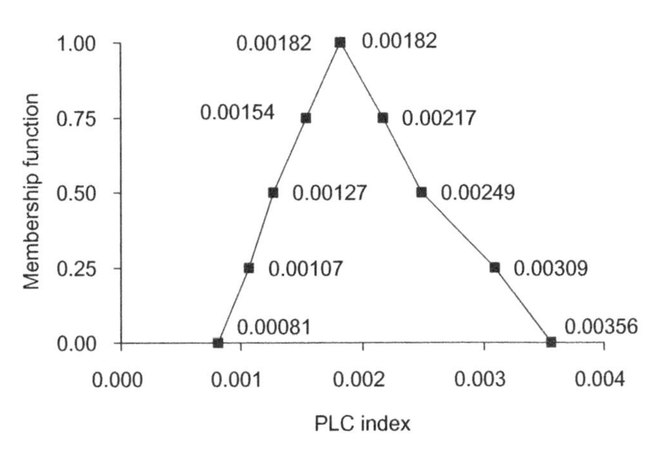

Figure 8.21. PLC membership function considering all the weather states.

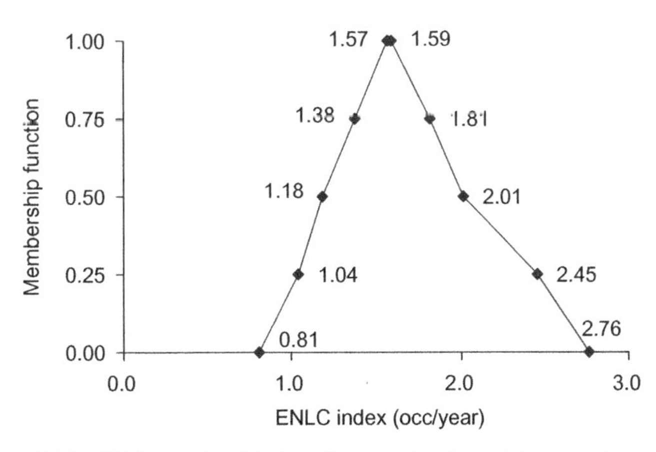

Figure 8.22. ENLC membership function considering all the weather states.

- The reliability indices associated with region 1 or 2 in adverse weather conditions do not have significant differences compared to those for the whole system exposed to the normal weather condition in this example. This is due to the fact that the 85-MW load at bus 3 is supplied through the two parallel lines L1 and L6 so that it will not be curtailed even if either one fails, although the adverse weather in region 1 increases the outage probabilities of the two lines and the 20-MW load at bus 2 is directly connected to generators at this bus and will not be affected by line outages.
- The system reliability becomes much worse in any weather state associated with region 3 in the adverse weather condition. This is mainly because there is a radial line L9 in region 3 and all the loads in this region are located at the end of the transmission system. When adverse weather causes high outage probabilities of the lines in this region, it will have a much greater effect on the reliability indices. By comparison between the indices in Tables 8.6 and 8.14, it can be seen that the contributions due to all the adverse weather states to the total indices account for 30–45%, depending on the index and the degree of fuzziness (the grade of membership functions).
- The membership functions of the three indices have a similar shape in this example. On the other hand, the shape of the membership function for the whole system exposed to normal weather is quite different from that for the whole system exposed to adverse weather, whereas the shape of the average membership function considering all the weather states is somewhere between these two situations.
- It can be observed that the membership functions of the reliability indices are no longer triangular, although the input data (outage rates and repair times) have triangular membership functions. This is a general conclusion.
- The membership functions in Figures 8.14–8.19 have only one peak value as the whole system is exposed to one weather state (either normal or adverse) and there is no impact of the fuzziness of the division between the different weather conditions, which is modeled using a rectangular membership function. The membership functions in Figures 8.20–8.22 have two peak values because of the effect of the rectangular membership function for the fuzziness of the division between the different weather conditions.
- The rectangular shape of the membership functions of the division ratios between the different weather conditions for line 3 and lines 2 and 7 does not have a significant impact on the shapes of membership functions of the reliability indices. In particular, this impact is not observed at the midpoint of the PLC membership function in Figure 8.21. However, this observation is true only for this example and cannot be generalized.
- A membership function represents the ranges of a reliability index at different fuzzy grades. This provides a broader insight into the uncertainty in the effects of weather conditions on line outages and system reliability. It is worth reemphasizing that this type of uncertainty is the fuzziness that cannot be characterized using a probability model.

TABLE 8.15. Comparisons between Fuzzy and Nonfuzzy Models in Indices for Individual Weather State Effects

	EENS (MWh/year)		PLC		ENLC (occ/year)	
Weather State	Nonfuzzy	Fuzzy	Nonfuzzy	Fuzzy	Nonfuzzy	Fuzzy
W0	131.8	131.7	0.00119	0.00119	1.05809	1.04304
W1	133.5	133.5	0.00123	0.00123	1.14253	1.13266
W2	133.0	135.2	0.00121	0.00122	1.19725	1.22139
W3	909.5	968.7	0.00808	0.00859	6.07870	6.43996
W12	145.9	151.2	0.00134	0.00137	1.42581	1.46508
W13	912.5	978.7	0.00815	0.00871	6.46432	6.90404
W23	914.3	977.0	0.00812	0.00866	6.86801	7.35244
W123	939.9	1004.2	0.00832	0.00888	7.44439	7.95244

TABLE 8.16. Comparisons among Three Methods in Total Mean Indices Considering All the Weather States

Weather Model	EENS (MWh/year)	PLC	ENLC (occ/year)
No weather model	156.0	0.00141	1.27622
Nonfuzzy weather model	202.8	0.00182	1.57435
Fuzzy weather model	209.0	0.00188	1.60627

8.6.3 Comparisons between Fuzzy and Traditional Models

In the traditional probabilistic method considering weather effects, it is assumed to have a single value of outage rate or repair time for an overhead line in a weather condition and a crisp boundary between different weather conditions. The reliability of the RBTS system was also evaluated using the method without considering any weather effect and the traditional nonfuzzy method for weather effects [10]. All other data were the same, except that the data for the fuzzy models is not needed. The mean values of the reliability indices are presented in Tables 8.15 and 8.16. The following observations can be made:

- The reliability indices obtained using the fuzzy and nonfuzzy weather models demonstrate differences for the weather states in which the adverse weather condition has relatively large effects (when region 3 is exposed to adverse weather) but are very close for the normal weather condition and the weather states in which the adverse weather condition has small effects (when region 1 or 2 is exposed to adverse weather). This is because the fuzzy ranges of outage parameters of components in the adverse weather condition are much wider than those in the normal weather condition.

- The total mean indices considering all the weather states obtained using the fuzzy and nonfuzzy weather models do not have significant differences. This is due to the fact that the weather states associated with region 3 in the adverse weather condition have low probabilities. On the other hand, the reliability indices considering the weather effects are much worse than those not considering the weather effects.
- In general, the fuzzy and nonfuzzy weather models are both acceptable for evaluation of the mean indices. The main advantage of using the fuzzy weather model is that it provides the membership functions of reliability indices. This is useful information, particularly for the impacts of individual weather states on system reliability, since it shows the ranges of reliability indices at different fuzzy grades but not just simple average values.

8.7 CONCLUSIONS

Dealing with the uncertainties of data in probabilistic transmission planning is a challenge for system planners. This chapter addressed this issue by presenting the fuzzy models for loads and system component outage parameters. A combined probabilistic–fuzzy method for transmission system analysis is also developed using the transmission reliability assessment as an example. The models and methods described can be extended to other system analyses in probabilistic transmission planning.

There exist two types of uncertainties (randomness and fuzziness) for loads and system component outage parameters in real life. The randomness can be characterized by a probabilistic model, whereas the fuzziness can be represented by a fuzzy model. The fuzzy model is required in the following two situations:

- A real fuzzy factor exists. For example, the outage parameters of system components are affected by weather states that are described in fuzzy terms and a noncrisp boundary between different weather conditions. The fuzziness in load forecast also exists for different reasons.
- A subjective judgment is needed or is more appropriate. For instance, in some cases, statistical data are insufficient for building an accurate probabilistic representation for a component outage parameter or peak load, but an interval estimate of the outage parameter or peak load can be established by engineers.

The mathematical essence of the models presented here is the establishment of combined probabilistic and fuzzy representations of random–fuzzy variables. Based on the probabilistic–fuzzy models, a hybrid method including membership functions and Monte Carlo simulation is developed to capture both randomness and fuzziness of system loads and component outage parameters. The interval estimates for establishing the fuzzy membership functions of input data can be either derived from statistical records or determined by experienced engineers. In the method presented here, Monte

Carlo simulation is an independent module and fuzzy modeling is performed from outside the Monte Carlo simulation process. This feature enables us to easily incorporate the method into the existing probabilistic evaluation techniques for transmission system analysis, particularly for reliability assessment.

It is important to appreciate that the purpose of introducing fuzzy models is not to replace probabilistic methods but to provide a new vehicle for handling the two types of uncertainty of data in probabilistic transmission planning.

9

NETWORK REINFORCEMENT PLANNING

9.1 INTRODUCTION

The basic task of network reinforcement planning is the decisionmaking (justification or rejection) on a project or alternative of equipment addition or network reconfiguration in a transmission system. This requires comprehensive comparisons among different alternatives and is associated with technical, economic, environmental, and societal assessments. Probabilistic network reinforcement planning focuses on a combined probabilistic reliability evaluation and economic analysis, whereas assessments in other aspects for selection of initial planning alternatives are performed as usual. Obviously, probabilistic planning is an important part of the whole planning process. The word *network* in this chapter refers to a transmission system but does not include individual substation configurations. The planning issues associated with substation configurations will be discussed in Chapter 11.

Transmission network planning issues can be classified into two major tasks: bulk power supply system planning and regional network planning. Bulk supply system planning is associated with the adequacy of power supply from generation and high-voltage transmission sources to each individual region, whereas regional network planning is related to the adequacy from power sources to local load points (buses) in a

Probabilistic Transmission System Planning, by Wenyuan Li

region. These two planning tasks are often considered separately in utilities' practice although coordination is necessary in some cases.

This chapter demonstrates two actual applications in the probabilistic network planning at BC Hydro of Canada. The first one in Section 9.2 is a bulk supply system planning issue where the adequacy of power supply to a large island region is a concern. The second one in Section 9.3 is a regional network planning issue in which the thermal constraint is the trigger for system reinforcement.

9.2 PROBABILISTIC PLANNING OF BULK POWER SUPPLY SYSTEM

More materials regarding this topic can be found in References 104–107.

9.2.1 Description of Problem

As shown in Figure 9.1, the Vancouver Island area is supplied by the two 500-kV transmission lines, two HVDC poles and a whole bunch of local generators. The HVDC poles 1 and 2 had been operated for 36 and 30 years respectively until 2006 and reached the end-of-life stage. It was decided to retire the HVDC subsystem in 2008 based on extensive planning studies and equipment status investigations. There are two options for the replacement of the existing HVDC subsystem: (1) a 230 kV AC transmission line; and (2) a new HVDC with the voltage source converter technology, which is called *HVDC light*.

The whole planning process was involved in economic, technical, societal, and environmental assessments [108]. In the technical aspect, traditional power flow and

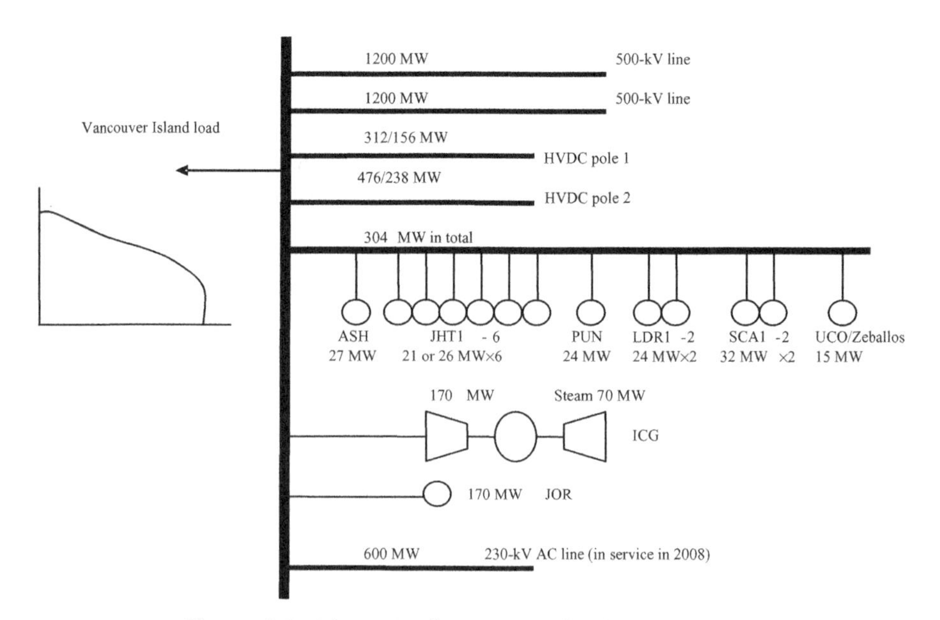

Figure 9.1. Schematic of Vancouver Island supply system.

transient stability studies were performed and remedial action schemes for special contingencies were investigated. Both options are feasible in societal and environmental assessments. The overall economic analysis and supply reliability become the crucial issues in making a final decision. Two main questions are

- Which option is more reliable and financially more competitive?
- What is the effect of the existing HVDC before and after the reinforcement project is in service?

9.2.2 Economic Comparison between Two Options

The economic analyses for both options are performed using the cash flow and present value methods given in Chapter 6. A comparison of the investment and operation costs between the two options (in million dollars) is summarized in Table 9.1 [108]. The term *project definition cost* refers to the present value of the expense incurred in the preproject stage, including feasibility studies and tests of new technologies. The *project implementation cost* is the present value of main capital investments. The *contingency cost* is the estimate of unexpected expenditures. The *present value of OMA* (operation, maintenance, and administration) is obtained using the useful lives of different equipment and the 6% discount rate. The *present value of taxes* is based on an approximate estimate. The power-flow-based loss studies indicate that the 230-kV AC line option creates less network losses than does the HVDC light option. The *indirect cost* item in the table includes a relative network loss cost by assuming that the loss cost for the 230-kV AC line option is zero and the loss cost for the HVDC light option has the cost due to the difference in the losses between the two options. The relative loss cost, which has been converted into the present value based on the useful lives of the 230-kV AC line and HVDC light and the 6% discount rate, is used only for purposes of relative comparison.

9.2.3 Reliability Evaluation Method

Reliability of the Vancouver Island supply system can be assessed using a generation reliability model with multiple power supply sources. As shown in Figure 9.1, the

TABLE 9.1. Comparison in Investment and Operation Costs between the Two Options (in M$)

Cost	230kV AC	HVDC Light	Difference
Project definition	0.0	24.5	24.5
Project implementation	208.0	311.0	103.0
Contingency	12.0	10.5	−1.5
Present value of OMA	2.5	13.5	11.0
Present value of taxes	27.5	27.5	0.0
Indirect cost	35.5	48.0	12.5
Total	285.5	435.0	149.5

500-kV transmission lines, HVDC poles, and local generators can all be modeled as independent power sources with their capacities and unavailability. The total demand in the Vancouver Island at different points of time during one year is represented using an annual load curve. This model ignores the local network in the island since it does not have any effect on the reliability of the supply system.

The reliability assessment includes the following steps [109]:

1. A multilevel load model is created using chronological hourly load records during one year (see Section 3.3.1).
2. System states at each load level are selected using Monte Carlo simulation techniques. This includes
   - The new single-pole HVDC light is modeled using a two-state random variable (up and down states), whereas the existing two-pole HVDC subsystem is modeled using a multistate random variable (full up, down, and derated states).
   - The generating units are modeled using multistate random variables or two-state random variables depending on the operational features of individual generators.
   - The AC transmissions are modeled using two-state random variables.

   The HVDC poles, transmission lines, and generating units are the power source components in the model. Take a three-state random variable as an example. A uniformly distributed random number is drawn in the interval [0,1] for each power source. The state of the jth power source is determined by

$$s_j = \begin{cases} 0 & \text{(up)} & \text{if} \quad R_j > (P_{\text{dr}})_j + (P_{\text{dw}})_j \\ 1 & \text{(down)} & \text{if} \quad (P_{\text{dr}})_j < R_j \le (P_{\text{dr}})_j + (P_{\text{dw}})_j \\ 2 & \text{(derated)} & \text{if} \quad 0 \le R_j \le (P_{\text{dr}})_j \end{cases} \tag{9.1}$$

   where R_j is the uniformed distributed random number for the ith power source, and $(P_{\text{dw}})_j$ and $(P_{\text{dr}})_j$ are the probabilities in the down and derated states of the jth power source. In the case of two-state random variables, the sampling concept is similar without considering the derated state by letting $(P_{\text{dr}})_j = 0.0$.
3. The capacity of each power source is determined according to its sampled state (up, down, or derated) so that the total supply system capacity can be obtained. For a given load level, the demand not supplied (DNS) in the kth sample is calculated by

$$\text{DNS}_k = \max\left\{0, \quad L_i - \sum_{j=1}^{m} G_{jk}\right\} \tag{9.2}$$

   where L_i is the load at the ith load level, G_{jk} is the available capacity of the jth power source in the kth sample, and m is the number of power sources supplying Vancouver Island.

4. Steps 2 and 3 are repeated for all the load levels on the load model.
5. The EENS (expected energy not supplied) index that represents the system supply reliability is calculated by

$$\mathrm{EENS} = \sum_{i=1}^{NL} \left(\frac{T_i}{N_i} \sum_{k=1}^{N_i} \mathrm{DNS}_k \right) \tag{9.3}$$

where NL is the number of the load levels in the multistep model of an annual load curve, T_i is the time length of the ith load level, and N_i is the number of samples at the ith load level.
6. The EDC (expected damage cost) is calculated by

$$\mathrm{EDC} = \mathrm{EENS} \times \mathrm{UIC} \tag{9.4}$$

where UIC is the unit interruption cost (in $/kWh), which can be estimated using one of the methods given in Section 5.3.1.

9.2.4 Reliability Comparison between Two Options

9.2.4.1 Data Preparation. The reliability of the Vancouver Island power supply system depends on two factors: the capacities and failure probabilities of power supply sources. The 230-kV AC line option has a capacity of 600 MW, whereas the HVDC light option has a capacity of 540 MW. The HVDC light option is composed of underground cables, submarine cables, and converter station equipment (thyristor valves, transformers, reactors, capacitors, and controls), whereas the 230-kV AC line consists of overhead lines, submarine cables, and a phase shifting transformer. All the other source components in the supply system are the same except for the two options. However, this never means that the reliability difference between the two supply systems with the 230-kV AC line and HVDC light link is simply the reliability difference between the 230-kV AC line and HVDC light link.

The failure data for HVDC components in the study are based on outage statistics in similar HVDC projects across the world. To capture the uncertainty in the statistics of HVDC cables, both optimistic and pessimistic failure data are estimated for the HVDC light. The failure data for the 230-kV AC line are based on outage statistics of 230-kV overhead lines in the BC Hydro system and an engineering estimate for the submarine cable (3 months of repair time due to extreme difficulties for repairing activities under water). The failure data of the phase shift transformer is based on historical failure records of a similar phase shift transformer in the BC Hydro system. The failure data for the 230-kV AC line and HVDC light options are presented in Tables 9.2, 9.3, and 9.4 respectively. In the two supply system alternatives, the 500-kV lines and local generators are identical and their data can be found in Reference 106. Note that the failure data for the 230-kV AC line option are given in failure frequency and repair time and those for the HVDC light option are given in unavailability and repair time due to the original forms in raw data collection. As mentioned in Chapter 7, any one of these three equipment indices can be calculated using the other two.

TABLE 9.2. Failure Data of 230-kV AC Line Option

Component	Failure Frequency (failures/year)	Repair Time (hours)
Overhead line (line-related)	0.2778	16.85
Overhead line (terminal-related)	0.2136	16.40
Submarine cable (HVAC)	0.1	2190
Phase shifting transformer	0.3333	3.06

TABLE 9.3. Failure Data of HVDC Light Option (Optimistic Estimates)

Component	Unavailability	Repair Time (hours)
Converter station facility	0.010769	49.01
Underground cable	0.000251	288
Submarine cable (HVDC)	0.019323	936
Whole HVDC light (equivalent)	0.030343	125.79

TABLE 9.4. Failure Data of HVDC Light Option (Pessimistic Estimates)

Component	Unavailability	Repair Time (hours)
Converter station facility	0.010769	49.01
Underground cable	0.005761	312
Submarine cable (HVDC)	0.054164	2190
Whole HVDC light (equivalent)	0.070694	268.85

9.2.4.2 EENS (Expected Energy Not Supplied) Indices. The EENS index is used to represent the system unreliability. Both the optimistic and pessimistic estimates of failure data for the HVDC light were used in the evaluation. The EENS indices of the supply system for the HVDC light and 230-kV AC line options in the 15-year period from 2008 to 2022 are shown in Table 9.5. It can be seen that the 230-kV AC line option provides better Vancouver Island supply reliability than does the HVDC light option. It is about 15–18% more reliable in terms of the EENS index if the optimistic data estimate for the HVDC light option is used and about 26–32% more reliable if the pessimistic failure data estimate for the HVDC light option is used.

In this example, it is unnecessary to convert the EENS indices into unreliability costs to obtain the total cost for comparison since the 230-kV line option not only is more reliable but also requires lower investment and operation costs compared to the HVDC light option. However, it should be pointed out that in a general case, if there is a conflict between reliability (EENS index) and economy (investment and operation costs) for different alternatives, the EENS index should be converted into unreliability costs so that the comparison can be made using the total costs of alternatives including investment, operation, and unreliability costs.

TABLE 9.5. EENS (MWh/year) of the Two Options

	230-kV AC Line	HVDC Light (Optimistic)		HVDC Light (Pessimistic)	
Year	EENS	EENS	Δ[a] (%)	EENS	Δ[a] (%)
2008	2,870	3,454	16.91	3,888	26.18
2009	2,779	3,349	17.02	3,767	26.23
2010	2,969	3,566	16.74	4,047	26.64
2011	3,085	3,693	16.46	4,211	26.74
2012	3,281	3,904	15.96	4,522	27.44
2013	3,523	4,193	15.98	4,824	26.97
2014	3,769	4,435	15.02	5,167	27.06
2015	3,991	4,719	15.43	5,468	27.01
2016	4,348	5,146	15.51	6,020	27.77
2017	4,692	5,650	16.96	6,716	30.14
2018	5,152	6,127	15.91	7,329	29.70
2019	5,710	6,746	15.36	8,059	29.15
2020	6,238	7,535	17.21	9,150	31.83
2021	6,989	8,493	17.71	10,299	32.14
2022	7,807	9,375	16.73	11,385	31.43

[a]Difference.

9.2.5 Effect of the Existing HVDC Subsystem

Investigation into the effects of the existing HVDC on the reliability of the Vancouver Island supply system was an important aspect in the reinforcement project, due to the following planning considerations:

- The existing HVDC was still one of the main power sources supplying the loads on Vancouver Island before the 230-kV AC line was placed in service. On the other hand, the performance of the existing HVDC had been greatly degraded because of its aging status. It was necessary to quantify the risks of the supply system before and after the 230-kV AC line was in service.
- The 230-kV AC line was expected to be in service in 2008, but there was a possibility of delay by one year. A contingency plan for the delay was needed.
- Was there any possibility of intentionally delaying the 230-kV AC line by continuing to use the existing HVDC? If possible, this would save the interest of the project investment cost by the deferral.

9.2.5.1 Comparison between Cases with and without the Existing HVDC. The cases of the supply system with and without the existing HVDC before (2006 and 2007) and after (2008–2010) the 230-kV AC line was placed in service were evaluated. The differences between the two cases reflect the benefit from the existing HVDC. The EENS indices for the comparison are shown in Tables 9.6 and 9.7. It can

TABLE 9.6. EENS (MWh/year) of Supply System before 230-kV Line Was Placed in Service

Year	Without HVDC	With HVDC	Difference
2006	13,016	4,850	8,166
2007	13,839	5,655	8,184
Total	26,855	10,505	16,350

TABLE 9.7. EENS (MWh/year) of the Supply System after 230-kV Line Was Placed in Service

Year	Without HVDC	With HVDC	Difference
2008	2870	1140	1730
2009	2779	1271	1508
2010	2969	1542	1427
Total	8618	3953	4665

be observed that before the 230-kV AC line was placed in service, the existing HVDC system could greatly reduce the EENS of the supply system, whereas after the 230-kV AC line was placed in service, the existing HVDC could provide only a limited improvement in the EENS. Also, the improvement would be decreased over the years. The results suggest that the existing HVDC should be retired as soon as its benefit cannot cover its operation and maintenance costs after the 230-kV AC line is placed in service.

The increasing trend in the EENS index is due to the load growth in the Vancouver Island area each year. It is noted that in the case without the existing HVDC, the EENS has a slight decrease in 2009. This is because a local generation capacity of 30 MW was added in 2009. However, the effect of the local generator is concealed by the HVDC for the case with the existing HVDC as the HVDC has a capacity much higher than 30 MW.

9.2.5.2 Effect of Replacing a Reactor of the Existing HVDC. A previous reliability study of the existing HVDC subsystem indicated that replacing the HVDC pole 2 reactor and using the old one as an on-site spare could be a measure to improve the availability of the existing HVDC subsystem and thus the reliability of the Vancouver Island power supply system [110]. The EENS indices of the supply system in the case of continuous use of the existing HVDC without any enhancement and with replacement of the pole 2 reactor are shown in Table 9.8. Note that the 230-kV AC line has been assumed not to be in service in this effect study in order to work out a contingency plan. It can be observed that the improvement (reduction) in the EENS due to replacing a reactor of the existing HVDC pole 2 and using the old one as an on-site spare is almost equivalent to delaying the degradation of Vancouver Island power supply reliability by one year before the new 230-kV AC line is placed in service. Therefore, this measure can be used as a contingency plan for a possible delay of the 230-kV AC line

TABLE 9.8. EENS (MWh/year) in the Case of Continuous Use of Existing HVDC without 230-kV AC Line

Year	Without Reactor Replaced	With Reactor Replaced	Difference
2007	5,655	5,002	653
2008	6,677	5,858	819
2009	7,261	6,207	1,054
2010	8,809	7,478	1,331
Total	28,402	24,545	3,857

TABLE 9.9. EENS (MWh/year) of Supply System for Using Existing HVDC versus 230-kV AC Line

Year	Existing HVDC	230-kV Line	Difference
2008	6,677	2,870	3,807
2009	7,261	2,779	4,482
2010	8,809	2,969	5,840
Total	22,747	8,618	14,129

addition project. Preparing a contingency plan is often a necessary consideration in transmission planning.

9.2.5.3 *Comparison between the 230-kV AC Line and Existing HVDC.*

As mentioned earlier, the 230-kV AC line project was expected to be in service in 2008. Was there any possibility of intentionally delaying the project to save the interest of the project investment cost? Such a question is popular in transmission planning. The economic analysis with quantified reliability worth assessment can help answer this question.

The EENS indices of the supply system for using the existing HVDC or new 230-kV AC line are shown in Table 9.9. It can be seen that the continuous use of the existing HVDC will result in a much higher risk for the Vancouver Island power supply than will using the new 230-kV AC line.

A further probabilistic economic analysis with reliability worth assessment was performed to compare the total costs between the 230-kV AC line addition and continuous use of the existing HVDC subsystem. The results are given in Table 9.10. The annual total cost is the sum of annual capital investment, OMA (operation, maintenance, and administration) and risk (unreliability) costs. The method of calculating the equivalent annual cost has been discussed in Chapter 6. The annual capital cost for the 230-kV AC line is $16.89 million, which is based on the present value of investment plus system operation cost (excluding the OMA), discount rate of 6%, and useful life of 50 years for the overhead line and useful life of 40 years for the submarine cable. It is assumed that the existing HVDC has reached the end of its useful life and no annual capital cost is left over. The OMA of the existing HVDC is about $5 million

TABLE 9.10. Probabilistic Economic Analysis for New 230-kV AC Line versus Existing HVDC

	Risk Cost (M$/year)		Annual Capital (M$/year)		OMA Cost (M$/year)		Total Cost (M$/year)		
Year	230-kV Line	Existing HVDC	230-kV Line	Existing HVDC	230-kV Line	Existing HVDC	230-kV Line	Existing HVDC	Difference
2008	10.93	25.44	16.89	0.00	0.16	5.00	27.98	30.44	2.46
2009	10.59	27.66	16.89	0.00	0.16	5.00	27.64	32.66	5.02
2010	11.31	33.56	16.89	0.00	0.16	5.00	28.36	38.56	10.20

per year. The OMA for the 230-kV AC line is only about $0.16 million per year. The risk cost equals the unit interruption cost times the EENS index, which increases over the years since the EENS increases over the years. It can be seen from the results that the 230-kV AC line should be in service as early as possible, not only because the annual total cost due to using the aged HVDC subsystem is higher than that required by adding the 230-kV AC line but also, more importantly, because this difference dramatically increases with the years (doubles each year). It should be noted that the risk costs in Table 9.10 are based on the conservative unit interruption cost of $3.81/kWh. The sensitivity studies with higher unit interruption costs were performed and can be found in References 104 and 107. The results with higher unit interruption costs greaterly favor the 230-kV AC line over the continuous use of the existing HVDC. Therefore the 230-kV AC line should be and was placed in service in 2008 as planned without any intentional delay.

9.2.6 Summary

This application demonstrated the probabilistic planning method for a bulk transmission system supplying a region using an actual example at the utility. The core of the method is the probabilistic reliability evaluation and economic analysis. The planning procedure includes comparison between different reinforcement alternatives in the total cost and reliability level, effect study of retired components or subsystem, contingency plan for possible delay of the reinforcement project, and probabilistic economic analysis for the in-service year.

The results indicate that in this particular example, the 230-kV AC line addition option not only requires the lower cost but also provides higher reliability than does a HVDC light option for the Vancouver Island supply after the existing HVDC subsystem is retired. However, it never implies the general conclusion that an AC line is always superior to an HVDC option. The existing HVDC subsystem, which reached its end-of-life stage and has a very high failure probability, should be retired as early as possible from both system reliability and overall economic analysis. If the new 230-kV AC line cannot be in service on time for some reason, replacing a reactor of the existing HVDC can be considered as a contingency plan, which can maintain the system reliability level (not to be degraded) for one more year.

9.3 PROBABILISTIC PLANNING OF TRANSMISSION LOOP NETWORK

The basic concept of probabilistic reinforcement planning method for transmission loop networks has been presented for years [6,111–113]. This section provides a new actual example [114] to demonstrate the application of the method.

9.3.1 Description of Problem

The single-line diagram of the central–southern Vancouver Island area at BC Hydro is shown in Figure 9.2. The load growth in this area would result in a thermal limit violation problem in the regional transmission system. Lines 1L115 and 1L116 are heavily loaded and are approaching the capacity limits during the winter peak even in the system's normal condition. In some single contingency events, either or both of 1L115 and 1L116 would be overloaded. Both lines are equipped with a special protection system (a remedial-action scheme) that is designed to open the two lines when overloading is detected. This leads to the situation in which more loads would be supplied from the VIT substation, which in turn creates the overloading on four 230/138-kV transformers at this substation. The regional system has to be reinforced to resolve the thermal limit violation problem on the 138-kV network and the four transformers at the VIT substation.

9.3.2 Planning Options

The following five reinforcement options were identified using traditional power flow studies and contingency analysis.

1. A new substation (indicated by the thick bold line in Figure 9.2) is built between the JPT and VIT substations, and two short 230 kV circuits (indicated by the dashed lines) are tapped to lines 2L123 and 2L128, and looped into the new substation.
2. Two phase shifting transformers are installed at the 138-kV level at the DMR substation to control or limit flows through lines 1L115 and 1L116. Two transformers (180 MVA each) at the VIT substation are replaced by two larger transformers (300 MVA each) in order to accommodate the higher flows that will result from installing the phase shifters.
3. The two 230-kV lines 2L123 and 2L128 from the DMR substation to the SAT substation are upgraded to the 500-kV level. Two 500/230-kV transformers are installed at the SAT substation. Two 180-MVA transformers at the VIT substation are replaced with two 300-MVA transformers.
4. The two 138-kV lines 1L115 and 1L116 are resized to increase the individual line capability up to 367 MVA for the normal rating. The two transformers at the VIT substation are replaced with 300-MVA transformers.
5. Two phase shifting transformers are installed at the 138 kV level at the DMR substation to control or limit the flows through lines 1L115 and 1L116. Two

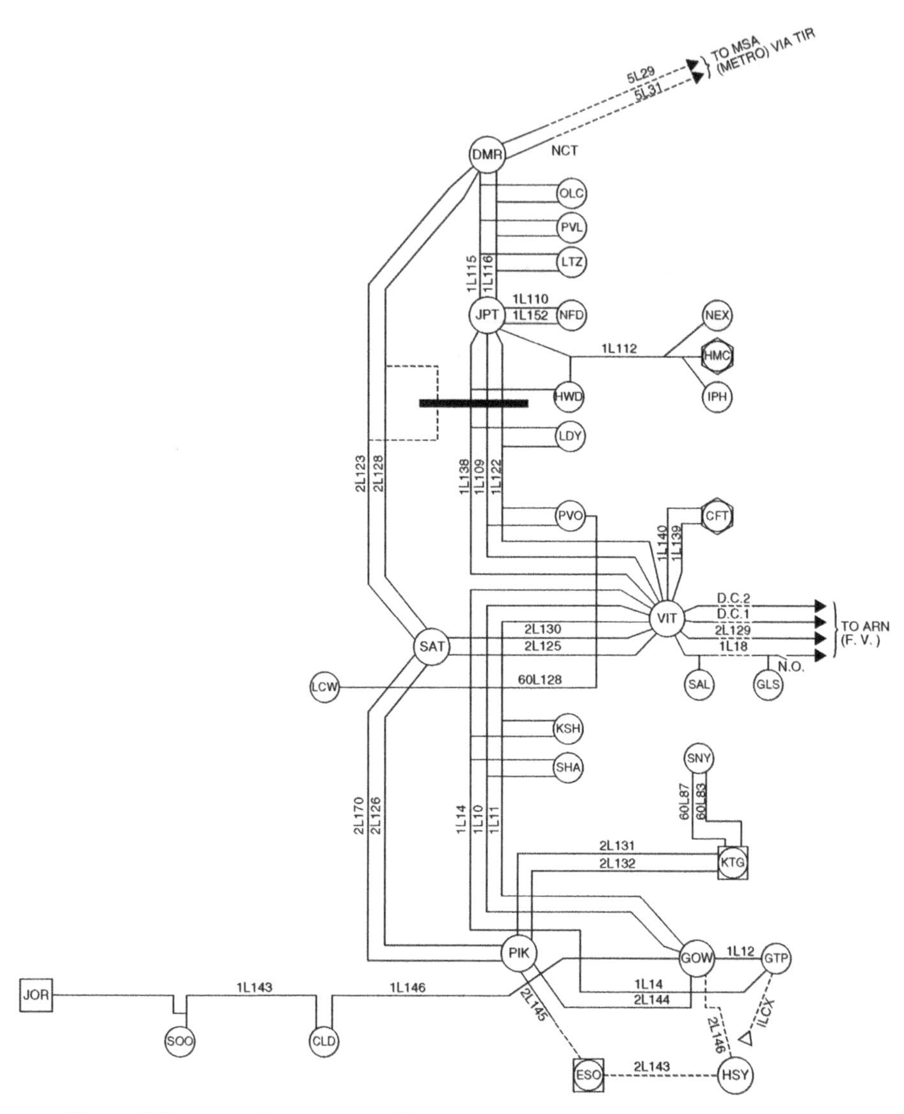

Figure 9.2. Single-line diagram of the central–southern Vancouver Island area.

230/138-kV transformers together with a 138-kV switchyard are added at the SAT substation. Three short 138-kV circuits (not shown in the figure) are built from the SAT substation to the three existing 138-kV lines 1L10, 1L11, and 1L14.

Each option can resolve the thermal limit violation problem but obviously with different reliability improvements and investment costs. The question that planners must answer is: Which option will be better in terms of overall system reliability and economic assessment? Apparently, the traditional $N-1$ criterion cannot answer this question.

9.3.3 Planning Method

9.3.3.1 Basic Procedure. The basic procedure includes the following steps:

1. Annual probabilistic unreliability costs are evaluated for the existing system and selected options over the planning timeframe using a reliability assessment tool for transmission systems [115]. The difference in the unreliability cost index between the existing system and each option is the reliability benefit in dollars due to the option.
2. The different options result in different system losses. Annual energy (MWh) losses are evaluated for the existing system and selected options over the same planning period using a power-flow-based computing tool considering an annual load curve [116]. The difference in the energy loss cost between the existing system and each option is the operational benefit in dollars due to the option.
3. The capital investments for the options are investigated and converted into an annual investment cash flow in the planning period.
4. The present values of the cash flows of annual investment costs and benefits (reductions in unreliability and loss costs) for the selected options are calculated.
5. A benefit/cost analysis is conducted to determine the best option.

9.3.3.2 Evaluating Unreliability Cost. The unreliability cost assessment includes the following aspects:

1. A multilevel load model is created using chronological hourly load records during one year. All the load levels are considered successively, and the resulting reliability indices for the load levels are weighted by their probabilities to obtain annual indices.
2. System states at each load level are selected using Monte Carlo simulation techniques. The generating units are modeled using multistate random variables, whereas transmission components (lines and transformers) are modeled using two-state random variables.
3. System contingency analyses are performed for each selected system state to identify overloading problems.
4. The following optimization model is used to reschedule generations, eliminate line overloads, and avoid load curtailments if possible or minimize the total load curtailment if unavoidable:

$$\min f = \sum_{i \in ND} W_i C_i \tag{9.5}$$

subject to

$$T_k = \sum_{i=1}^{N} A_{ki}(PG_i - PD_i + C_i) \qquad (k \in L) \tag{9.6}$$

$$\sum_{i \in NG} PG_i + \sum_{i \in ND} (C_i - PD_i) = 0 \tag{9.7}$$

$$PG_i^{\min} \le PG_i \le PG_i^{\max} \qquad (i \in NG) \tag{9.8}$$

$$0 \le C_i \le PD_i \qquad (i \in ND) \tag{9.9}$$

$$-T_k^{\max} \le T_k \le T_k^{\max} \qquad (k \in L) \tag{9.10}$$

where f is the weighted total load curtailment; C_i is the load curtailment variable (MW) at bus i; W_i is the weighting factor reflecting bus load importance; PG_i and PD_i are the generation variable and load demand at bus i, respectively; T_k is the real power flow on Line k; $PG_i^{\min}$, $PG_i^{\max}$ and $T_k^{\max}$ are the limit values, respectively, of PG_i and T_k; A_{ki} is an element of the relation matrix between real power flows and power injections at buses; ND, NG, and L are the sets of load buses, generator buses, and lines, respectively; and N is the number of buses in the system. The objective of the optimization model is to minimize the total load curtailment while satisfying the power balance, linearized power flow relationships, and limits on branch flows and generation outputs. The model can be solved using a linear programming algorithm.

5. The EENS index is calculated by using the probabilities and load curtailments for all sampled system states, which are obtained in Monte Carlo simulation and the optimization model, respectively.
6. The expected damage cost (EDC) index is calculated by multiplying the EENS by the unit interruption cost in ($/kWh), which is based on the customer survey and customer composition in the region of interest.

Obviously, steps 1, 2, 5, and 6 are similar to the steps described in Section 9.2.3, whereas steps 3 and 4 are associated with much more complex modeling techniques for a looped transmission network.

9.3.3.3 Evaluating Energy Loss Cost. The term *annual energy loss* refers to the MWh loss in one year. The traditional power flow can be used to only calculate the MW loss at a load level. The annual energy loss evaluation requires the automatic incorporation of an annual load curve and load coincidence factors of all buses at each level of the load curve. The energy loss cost evaluation includes the following two steps:

1. The annual energy losses are calculated using a power-flow-based method with incorporating an annual load curve model [116]. If necessary, a probabilistic power-flow-based method can be used to capture random factors impacting energy losses. The probabilistic power flow method has been discussed in Section 4.3.
2. The energy loss cost is calculated by multiplying the energy losses by the energy price in $/MWh.

9.3.3.4 *Evaluating Annual Investment Cost.* The cash flow of annual investment cost is estimated using the capital return factor (CRF), which has been discussed in Section 6.3.4.3:

$$\text{AI} = V \cdot \text{CRF} \tag{9.11}$$

$$\text{CRF} = \frac{r(1+r)^n}{(1+r)^n - 1} \tag{9.12}$$

Here, AI is the equivalent annual investment cost, V is the actual investment in the initial year, r is the discount rate, and n is the useful life (years) of the investment V.

9.3.3.5 *Calculating Present Values of Costs.* The cash flows of unreliability and energy loss costs are calculated through year-by-year evaluations, and the equal annuity cash flow of investment cost is obtained using Equations (9.11) and (9.12). The present value of the investment, unreliability cost, or energy loss cost is calculated using the following formula:

$$\text{PV} = \sum_{j=1}^{m} \frac{A_j}{(1+r)^{j-1}} \tag{9.13}$$

where PV is the present value, A_j is the annual cost in year j, r is the discount rate, and m is the number of years considered in the system planning period. Note that the reference year in Equation (9.13) is year 1 and the reference year in the corresponding equations of Chapter 6 is year zero.

In the benefit/cost analysis, the present value of the reduction in the unreliability cost or energy loss cost due to each planning option is the benefit and the present value of the investment is the cost.

9.3.4 Study Results

9.3.4.1 *Unreliability Costs.* The reliability evaluation method in Section 9.3.3.2 was used to assess the expected damage cost (EDC) indices of the existing system and five reinforcement options. The difference in the EDC between the existing system and each option is the reduction of unreliability cost due to the option. The planning timeframe was the 11 years from 2010 to 2020. The system loads were based on the load forecast for the 11 years. The failure frequencies and repair times of all transmission lines and transformers in the system were based on historical outage records in the previous 10 years, which were obtained from an outage database called the *reliability decision management system* (RDMS) [93] of the utility. The local generator at the JOR station is a run-of-river hydro unit, and its unavailability is dominated by water conditions but not by physical failures. The historical power output records at this generator were used to estimate the availability index of the generator.

The EENS reductions due to the five reinforcement options from 2010 to 2020 are shown in Table 9.11. Note that the fiscal years (from April 1 in one year to March 31 in the following year) are used in the evaluation.

TABLE 9.11. EENS Reductions (MWh/year) Due to the Five Reinforcement Options

Year	Option 1	Option 2	Option 3	Option 4	Option 5
2010/11	2490	1844	1889	1828	1874
2011/12	2882	2218	2269	2206	2255
2012/13	3184	2518	2564	2497	2548
2013/14	3658	2975	3029	2957	3010
2014/15	4151	3453	3516	3438	3506
2015/16	4654	3995	4059	3986	4042
2016/17	5217	4542	4613	4538	4594
2017/18	6025	5334	5410	5338	5407
2018/19	6999	6286	6373	6302	6373
2019/20	8035	7300	7400	7329	7401
2020/21	9305	8553	8665	8599	8672

TABLE 9.12. Unit Interruption Costs of Customer Sectors ($/kWh)

Duration (min)	Residential	Commercial	Industrial
10	1.2	68.4	33
40	0.9	39.6	12.9
90	1.9	26.7	13.1
180	1.7	24.2	11.2
360	1.2	24.6	8.6
Average	1.38	36.7	15.76

The EDC reductions due to the five options were calculated by multiplying the EENS reductions by the unit interruption cost (UIC). The composite UIC for each substation can be obtained from the UICs of customer sectors presented in Table 9.12 and the customer sector compositions at each substation in the region. The customer load compositions and composite UICs (in $/kWh) are shown in Table 9.13. The monetized reliability benefits (EDC reductions) due to the five reinforcement options are presented in Table 9.14.

9.3.4.2 *Energy Loss Costs.* The energy loss cost evaluation method in Section 9.3.3.3 was used to assess the energy loss costs of the existing system and the five reinforcement options. The difference in the energy loss cost between the existing system and each option is the reduction of energy loss cost due to the option. The seasonal load coincidence factors (LCFs) of each substation were used to enhance accuracy in the energy loss assessment. The concept of LCF has been described in Section 7.2.4. The energy loss reductions due to the five reinforcement options from 2010 to 2020 are shown in Table 9.15. The energy loss cost reductions were calculated by multiplying the energy loss reductions by the energy price, which is $88/MWh for the region studied. The monetized benefits in energy loss cost reduction due to the five reinforcement options are presented in Table 9.16.

TABLE 9.13. Customer Load Compositions and Composite Unit Interruption Costs (UIC) for Substations in Central-Southern Vancouver Island Region

Substations	Residential (%)	Commercial (%)	Industrial (%)	Composite UIC ($/kWh)
PVO	80	13	7	6.98
LDY	71	12	17	8.06
HWD	67	20	13	10.31
KTG	84	12	4	6.19
NFD	75	21	4	9.37
PVL	81	16	3	7.46
KSH	60	21	19	11.53
SNY	76	18	6	8.60
SHA	88	10	2	5.20
GOW	84	16	0	7.03
GTP	80	19	1	8.23
HSY	60	39	2	15.46
CLD	80	17	3	7.82
SOO	91	7	2	4.14
SAL	88	10	2	5.20
LCW	83	14	3	6.76
GLS	91	8	1	4.35
LTZ	76	23	1	9.65
QLC	88	9	3	4.99
Average				9.04

TABLE 9.14. EDC Reductions (M$/year) Due to the Five Reinforcement Options

Year	Option 1	Option 2	Option 3	Option 4	Option 5
2010/11	22.51	16.67	17.08	16.53	16.94
2011/12	26.05	20.05	20.51	19.94	20.39
2012/13	28.79	22.77	23.18	22.57	23.03
2013/14	33.06	26.90	27.38	26.73	27.21
2014/15	37.53	31.21	31.78	31.08	31.69
2015/16	42.07	36.11	36.69	36.04	36.54
2016/17	47.16	41.06	41.70	41.02	41.53
2017/18	54.47	48.22	48.91	48.25	48.88
2018/19	63.27	56.82	57.61	56.97	57.61
2019/20	72.63	65.99	66.90	66.26	66.90
2020/21	84.12	77.32	78.33	77.74	78.39

9.3.4.3 Cash Flows of Annual Investments. The capital investments of the five reinforcement options were estimated through comprehensive engineering and financial assessments and are shown in Table 9.17. The capital return factor method in Section 9.3.3.4 was used to calculate the equivalent annual investments of the five reinforcement options. The discount rate of 6% and the useful life of 40 years were

TABLE 9.15. Annual Energy Loss Reductions (MWh/year) Due to the Five Reinforcement Options

Year	Option 1	Option 2	Option 3	Option 4	Option 5
2010/11	25,367	11,255	55,217	31,043	11,175
2011/12	26,007	11,526	62,046	32,439	12,140
2012/13	26,252	11,744	57,977	32,913	11,840
2013/14	26,638	12,027	58,776	33,971	13,957
2014/15	27,394	12,502	66,262	35,196	15,464
2015/16	27,207	12,701	60,753	35,805	14,095
2016/17	27,944	12,868	64,265	36,643	13,132
2017/18	28,687	12,994	67,138	37,402	14,851
2018/19	29,431	13,101	72,057	38,155	13,240
2019/20	29,679	13,217	75,281	38,950	13,010
2020/21	32,331	13,351	77,247	39,783	13,095

TABLE 9.16. Energy Loss Cost Reduction (M$/year) Due to the Five Reinforcement Options

Year	Option 1	Option 2	Option 3	Option 4	Option 5
2010/11	2.23	0.99	4.86	2.73	0.98
2011/12	2.29	1.01	5.46	2.85	1.07
2012/13	2.31	1.03	5.10	2.90	1.04
2013/14	2.34	1.06	5.17	2.99	1.23
2014/15	2.41	1.10	5.83	3.10	1.36
2015/16	2.39	1.12	5.35	3.15	1.24
2016/17	2.46	1.13	5.66	3.22	1.16
2017/18	2.52	1.14	5.91	3.29	1.31
2018/19	2.59	1.15	6.34	3.36	1.17
2019/20	2.61	1.16	6.62	3.43	1.14
2020/21	2.85	1.17	6.80	3.50	1.15

TABLE 9.17. Capital Investments (M$) of the Five Reinforcement Options

	Option 1	Option 2	Option 3	Option 4	Option 5
Total investment	82.2	114.7	153.0	169.5	78.0

assumed in the calculation. The cash flows of annual capital investment for the five reinforcement options from 2010 to 2020 are shown in Table 9.18.

9.3.4.4 *Benefit/Cost Analysis.* The total benefit of each option is the sum of the benefits in the reliability improvement and energy loss reduction. The monetized benefits of the five options shown in Tables 9.14 and 9.16 are added to obtain the total annual benefit cash flows, which are presented in Table 9.19, whereas the annual investment cost cash flows of the five options have been given in Table 9.18. The present

TABLE 9.18. Cash Flows of Annual Capital Investment (M$/year) for the Five Reinforcement Options

Year	Option 1	Option 2	Option 3	Option 4	Option 5
2010/11	5.46	7.62	10.17	11.26	5.18
2011/12	5.46	7.62	10.17	11.26	5.18
2012/13	5.46	7.62	10.17	11.26	5.18
2013/14	5.46	7.62	10.17	11.26	5.18
2014/15	5.46	7.62	10.17	11.26	5.18
2015/16	5.46	7.62	10.17	11.26	5.18
2016/17	5.46	7.62	10.17	11.26	5.18
2017/18	5.46	7.62	10.17	11.26	5.18
2018/19	5.46	7.62	10.17	11.26	5.18
2019/20	5.46	7.62	10.17	11.26	5.18
2020/21	5.46	7.62	10.17	11.26	5.18

TABLE 9.19. Total Annual Benefits (M$/year) of the Five Reinforcement Options

Year	Option 1	Option 2	Option 3	Option 4	Option 5
2010/11	24.74	17.66	21.94	19.26	17.92
2011/12	28.34	21.06	25.97	22.79	21.46
2012/13	31.10	23.80	28.28	25.47	24.07
2013/14	35.40	27.96	32.55	29.72	28.44
2014/15	39.94	32.31	37.61	34.18	33.05
2015/16	44.46	37.23	42.04	39.19	37.78
2016/17	49.62	42.19	47.36	44.24	42.69
2017/18	56.99	49.36	54.82	51.54	50.19
2018/19	65.86	57.97	63.95	60.33	58.78
2019/20	75.24	67.15	73.52	69.69	68.04
2020/21	86.97	78.49	85.13	81.24	79.54

value method in Section 9.3.3.5 was used to conduct benefit/cost analysis for the five reinforcement options. The *benefit/cost ratio* is defined as the present value of the annual benefit cash flow divided by the present value of the annual cost cash flow. The *net benefit present value* is defined as the present value of the annual benefit cash flow minus the present value of the annual investment cost cash flow. The present values of benefits, costs, and net benefits as well as benefit/cost ratios for the five reinforcement options are shown in Table 9.20.

The following observations can be made:

- All five options are economically justifiable since their benefit/cost ratios are all greater than 3.0 and the net benefits are all above 200 M$.
- Option 1 (links between 230 and 138 kV) has the highest benefit/cost ratio and the largest net benefit, followed by option 5 (phase shifters and 230/138 kV transformation at the SAT substation).

TABLE 9.20. Present Values of Benefits, Costs, and Net Benefits as Well as Benefit/Cost Ratios for the Five Reinforcement Options

Index	Option 1	Option 2	Option 3	Option 4	Option 5
PV of benefits (M$)	319.93	267.10	303.24	281.00	271.13
PV of costs (M$)	38.33	53.49	71.39	79.04	36.36
PV of net benefit (M$)	281.60	213.61	231.85	201.96	234.77
Benefit/cost ratio	8.35	4.99	4.25	3.56	7.46

- Option 5 is less competitive than option 1, although it requires the lowest investment.
- Some options can provide a greater energy loss reduction. However, the energy loss cost reduction makes much fewer contributions to the total benefit compared to the unreliability cost reduction in this example.
- The benefit/cost ratio and net benefit approaches provide slightly different ranking orders. Option 3 ranks higher than option 2 in terms of the net benefit present value, but their ranking order is reversed if the benefit/cost ratio is used. Nevertheless, this does not affect the final decisionmaking on the best option 1.

9.3.5 Summary

This application illustrated the probabilistic planning method for a transmission system with thermal constraints due to load growth using an actual regional network at the utility. The core of the method is still the probabilistic reliability evaluation and overall economic assessment. The reliability model in this case is more complex than that used for the bulk supply system in the first example. The economic analysis includes the assessments of investment, operation, and unreliability costs. A benefit/cost analysis is required since no planning alternative can lead to the lowest values for all three costs. In the analysis, the benefit is the reduction in the unreliability and energy loss costs and the cost is the capital investment.

The five representative reinforcement options were selected using the traditional power flow and contingency studies and have been compared using the probabilistic benefit/cost analysis method. These options include the network reconfiguration by bridging between 230- and 138-kV lines (option 1), addition of phase shifters plus transformer capacity enhancement (option 2), line voltage upgrading from 230 to 500 kV (option 3), resizing of the 138-kV lines constrained by thermal limits (option 4), and addition of phase shifters plus new transformation between 230 and 138 kV at a substation (option 5). The results indicate that option 1 has the highest benefit/cost ratio and the largest net benefit. This alternative was finally selected by planners and decisionmakers.

9.4 CONCLUSIONS

The drivers for network reinforcement include load growth, equipment aging, interconnection of new generation or load customers, and increased energy exports. The basic task of network reinforcement planning is to make a decision on an equipment addition project in the network, which requires comprehensive comparisons among different alternatives. As an important part of the whole planning process, probabilistic reinforcement planning focuses on the quantified reliability evaluation and the overall economic analysis, which includes the assessments of investment, operation (loss) and unreliability costs.

This chapter discussed the concepts, methods, and procedures of probabilistic planning for transmission network reinforcement through two actual applications: (1) bulk system planning for supplying a region in which local network constraints are not needed for consideration and (2) regional system planning, in which the thermal constraint problem of local network is the trigger for reinforcement. These are two typical transmission network reinforcement issues. A simplified generation–demand reliability model can be used in the first application, whereas a composite generation and transmission reliability assessment has to be performed in the second application. It should be appreciated that the reliability model used in the second application is general and can be applied to any case (a regional or overall system) in which power flow equation constraints and optimal load curtailments need to be taken into consideration. In economic assessment, it is unnecessary to conduct a benefit/cost analysis in the first example because the 230-kV AC line alternative is superior to the HVDC light alternative in both reliability indices and investment/operation costs. However, the benefit/cost analysis must be performed for the five potential reinforcement options in the second example since no option can have consistent superiority in all three costs of investment, operation, and unreliability. In such a general case, the reduction in unreliability and energy loss costs is the benefit and the capital investment is the cost. The unreliability cost reduction is the dominant factor in the benefit over the energy loss cost reduction in the second example. This is frequent situation in regional system planning in which thermal constraints are the main problems to be resolved. However, it should be appreciated that this may not be true for other system planning issues, particularly for a bulk system with a long-distance transmission network.

There are two benefit/cost analysis approaches: benefit/cost ratio and net benefit. The two approaches may or may not provide the same ranking order of alternatives. As explained in Chapter 6, the benefit/cost ratio represents the relative economic return, whereas the net benefit reflects an absolute economic profit. As long as both approaches rank the same alternative as the first one, which is the case in the second application, there is no conflict in decisionmaking. Otherwise, more engineering and financial analyses are required.

10

RETIREMENT PLANNING OF NETWORK COMPONENTS

10.1 INTRODUCTION

The retirement and replacement planning of network components is one of the basic tasks in transmission planning, although most planning activities are associated with component additions or system reconfiguration. Traditionally, equipment retirement or replacement is treated as an issue of asset management and any decision is based only on equipment conditions. The following three strategies have been used in utilities:

- A piece of aged equipment is continuously used until it dies, or ceases to function. The problem with this strategy is that when a major component of transmission system equipment (such as a cable or transformer) dies as a result of end-of-life failure, it will take more than one year to complete the whole replacement process, including the purchase, transportation, installation, and commissioning of new equipment. Obviously, it is unacceptable for any transmission system to function without a key component for a long time since the system may experience a very high risk of failure or may not even be able to satisfy security criteria during the period of more than one year.

Probabilistic Transmission System Planning, by Wenyuan Li

- A piece of aged equipment is continuously used with close field monitoring. The process of purchasing new equipment for replacement starts whenever a prephenomenon of fatal failure occurs. Unfortunately, some equipment cannot be monitored in such a way. For example, it is almost impossible to monitor the aging process of a cable. Apparently, the method of sampling a section of cable does not work since one section cannot represent the status of the whole cable. There have been many real-life examples where an aged cable experiences sudden multiple leakages simultaneously leading to a fatal failure. For a power transformer, although oil sampling can be performed to partially monitor the status of its wearout, a replacement may still not be able to be completed in time because it is impossible to take oil samples in such a way that its end-of-life failure could be foreseen prior to more than one year, which is needed for the process of purchase, transportation, and installation of a new transformer.
- The third strategy is to set a retirement age, which is normally around the estimated mean life of equipment. Once a piece of equipment reaches this age, it will be forced to retire. The purchase and transportation can proceed in advance. The concern associated with this strategy is the fact that the survival of one specific piece of equipment may be shorter or longer than the specified retirement age. If it is shorter, this will cause the same problem as described above for the other two strategies. If the component can survive longer, its early retirement will result in a waste of capital because of unnecessary earlier investment.

All three strategies missed an important point in that any piece of equipment is a component installed in a transmission system; therefore, any decision on its retirement or replacement should be based not only on its own status but also on its impact on the system. From this viewpoint, the retirement or replacement of a network component is essentially a system planning issue, which requires answers to the following questions:

- Should a piece of equipment be retired from a system risk perspective?
- If a piece of equipment is retired, should it be replaced, or should an alternative reinforcement be implemented to ensure system reliability and security?
- If the replacement of equipment is considered, when should it be replaced?

This chapter addresses this issue using probabilistic analysis techniques through two applications. The first one, in Section 10.2, is determination of retirement time of an aged AC cable in a transmission system. The second one, discussed in Section 10.3, entails decisionmaking on the replacement strategy of a damaged HVDC cable in a bulk power supply system.

10.2 RETIREMENT TIMING OF AN AGED AC CABLE

In most cases, a replacement is needed when a piece of equipment is retired from a transmission network. This application demonstrates how to determine the retirement

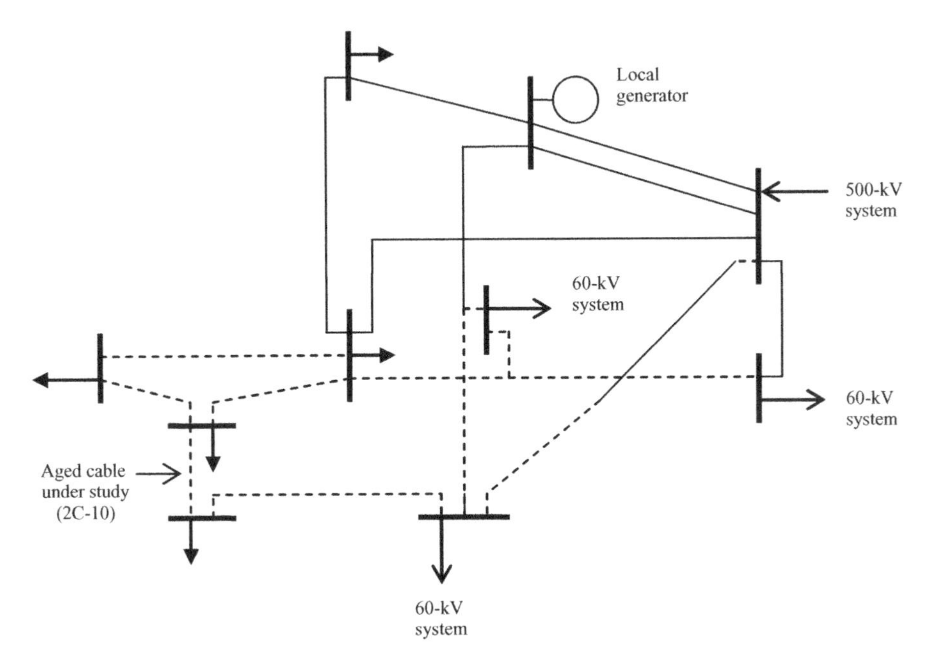

Figure 10.1. Single-line diagram of a utility system.

and replacement timing of equipment in a transmission network using an aged AC cable as an example [117].

10.2.1 Description of Problem

The single-line diagram of an actual 230-kV system is shown in Figure 10.1. A 500-kV system provides the main power supply to this area in addition to a local generating plant. Three 60-kV subsystems with considerable loads are connected to the 230-kV system. Both the 500-kV and 60-kV systems are considered in the analysis, but their network connection details are omitted from the single-line diagram for simplicity.

The south portion of the system is composed of underground cables that are expressed by dashed lines in the figure, whereas its north portion is composed of overhead lines, which are represented by solid lines. The aged cable 2C-10 under the study, which is labeled in the figure, had been operating for 45 years as of 2005 and reached the end-of-life stage. The issues that the utility was facing were whether this cable should be retired right away (in 2005) or continue to operate and, if it continued to operate, when it should be retired and replaced.

10.2.2 Methodology in Retirement Planning

10.2.2.1 Basic Procedure. The aging status of cable 2C-10 is the essential driver of the problem, and therefore its aging failure probability must be modeled. The basic idea is to evaluate the expected damage cost caused by end-of-life failure of the aged cable and compare the damage cost with the saving in capital investment due to delaying its replacement. The expected damage cost depends not only on the consequence of end-of-life failure but also on the probability of occurrence of the failure. The method presented includes the following steps:

- Estimating the scale and shape parameters in the Weibull distribution model for end-of-life failure of equipment
- Estimating the unavailability of aged equipment due to its end-of-life failure using the Weibull model
- Quantifying the expected system damage cost caused by the end-of-life failure
- Performing an economic comparison analysis

10.2.2.2 Evaluating Parameters in the Weibull Model. The method of estimating the scale and shape parameters in the Weibull distribution model is summarized as follows [118]:

1. The data of same type of equipment (underground AC cables in this example) under similar operation conditions are prepared. Most equipment pieces are in the status of being used and their in-service years are needed. For previously retired equipment, both in-service and retired years are needed.
2. A table containing the following two columns for all pieces of equipment is created: column 1—age year; column 2—survival probability of equipment corresponding to the age year. For a piece of previously retired equipment, its age is the difference between the retired year and in-service year. For a piece of surviving equipment, its age is the difference between the current year and in-service year. The number of exposed equipment pieces and the number of retired equipment pieces for each age year can be easily obtained from the data in step 1. The discrete failure density probability for each age year is the number of retired equipment pieces divided by the number of exposed equipment pieces for that year. The survival probability for each age year equals 1.0 minus the cumulative failure probability up until that year.
3. The Weibull distribution model has the following reliability function [119]

$$R = \exp\left[-\left(\frac{t}{\alpha}\right)^{\beta}\right] \tag{10.1}$$

where α and β are the scale and shape parameters, R is the survival probability, and t is the age year. Using the data in the table prepared in step 2, each pair of R and t values can apply to Equation (10.1) with an error, and the sum of squares of all the errors corresponding to M pairs of R and t values can be obtained from the following function:

$$L = \sum_{i=1}^{M}\left[\ln R_i + \left(\frac{t_i}{\alpha}\right)^{\beta}\right]^2 \tag{10.2}$$

The best estimates of α and β can be obtained when L reaches its minimum. An optimization technique can be used to conduct the minimization. An initial estimate of the shape and scale parameters is needed in the optimization and can be obtained from an initial estimate of the mean life and standard deviation of the equipment [118]. It can be seen from the model that not only previously retired but also surviving equipment contributes to the estimate of the parameters.

10.2.2.3 Evaluating Unavailability of System Components. In the estimation of expected system damage cost, the unavailability due to repairable failure of system components and the unavailability due to end-of-life failures of aged equipment are basic input data.

The unavailability of a system component due to its repairable failure is calculated by

$$U_r = \frac{f * \text{MTTR}}{8760} \tag{10.3}$$

where f is the average failure frequency (failures/year) and MTTR is the mean time to repair (hours/repair).

The unavailability of aged equipment due to its end-of-life failure depends on its age and a subsequent period to consider. By denoting its age and the subsequent period by T and t, respectively, and dividing the t into N equal intervals with an interval length D, the unavailability due to its end-of-life failure using a Weibull distribution model can be calculated by [120]:

$$U_a = \frac{1}{t}\sum_{i=1}^{N} P_i \cdot \left[t - \frac{(2i-1)D}{2}\right] \tag{10.4}$$

$$P_i = \frac{\exp\{-[T+(i-1)D]/\alpha\}^{\beta} - \exp\{-[T+iD]/\alpha\}^{\beta}}{\exp[-(T/\alpha)]^{\beta}} \quad (i = 1,2,\ldots,N) \tag{10.5}$$

Here, α and β are the scale and shape parameters, which have been estimated using the approach in Section 10.2.2.2. The P_i in Equation (10.5) is essentially the failure probability in the ith interval. If N is large enough, the equation will provide a sufficiently accurate estimate. The unavailability due to the end-of-life failure is a conditional probability given that the aged equipment has survived for T years and increases as T advances.

The total unavailability due to the two failure modes is obtained using a union concept:

$$U_t = U_r + U_a - U_r U_a \tag{10.6}$$

10.2.2.4 Evaluating Expected Damage Cost Caused by End-of-Life Failure. A basic principle in reliability economics states that the value of a component in a system does not depend on its own status but on impacts created by its absence from the system. The value of a piece of equipment should be measured using the system damage cost caused by its unavailability. In other words, if the possible failure of a piece of aged equipment does not create a severe system risk, it may not be necessary to replace it right away. Otherwise, if its failure leads to a high system risk of failure, its replacement should be considered as early as possible.

The expected damage cost (EDC) due to end-of-life failure of aged transmission system equipment can be quantified using a system unreliability cost (reliability worth) assessment method. Two cases are evaluated. In the first one, the unavailability values due to only repairable failures of all transmission components (including the aged equipment considered for retirement) are modeled. In the second one, the unavailability due to the end-of-life failure of the aged equipment for retirement is also included. The difference in the expected system damage cost between the two cases is that caused by the end-of-life failure of the aged equipment.

The system unreliability cost assessment method described in Section 9.3.3.2 can be used to estimate the expected damage cost (EDC) for each case. One of the following three approaches for calculating the EDC can be applied depending on the form of unit interruption cost (UIC):

1. Only an average UIC for the whole system or a region is known, but individual unit interruption costs for every bus are not available. In this case, the W_i in the objective function expressed by Equation (9.5) can be the weighting factors representing the relative importance of different bus loads. The optimization model given in Section 9.3.3.2 is used to calculate the EENS index first, and then the EDC is calculated by the EENS times the average unit interruption cost in \$/kWh for the whole system or a region considered. This can be expressed as

$$\mathrm{EDC} = b \cdot \sum_{j \in V} \sum_{i \in ND} C_{ij} F_j D_j \tag{10.7}$$

   where b is the average unit interruption cost in \$/kWh for a whole system or region; F_j and D_j are the frequency (occurrences/year) and duration (hours) of sampled system state j, respectively, which can be determined in the Monte Carlo simulation; C_{ij} is the load curtailment at bus i in system state j; ND is the set of load buses; and V is the set of all the sampled system states that require load curtailments. Obviously, the right side of Equation (10.7) except b is nothing else than the total EENS. In this case, no change is needed for the objective function expressed by Equation (9.5).

2. The individual unit interruption costs for every bus are available. In this case, the W_i in the objective function expressed by Equation (9.5) are replaced by the unit interruption costs (in \$/kWh) for each bus. The EDC is calculated by

$$\mathrm{EDC} = \sum_{j \in V} \sum_{i \in ND} b_i C_{ij} F_j D_j \tag{10.8}$$

   where b_i is the unit interruption cost in \$/kWh for bus i. In this case, C_{ij}, F_j, and D_j in each sampled system state j must be separately recorded in order to calculate the EDC.

3. The individual unit interruption cost for each bus is not a single value but a function of outage duration. In this case, the objective function in Equation (9.5) is changed to directly use the customer damage functions (CDFs) in \$/kW at all buses

$$\min f = \sum_{i \in ND} W_i(D_j) \cdot C_i \tag{10.9}$$

where $W_i(D_j)$ is the CDF at bus i, which is a function of the duration D_j of system state j. See Section 5.3.2 for more information on CDF. The EDC is calculated by

$$\mathrm{EDC} = \sum_{j \in V} \sum_{i \in ND} C_{ij} F_j W_i(D_j) \tag{10.10}$$

The first approach is the easiest for implementation since it has the lowest requirement on the data. Application of the third approach requires the customer damage functions (in \$/kW) for all load buses to be available, which is usually a difficult task. It has rarely been used in the practice of utilities so far.

10.2.2.5 Economic Analysis Approach. The expected system damage cost caused by the end-of-life failure of the aged equipment is calculated for each year over a timespan in the future starting from the current year. Apparently, this damage cost will increase over the years as the unavailability due to the end-of-life failure increases with the age of equipment. The expected system damage cost for each year represents the incremental risk cost that the system will experience if the aged equipment is not retired and replaced by new equipment. On the other hand, the replacement of aged equipment requires a capital investment. The replacement will be carried out sooner or later. The capital saved by delaying the retirement of the aged equipment by one year is the interest of the investment in that year, which can be calculated as the investment multiplied by an interest rate.

The current year is used as a reference (the first year). The economic comparison can be expressed using the criterion of finding the minimum n in the following inequality:

$$\sum_{i=1}^{n} E_i > n \cdot rI \tag{10.11}$$

where E_i is the expected system damage cost caused by the end-of-life failure of the aged equipment in year i, I is the capital investment required for replacing the aged equipment by new equipment, r is the interest rate, and n is the year to be determined in which the aged equipment should be retired and replaced.

If a compound interest is considered, Equation (10.11) becomes

$$\sum_{i=1}^{n} E_i > \sum_{i=1}^{n} (1+r)^{i-1} \cdot rI \tag{10.12}$$

The economic comparison process using Equation (10.11) or (10.12) is explained as follows:

- When $n = 1$, the inequality indicates that if the expected damage cost caused by the end-of-life failure of the aged equipment in the first year is larger than the saving of the investment interest due to delaying the retirement by one year, the aged equipment should be retired in the first year; otherwise, the equipment should not be retired in the first year. Go to the next step to examine the second year.
- When $n = 2$, the inequality indicates that if the sum of expected damage costs caused by the end-of-life failure of the aged equipment in the first 2 years is larger than the total saving of the investment interests due to delaying the retirement by 2 years, the aged equipment should be retired in the second year; otherwise, the third year is examined.

The examination proceeds until the n is found so that the total expected system damage cost caused by the end-of-life failure of the aged equipment in the first n years is just larger than the total saving of the investment interests due to delaying retirement by n years.

10.2.3 Application to Retirement of the Aged AC Cable

The method presented above was applied to retirement of the AC cable 2C-10 shown in Figure 10.1.

10.2.3.1 α and β in the Weibull Model. There were 18 cables in the system shown in Figure 10.1 and its neighboring subsystems, which are of the same type, at the same voltage level and under similar operation conditions. Four of the cables were retired in the past. Table 10.1 presents the data of the cables. The aged cable under study is the cable with the ID of 2C-10. The year 2005 was used as a reference year in calculating the ages of the cables. Using the estimation method given in Section 10.2.2.2, the scale and shape parameters in the Weibull distribution model were estimated as $\alpha = 47.30$ and $\beta = 27.96$, respectively. The corresponding mean life and standard deviation are $\mu = 46.34$ years and $\sigma = 2.55$ years. It is worth mentioning that the mean life estimated using the presented method is longer than the average age of the four previously retired cables since not only the retired cables but also the surviving cables contribute to the estimate in the model.

10.2.3.2 Unavailability Due to End-of-Life Failure. With the parameters α and β, Equations (10.4) and (10.5) were used to estimate the unavailability values of cable 2C-10 due to its end-of-life failure from 2005 to 2010. Cable 2C-10 was in service in 1960, and the T in Equation (10.5) for each year from 2005 to 2010 could be easily calculated. The t in Equation (10.4) was one year. The unavailability values due to the end-of-life failure of cable 2C-10 are given in Table 10.2. For comparison, the unavailability due to repairable failure of cable 2C-10, which is based on the average failure

TABLE 10.1. Cable Data

Cable ID	In-Service Year	Retired Year
2C-1	1981	—
2C-2	1974	—
2C-3	1974	—
2C-4	1957	2001
2C-5	1957	2003
2C-6	1962	—
2C-7	1957	2003
2C-8	1962	—
2C-9	1975	—
2C-10	1960	—
2C-11	1965	2004
2C-12	1969	—
2C-13	1980	—
2C-14	1980	—
2C-15	1980	—
2C-16	1980	—
2C-17	1982	—
2C-18	1987	—

TABLE 10.2. Unavailability of Cable 2C-10

Year	Age	Unavailability (Repairable Failure)	Unavailability (End-of-Life Failure)
2005	45	0.01517	0.08879
2006	46	0.01517	0.15204
2007	47	0.01517	0.24829
2008	48	0.01517	0.37999
2009	49	0.01517	0.53453
2010	50	0.01517	0.68262

frequency and repair time, is also presented in Table 10.2. It can be seen that the unavailability due to the repairable failure is a constant value, whereas the unavailability due to the end-of-life failure increases dramatically over the years after the age of cable 2C-10 exceeds the mean life.

10.2.3.3 Expected Damage Costs. The expected system damage cost caused by the end-of-life failure of cable 2C-10 was evaluated using the method described in Section 10.2.2.4. The following two cases were assessed:

1. The unavailability values due to only repairable failures of all the transmission components in the system including cable 2C-10 were considered.
2. The unavailability due to the end-of-life failure of cable 2C-10 was also included on top of case 1.

TABLE 10.3. Expected System Damage Costs (k$)

Year	Case 1	Case 2	EDC Due to End-of-Life Failure of 2C-10 (Case 2 – Case 1)
2005	273.2	512.9	239.7
2006	287.5	830.1	542.6
2007	303.7	1516.2	1212.5
2008	321.5	2649.1	2327.6
2009	342.5	4320.8	3978.3
2010	368.0	6785.0	6417.0

The unavailability values due to repairable failures were calculated using Equation (10.3). The average failure frequency and repair time of each transmission component were obtained from historical statistics in the previous 10 years. It was assumed that each year had a load growth of 3% in the system. The equivalent power injection source from the 500 kV system and the local generator were assumed to be 100% reliable in the evaluation since their failures have no effect on the expected damage cost caused by the end-of-life failure of cable 2C-10, which is the difference between the two cases. The results are shown in Table 10.3.

It can be seen that the expected system damage cost considering only repairable failures of all the transmission components will slightly increase with the years. This is caused by the load growth. The expected system damage cost including the end-of-life failure of cable 2C-10 will greatly increase over the years mainly because of the dramatically increased unavailability due to the end-of-life failure of cable 2C-10. Obviously, the end-of-life failure of cable 2C-10 dominates the expected system damage cost as time advances.

10.2.3.4 Economic Comparison. The capital investment of using a new cable to replace the cable 2C-10 was estimated to be $30 million in the 2005 dollar value. The interest rate of 5% was used. With the capital investment, interest rate, and results in Table 10.3, the cumulative expected system damage cost (incremental system risk cost) due to the end-of-life failure of cable 2C-10 and cumulative interest saving due to delaying the retirement of cable 2C-10 up to each year from 2005 to 2010 were calculated using Equations (10.11) and (10.12) and are shown in Table 10.4. There are two columns for the cumulative interest saving: the one corresponding to the noncompound interest and the other to the compound interest. It can be seen that both the system risk cost and capital interest saving due to delaying the retirement of cable 2C-10 would increase over the years. However, the system risk cost increases more than does the capital interest saving. The results in Table 10.3 show that at the time of conducting this study (in 2005), the cumulative system risk cost was expected to exceed the cumulative capital interest saving by 2009. Therefore it was decided that cable 2C-10 should be retired and replaced in 2009, when it reached 49 years old. It is worth noting that the calculations should be updated every subsequent year since a few factors change with time, including the interest rate and the α and β parameters in the Weibull model

TABLE 10.4. Cumulative System Risk Cost and Capital Interest Saving Due to Delaying Retirement of 2C-10 Cable

Year	System Risk Cost (k$)	Capital Interest Saving (k$)	
		Noncompound	Compound
2005	239.7	1,500	1,500.0
2006	782.3	3,000	3,075.0
2007	1,994.8	4,500	4,728.8
2008	4,322.4	6,000	6,465.2
2009	8,300.7	7,500	8,288.5
2010	14,717.7	9,000	10,202.9

that depend on the ages of surviving cables. The leading time of 1.0–1.5 years before a replacement is usually needed for a final decision.

10.2.4 Summary

This application presents a probabilistic analysis method for decisionmaking on retirement and replacement timing of aged equipment in transmission systems. The basic idea is to quantify and compare the expected system damage cost and capital saving due to delaying the retirement and replacement of aged equipment. The method uses the Weibull distribution to model the unavailability due to the end-of-life failure of aged equipment, evaluates the expected system damage cost caused by the end-of-life failure using a system reliability assessment technique, and performs an economic comparison analysis between the incremental system risk cost and capital saving.

An actual example of an aged underground AC cable in a utility regional system has been used to demonstrate the application of the approach presented above step by step. The results in the study conducted in 2005 indicate that the retirement and replacement of the aged cable should be delayed to 2009 although it would reach the mean life in 2006. It can be seen from Table 10.4 that the 3 years' delay (from 2006 to 2009) could save $4.5 to $5.2 million.

10.3 REPLACEMENT STRATEGY OF AN HVDC CABLE

This section discusses the replacement strategy of an aged HVDC cable [121]. An HVDC link is more complex than an AC circuit since it consists of not only overhead lines or underground/submarine cables but also a variety of converter station equipment, including thyristor valves, converter transformers, smoothing reactors, filters, and auxiliary protection and control devices. It is actually a subsystem with multiple components.

10.3.1 Description of Problem

The Vancouver Island region in the BC Hydro system was supplied through two 500-kV lines, a bipolar HVDC link, and several local generators. The schematic diagram of the

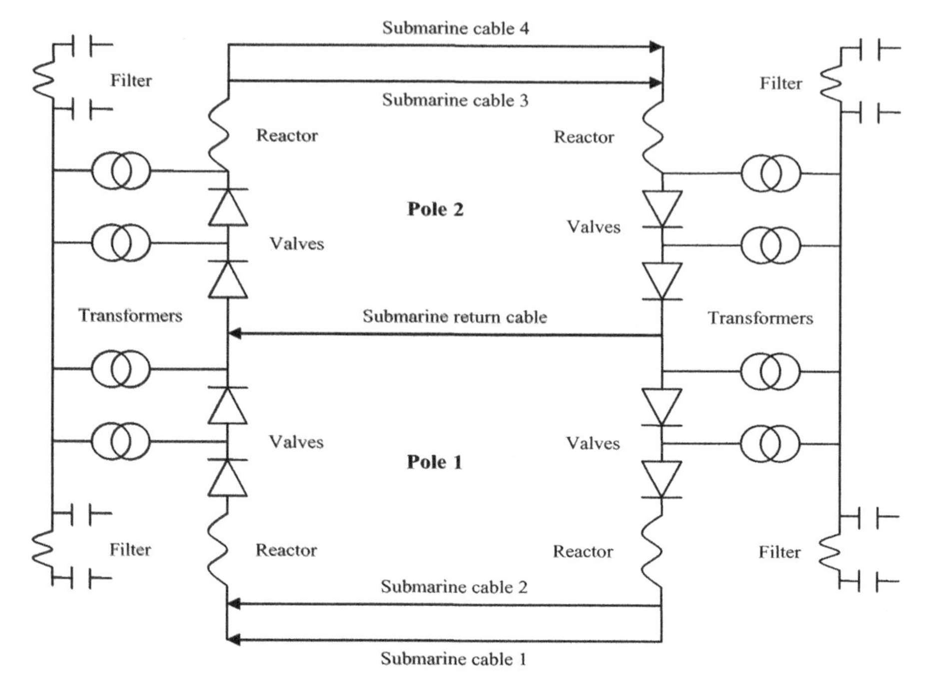

Figure 10.2. Schematic diagram of HVDC system.

island supply system is shown in Figure 9.1. The HVDC link was an aged system with pole 1 in service for 36 years and pole 2 for 30 years. The schematic diagram of the HVDC system is shown in Figure 10.2. As discussed in Section 9.2, the separate probabilistic system planning study justified that a new 230-kV AC line would be added to the power supply system in 2008 to replace the aged HVDC system. On the other hand, the existing HVDC system must be available before the new 230-kV AC line is in service and would be kept for a few years longer until the 230-kV AC line achieves a stable operation condition. Unfortunately, a field inspection in 2005 found that a section of cable 1 of HVDC pole 1 had some armor damages with three broken wire strands. Cable experts estimated that cable 1 still could be used but the damaged section (5 km) would have a very high possibility of fatal failure within a couple of years before the new 230-kV line was placed in service. The questions facing the utility were: Should the damaged section be replaced? If yes, should it be replaced before or after it totally fails? In other words, there were the following three replacement options:

- Not replacing it at all
- Replacing it before its fatal failure
- Replacing it after its fatal failure

Apparently, the questions in this example are somewhat different from those in the first application, although both are the issues associated with replacement. In the first application, the AC cable has to be replaced by a new one and the issue is simply when it should be replaced. In this example, not replacing the damaged section of the HVDC cable 1 is a possible option if the increased system risk is acceptable since the whole HVDC system will be out of service several years later. Besides, the HVDC reliability must be modeled in a more detailed way so that the impact of the damaged cable section can be quantified.

10.3.2 Methodology in Replacement Strategy

10.3.2.1 Basic Procedure. As mentioned earlier, the value of a component in a transmission system depends on the variation of system risk caused by its absence from the system. If the absence of a component creates very marginal degradation in system reliability, the benefit of replacing it becomes minor. This situation may not occur often since each transmission system component is installed for a specific purpose that contributes to the reliable delivery of power. However, when the transmission system configuration is changed or a system enhancement is performed, some equipment may become less important to the system. The example to be discussed falls within this situation.

Obviously, the three replacement options have varied impacts on the total system risk. The subject considered for replacement is a cable in the HVDC subsystem that consists of multiple components and can be operated at different capacity levels. It is difficult to directly evaluate the system reliability of a transmission system containing HVDC subsystems. The basic idea is to calculate the capacity probability distribution of the HVDC subsystem first and then to incorporate it into the reliability evaluation of the whole transmission system as an equivalent component with multiple states. The presented approach includes the following steps:

1. Estimating the average unavailability values of system components, particularly for individual HVDC components, including both repairable and end-of-life failure modes
2. Calculating the capacity levels and capacity probability distributions of the HVDC subsystem for the three cases: the first one with all existing components, including the partially damaged cable; the second one, with the damaged cable to be replaced; and the third one, with the damaged cable out-of-service without replacement
3. Evaluating the unreliability or risk costs of the whole transmission system containing the HVDC subsystem for the three cases
4. Performing the benefit/cost analysis of the three replacement options for the cable under investigation

One purpose in the first two steps is to obtain an equivalent component of the HVDC subsystem in all three cases. For a system without HVDC link, the procedure would be simpler. An AC component generally can be represented using a two-state model (up and down) and only the unavailability values of AC components are prepared. The three cases to be evaluated for an AC component under investigation for replacement strategy are the same.

10.3.2.2 Evaluating Capacity State Probability of HVDC Subsystem.

The approaches to estimating the average unavailability values of system components, including individual HVDC components, are the same as those in Section 10.2.2.3. The unavailability due to repairable failure is calculated using Equation (10.3). The unavailability due to end-of-life failure is modeled by the Weibull distribution and estimated using Equations (10.4) and (10.5). The parameters in the Weibull model can be estimated using the method in Section 10.2.2.2. The total unavailability is calculated using Equation (10.6).

The unavailability is sufficient for modeling a two-state model for AC components, whereas an equivalent multiple capacity state model is needed for a HVDC pole. Once the unavailability of each HVDC component is obtained, different methods can be used to calculate the multicapacity probability distribution of each HVDC pole [122,123]. The following is one of the methods.

The HVDC pole has its full capacity when all its components are available. The probability at the full capacity is calculated by

$$P_{\text{full}} = \prod_{i=1}^{m}(1-U_i) \tag{10.13}$$

where U_i is the total unavailability of component i, including both repairable and end-of-life failure modes; and m is the number of components in the HVDC pole.

The failure of one or more components leads to a derated state, which is often called the *half-pole operation mode*. The probability of derated capacity level is calculated by

$$P_{\text{dr}} = \sum_{j=1}^{M} \frac{\prod_{k=1}^{N_j} U_k}{\prod_{k=1}^{N_j}(1-U_k)} P_{\text{full}} \tag{10.14}$$

where M is the number of the failure states that lead to the derated capacity level and N_j is the number of failed components in the jth failure event. It is noteworthy that the probability of each failure state defined in Equation (10.14) includes the probabilities of both failed and non failed components, and thus the failure states are mutually exclusive.

The probability of the entire HVDC pole being down (at the zero capacity) is

$$P_{\text{dw}} = 1 - P_{\text{full}} - P_{\text{dr}} \tag{10.15}$$

Obviously, multiple derated states of an HVDC pole can be modeled in a similar way if necessary.

10.3.2.3 Evaluating Reliability of Overall System.

The equipment whose replacement strategy is investigated is cable 1 of HVDC pole 1 in the example. Once the capacity probability distribution of each HVDC pole is obtained, each pole can be represented as an equivalent three-state component in overall transmission system reliability evaluation. The three cases defined in step 2 of Section 10.3.2.1 have different capacity probability distributions and therefore have varied impacts on the overall system risk.

For a general transmission loop network, the system reliability evaluation method described in Section 9.3.3.2 can be used to calculate the EENS index for different cases. In the given example, the overall system containing the HVDC subsystem with the damaged cable is a power supply system from power sources to an island, as described in Section 9.2, and therefore the simpler reliability evaluation method given in Section 9.2.3 can be used for this purpose.

The uncertainty of load can be considered. In this case, the load level L_i in Equation (9.2) should be replaced by a sampled load value considering the uncertainty. The L_i is used as the mean with the uncertainty represented by a standard deviation σ_i. Note that the unit of both the load level L_i and σ_i is in MW. A standard normal distribution random number X_k in the kth sample is created using the approximate inverse transformation method (see Section A.5.4.2 in Appendix A). This sampling is added to step 3 in Section 9.2.3. The sampled value of the load in the kth sample is given by

$$L_{\sigma i} = X_k \sigma_i + L_i \tag{10.16}$$

The $L_{\sigma i}$ is used to replace the L_i in Equation (9.2) in order to capture the uncertainty of load.

10.3.2.4 Benefit/Cost Analysis of Replacement Strategies. Different replacement strategies lead to varied system risks while they require different costs. The costs of replacement are straightforward. With the EENS indices for the three cases, which are obtained from the reliability evaluation, the system risk costs due to different replacement strategies can be further assessed. Therefore they can be compared using a benefit/cost analysis approach. The details of the benefit/cost analysis are illustrated through the actual example, which will be given in Section 10.3.3.4. The analysis process may be slightly different for other similar applications.

10.3.3 Application to Replacement of the Damaged HVDC Cable

10.3.3.1 Study Conditions. The main study conditions include:

- A new 230-kV AC line was expected to be in service in 2008. The HVDC subsystem had a much smaller effect on the reliability of the island supply system after the 230-kV line was placed in service than before. This has been shown in Section 9.2.5.1.
- The HVDC subsystem was an aged system. Once the 230-kV line was in service, the HVDC subsystem would be kept for a transition period and possibly retired around 2010 when the cost of maintenance was expected to exceed the benefit. The timeframe in the study was the 5 years from 2006 to 2010.

- The replacement of the damaged cable section would take about one year because the submarine work could be performed only under fair weather and surrounding conditions. Preparation and replacement under the submarine condition would also take long time to complete.
- The annual peak loads in the island region from 2006 to 2010 were based on the recent load forecast. It was assumed that the annual load curves for all 5 years would follow the same shape that was based on the hourly load records in 2005.
- Poles 1 and 2 of the HVDC subsystem were both modeled using three capacity states (up, derated, and down states). According to the HVDC configuration (see Figure 10.2) and operation criteria, if cable 1 of pole 1 fails with no replacement, the maximum capacity of HVDC pole 1 would be derated to 156 MW from 312 MW, whereas the maximum capacity of pole 2 would be derated to 336 MW from 476 MW.
- Both repairable and end-of-life failure modes of all components in the HVDC subsystem were modeled, whereas only repairable failure modes for AC transmission components and local generators were considered. The repairable failure data were obtained from historical records, and the end-of-life failure was represented using the Weibull distribution model.

10.3.3.2 Capacity Probability Distributions of HVDC Subsystem. The state capacity probabilities of the HVDC poles in the three cases (the existing HVDC subsystem, the HVDC subsystem with the replacement of the damaged cable section, and the HVDC subsystem with the damaged cable out of service) were evaluated using the methods given in Section 10.3.2.2. The results are shown in Table 10.5–10.10. The following observations can be made:

TABLE 10.5. Capacity State Probabilities of Pole 1 for Existing HVDC Subsystem

Year	At 312 MW	At 156 MW	At 0 MW
2006	0.106243735	0.152434503	0.741321762
2007	0.075725132	0.124754433	0.799520435
2008	0.051009050	0.097306577	0.851684374
2009	0.032753449	0.072326656	0.894919895
2010	0.019887959	0.050931581	0.929180460

TABLE 10.6. Capacity State Probabilities of Pole 2 for Existing HVDC Subsystem

Year	At 476 MW	At 238 MW	At 0 MW
2006	0.554333069	0.216997424	0.228669507
2007	0.512838492	0.217244321	0.269917187
2008	0.463541606	0.218515517	0.317942876
2009	0.413689862	0.216221708	0.370088431
2010	0.362198344	0.211159543	0.426642113

TABLE 10.7. Capacity State Probabilities of Pole 1 for Damaged Cable 1 Section Replaced

Year	At 312 MW	At 156 MW	at 0 MW
2006	0.106944494	0.152709123	0.740346383
2007	0.076228654	0.125058682	0.798712664
2008	0.051351387	0.097602359	0.851046254
2009	0.032975628	0.072585300	0.894439072
2010	0.020024502	0.051138621	0.928836877

TABLE 10.8. Capacity State Probabilities of Pole 2 for Damaged Cable 1 Section Replaced

Year	At 476 MW	At 238 MW	At 0 MW
2006	0.557989321	0.214615684	0.227394995
2007	0.516248523	0.215131435	0.268620042
2008	0.466652574	0.216735362	0.316612064
2009	0.416496079	0.214758473	0.368745447
2010	0.364685055	0.210011597	0.425303347

TABLE 10.9. Capacity State Probabilities of Pole 1 for Cable 1 Out of Service

Year	At 156 MW	At 78 MW	At 0 MW
2006	0.122508347	0.147066434	0.730425219
2007	0.087353876	0.123346221	0.789299902
2008	0.059378386	0.098648047	0.841973567
2009	0.038348715	0.074954605	0.886696679
2010	0.023438967	0.053882509	0.922678524

TABLE 10.10. Capacity State Probabilities of Pole 2 for Cable 1 Out of Service

Year	At 336 MW	At 168 MW	At 0 MW
2006	0.578098707	0.201516114	0.220385180
2007	0.535003695	0.203510560	0.261485745
2008	0.483762895	0.206944509	0.309292596
2009	0.431930277	0.206710684	0.361359039
2010	0.378361967	0.203697894	0.417940139

- The HVDC pole 1 has extremely high failure probability since its age has greatly exceeded its mean life. The failure probability of HVDC pole 2 is also high because the ages of its major components are close to or beyond the mean life.
- Replacing the damaged cable 1 can slightly increase the probabilities of both poles operating at the maximum capacity levels. However, the increase is very

marginal because only one cable section of 5 km is replaced and the remaining portion (27.5 km) is still in the aging state. This indicates that the impact of replacing cable 1 on the capacity probability distribution of the whole HVDC is minimal.

- The out-of-service condition of cable 1 leads to slightly higher probabilities of the two poles being operated at the reduced maximum capacity compared to the probabilities of the two poles with cable 1 in service at the original maximum capacity. This is because all the cables in the HVDC subsystem must be available for the HVDC poles operating at maximum capacity, which means that all the cables are logically in series in the reliability model. One basic concept in reliability evaluation is that removing one component from a series reliability network leads to a higher success (at the maximum capacity) probability and a lower failure (at the derated and zero capacity states) probability. The impact of the cable 1 out-of-service condition on the HVDC subsystem reliability includes the two factors of reduced capacities and changed capacity probability distributions. It can be observed from the tables that the maximum and derated capacities for both the poles are greatly reduced when cable 1 is out of service. It will be seen in the following EENS indices that the impact of the reduced capacities on the HVDC poles 1 and 2 is a dominant factor in this example.

10.3.3.3 EENS Indices of Supply System. The risk of the transmission system supplying the Vancouver Island region was assessed for the three cases with the existing cable 1, with the damaged cable 1 section replaced and with the cable 1 out of service using the reliability evaluation method. The EENS index is used as the indicator of system risk. The EENS indices for the three cases from 2006 to 2010 are shown in Table 10.11. It can be seen that the EENS indices for using the existing cable 1 with the damaged section and replacing the damaged section are almost the same because there are the same state capacities but very minor differences in the capacity probability distributions for the two cases. The EENS indices for the cable 1 out-of-service case are larger than those in the other two cases. Note that the EENS indices have big drops in 2008 because the 230-kV AC line was expected to be in service in that year (see Section 9.2). Other than the drops in 2008, the EENS indices increase with the years because of the load growth and increased end-of-life failure probabilities of aged HVDC components.

TABLE 10.11. EENS Indices for Vancouver Island Supply System (MWh/year)

Year	With Existing Cable 1	With Replaced Cable 1	Without Cable 1
2006	4850	4843	6097
2007	5655	5642	6881
2008	1140	1138	1406
2009	1271	1268	1504
2010	1542	1541	1755

10.3.3.4 Strategy Analysis of Three Replacement Options. Using the results in Table 10.11, a replacement strategy analysis for cable 1 can be performed. The following three replacement options are considered for comparison:

1. Replacing the damaged section of cable 1 right away (in 2006) before it fails
2. Replacing the damaged section of cable 1 after it fails
3. Not replacing the damaged section of cable 1 at all (using it until it fails and operating the HVDC subsystem without it after its failure)

As mentioned in the study conditions, the replacement duration is assumed to be 1 year and the period of 5 years from 2006 to 2010 is considered in the analysis. If the replacement duration is less than 1 year, the EENS indices need to be evaluated in a shorter timeframe such as on a seasonal or monthly basis but the analysis process remains the same. The analysis approach can be described as follows (timeframes expressed in present tense at time of writing):

1. If the damaged section of cable 1 is replaced in 2006 before it fails, the HVDC subsystem will be operated without cable 1 during the replacement in 2006 and with it after the replacement from 2007 to 2010. The total EENS for the 5-year period is (see the values in Table 10.11): 6097 + 5642 + 1138 + 1268 + 1541 = 15,686 MWh.
2. If the damaged section of cable 1 is replaced after it fails, there will be different possibilities since it can fail in any year from 2006 to 2010. If it fails in the first year (2006) and is replaced right away, the total EENS for the 5-year period is the same as that for option 1 (replacing it in 2006). If it fails in a later year and the replacement begins immediately after its failure, the HVDC will be operated without cable 1 for that year, with the existing cable 1 for the years before that year, and with the replaced cable 1 for subsequent years after that year. For example, if it fails in 2007, the total EENS for the 5-year period is: 4850 + 68 81 + 1138 + 1268 + 1541 = 15,678 MWh. The total EENS indices for the replacement after the failure in the 5-year period for the different failure years are summarized in Table 10.12 (option 2).

TABLE 10.12. Total EENS (MWh) in the 5-Year Period for Options 2 and 3

Failure Year of Cable 1	Option 2	Option 3
In 2006	15,686	17,643
In 2007	15,678	16,396
In 2008	14,720	15,170
In 2009	14,690	14,904
In 2010	14,671	14,671

3. If the section of cable 1 is never replaced, the Vancouver Island supply risk also depends on the year in which it fails. The later it fails, the lower the risk is. For example, if it fails in 2008, the total EENS for the 5-year period is: 4850 + 5655 + 1406 + 1504 + 1755 = 15,170 MWh. The total EENS indices for not replacing the damaged section of cable 1 even after its failure in the 5-year period for the different failure years are also summarized in Table 10.12 (option 3). Note that if the section of cable 1 fails in early 2010, the total EENS without replacement in the 5-year period is the same as that with replacement because the replacement is assumed to take one year and therefore the HVDC will still be operating without cable 1 during the replacement. Performing the replacement in 2010 will benefit the island reliability only after 2010, which will be minimal. As mentioned in the study conditions, according to previous planning studies (see Section 9.2), the HVDC subsystem will be retained for only a few years before its complete retirement once the new 230-kV line is in service.

By comparing the EENS indices between options 1 and 2 (the value calculated in item 1 and the values in the option 2 column of Table 10.12), it can be seen that replacing the damaged section of cable 1 after its failure has a lower risk (the smaller EENS), except that it fails and is replaced in 2006, resulting in the same EENS as that for option 1. The later its failure occurs, the lower the risk is. Between options 2 and 3, it is necessary to compare the reduced risk due to replacing the damaged section of cable 1 against the cost required to replace it. The reduced EENS index for each failure year is the difference in EENS index between the two options. The reduced risk cost is the product of the reduced EENS and unit interruption cost. The unit interruption cost is $3.81/kWh, which is the same as that used in Section 9.2.5.3. The reductions in the EENS and risk cost due to replacing the damaged section of cable 1 for different failure years are given Table 10.13.

The cost required by replacing the damaged section (5 km) of cable 1 is estimated to be $8 million. The reduction of risk cost due to the replacement is the benefit. The benefit/cost ratios of replacing the damaged section of cable 1 for different failure years are summarized in Table 10.14. It can be seen that the benefit/cost ratio for any year in which cable 1 may fail is less than 1.0. This indicates that not replacing cable 1 is more cost-effective than replacing it in this example.

TABLE 10.13. Reductions in EENS (MWh) and Risk Cost (M$) Due to Replacing the Damaged Section of Cable 1

Failure Year of Cable 1	Reduction of EENS (MWh)	Reduction of Risk Cost (M$)
In 2006	1957	7.456
In 2007	718	2.736
In 2008	450	1.715
In 2009	214	0.815
In 2010	0	0.000

TABLE 10.14. Benefit/Cost Ratios for Replacement of Damaged Section of Cable 1

Failure Year of Cable 1	Benefit/Cost Ratio
In 2006	0.932
In 2007	0.342
In 2008	0.214
In 2009	0.102
In 2010	0

10.3.4 Summary

The system risk-evaluation-based approach to the replacement strategy of aged HVDC components is presented in this application. The approach includes the estimation of average unavailability values of individual HVDC components due to both repairable and end-of-life failure modes, calculations of capacity levels and capacity probability distributions of HVDC poles, quantified risk assessment of the transmission system containing HVDC poles, and a benefit/cost analysis for different replacement options.

The replacement strategy for an aged submarine cable of the HVDC link in a power supply system at BC Hydro of Canada has been analyzed as an example to demonstrate the actual application of the approach presented. The analysis procedure has been explained in detail. The results show that not replacing the damaged section of cable 1 in the HVDC subsystem is the most cost-effective option in this example. It should be emphasized that this is not a universal conclusion but applies only to this particular case. Different results and conclusions may be obtained in other cases. However, the approach and analysis process given in this application are general.

10.4 CONCLUSIONS

This chapter discussed retirement and replacement planning issues of aged components in a transmission system. Determining the retirement time and replacement strategy of aged equipment in transmission systems has been a challenge in the utility industry for many years. This is due mainly to the following facts: (1) it is difficult or even impossible to accurately monitor the aging process of equipment and predict the time of end-of-life failure; (2) early replacement leads to a waste of capital investment, whereas a too late retirement may result in a huge system damage cost caused by a sudden fatal failure of aged equipment; and (3) other new equipment additions in system planning studies may change the importance of some existing equipment, and a direct replacement at the original location may or may not be the best option from a viewpoint of overall transmission system reliability and security.

Traditionally, a decision on retirement and replacement of equipment is based only on the equipment's own status and condition. In contrast with the traditional decision process, the most essential idea in the method presented is to focus on impacts of equipment retirement and replacement on the system risk. The unavailability models due to end-of-life failures of aged equipment, system risk evaluation techniques, and economic analysis methods for retirement – replacement timing or different replacement options have been developed and integrated into a decisionmaking process. Obviously, the retirement and replacement of a system component is no longer associated only with the asset management of individual equipment, but becomes a system planning task.

There are two issues: (1) A piece of equipment in a system must be replaced after its retirement. In this case, the basic question is when it should be retired and replaced. This issue is addressed in Section 10.2 using an application to determining the retirement timing of an aged AC cable in a transmission system. (2) Not replacing a piece of aged equipment may be a potential option due to other system reinforcement. In this case, the best replacement strategy should be selected through comparisons of different options. This issue is illustrated in Section 10.3 using an application to determining the replacement strategy of an aged HVDC cable in a bulk power supply system. The latter example also demonstrated how to create a capacity probability distribution model of HVDC subsystems and incorporate the model into the reliability evaluation of an overall transmission system. The method presented in the second application is not limited to the replacement strategy of aged HVDC components but can also be applied to any other transmission system component.

11

SUBSTATION PLANNING

11.1 INTRODUCTION

Substation planning covers a wide range of topics. The most important topic is selection of substation configuration. It is commonly recognized that configurations with different layouts result in differences in substation reliability, operational flexibility, and economy of investment. In simple cases, a qualitative observation can help determine which configuration is more reliable. In many cases, however, an observation cannot provide a correct judgment. A phenomenon called *noncoherence* may exist in substation configurations. The noncoherence is defined as a case where the addition of more components in a substation layout leads to deterioration of substation reliability or an increase in one or more unreliability indices. Such a case is opposite to intuition and cannot be identified by a qualitative judgment of planners. Quantified reliability evaluation is needed in order to conduct an overall comparison between different substation arrangements, including reliability and investment costs.

Another important topic in substation planning is adequacy of transformers in a substation. This is associated with determination of the number and capacity of transformers. There are two strategies for transformer adequacy. The traditional one is to apply the N–1 principle in each individual substation. According to this principle, when

Probabilistic Transmission System Planning, by Wenyuan Li

any transformer fails, the remaining transformer(s) in a substation must be able to transfer full loads. When this criterion is no longer satisfied because of load growth, a new transformer is added even if overloading due to a transformer failure is very marginal. The *N*–1 principle is generally secure but expensive since each substation has to satisfy the principle individually. This strategy usually is applied to important substations. There still exists a risk of load curtailment for the *N*–1 principle since it does not cover the cases in which two or more transformers fail at the same time. The other strategy is the use of transformer spares shared by multiple substations. When a transformer fails, it is replaced temporarily by a spare transformer until the failed transformer is repaired. As long as the number of spares is sufficient, events of simultaneous multiple transformer failures can be also handled with an acceptable overall reliability level. There is a possibility of overloading during the installation of spare. However, the probability of possible overloading is very low because it can occur only when a transformer fails at the time of annual peak load, which is very short. The second strategy is often applied to a group of *substations* with less importance. The group can be composed of different substations in which some have more than one transformer and some have only one single transformer. A reliability-based approach is required to determine the number and timing of transformer spares shared by the substation group.

This chapter provides two actual applications in reliability-based substation planning. The first one is associated with selection of switching station configuration and is described in Section 11.2. The second one is associated with determination of spare transformers for a substation group and is presented in Section 11.3. In this chapter, the term substation refers to either a substation with voltage transformation or a switching station at an independent power producer (IPP).

11.2 PROBABILISTIC PLANNING OF SUBSTATION CONFIGURATION

Similar to probabilistic network planning, probabilistic substation (or switching station) configuration planning requires two basic assessments: reliability evaluation and economic analysis. An example at an actual utility is used to illustrate the probabilistic planning process for substation configuration.

11.2.1 Description of Problem

A radial supply system consists of two long overhead transmission circuits connected in series to supply loads at several substations [124]. On the basis of the radial supply structure, the loads at all substations can be aggregated into an equivalent load located at the left end as shown in Figures 11.1 and 11.2. Circuit 1 (197 km) at the right side is connected to the power source, and circuit 2 (129 km) at the left side is connected to the equivalent load. One new generating station owned by an independent power producer (IPP) will be inserted into circuit 2, and thus circuit 2 is sectionalized into two parts designated as circuits 2a (61.8 km) and 2b (67.2 km). Two different station configurations are considered for comparison. The one in Figure 11.1 contains three breakers and is called the *three-breaker ring configuration*, and the one in Figure 11.2

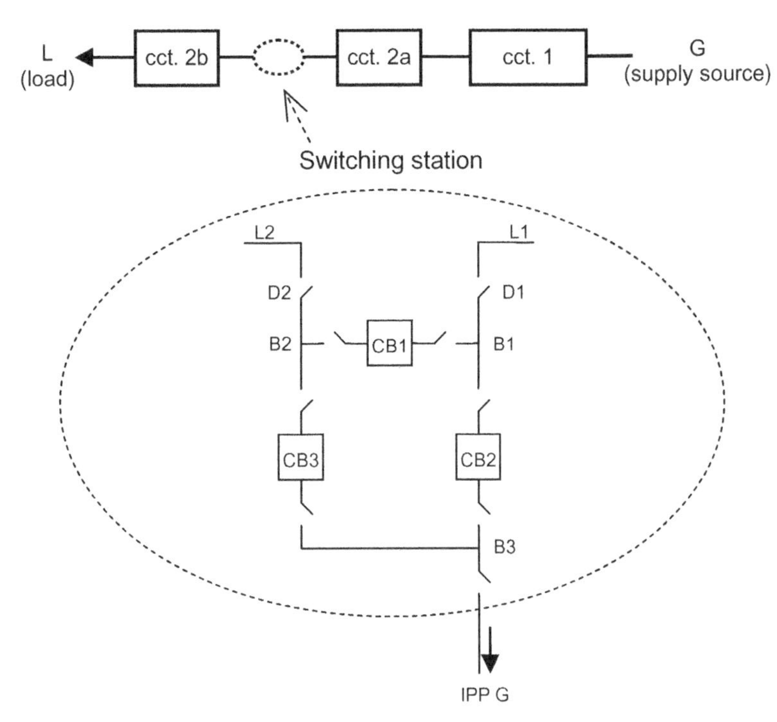

Figure 11.1. Block diagram of the radial supply system with the IPP switching station of three-breaker ring configuration (cct. = circuit).

contains two breakers and is called the *two-breaker configuration*. The transformer and breaker at the side of the IPP generator are excluded from modeling since their impacts are the same for both configurations. The access of the IPP generator to the system was justified in a previous IPP interconnection planning study. The purpose of this planning study is to select a better switching station configuration.

11.2.2 Planning Method

11.2.2.1 Simplified Minimum Cutset Technique for Reliability Evaluation. The system state enumeration technique for substation reliability evaluation was presented in Section 5.4 as a general method. In some cases, a simpler technique is applicable. An approximate method based on failure event enumeration, which can be viewed as a simplified minimum cutset technique, is applied to this example. The method includes the following steps:

1. Failure events are enumerated first. A *failure* is defined as any outage leading to load curtailment. Minimum failed components resulting in an outage event compose a minimum cutset of the failure event. For a complex substation configuration, a minimum cutset search technique is needed to enumerate all

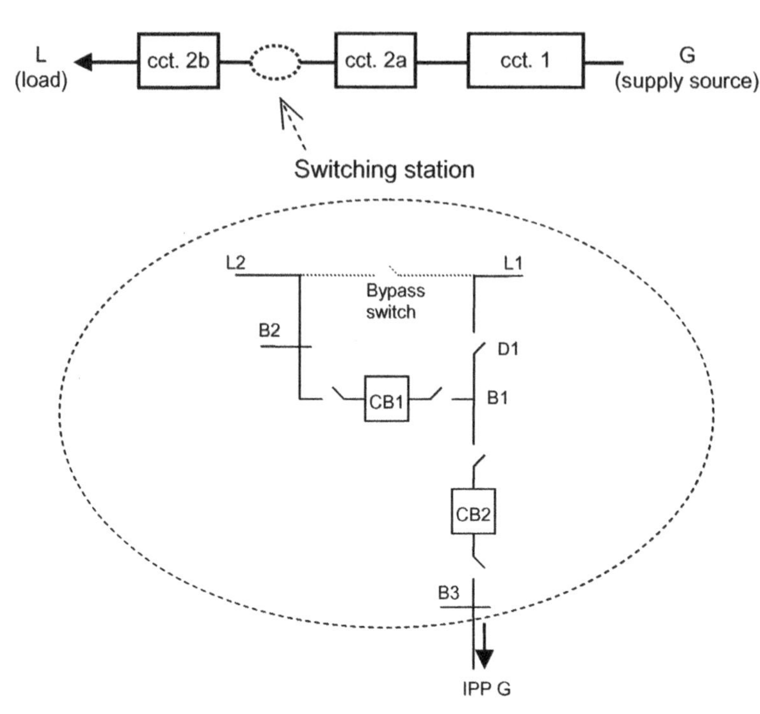

Figure 11.2. Block diagram of the radial supply system with the IPP switching station of two-breaker configuration (cct. = circuit).

minimum cutsets. A challenging difficulty in developing a general search technique of minimum cutsets is how to deal with dependent failures between components, multiple failure modes including breaker stuck conditions, breaker switching, and protection coordination. In these situations, the physical network of substation configuration cannot be directly used as a topology structure for searching minimum cutsets [62]. Fortunately, many substation configurations in actual systems are relatively simple, and their failure events can be identified by direct observation or judgment. This is the case in the given example.

2. The failure frequency, recovery duration, and unavailability due to each failure event are calculated. For a first-order failure event in which only one component fails, the failure frequency and duration are simply the failure data of the failed component. For a second-order failure event in which either two components fail, or one component fails with a breaker stuck condition or bypass switching action, simple calculations are required. The similar concept can be easily extended to a third-order failure event which is associated with three failure components. In the following equations, f, r, and U represent the failure frequency (failure/year), recovery duration (hours/recovery) and unavailability (hours/year), respectively; the subscript i represents the failure event i; the subscript 1, 2, or 3 represents component 1, 2, or 3; f_p and f_a are the passive and

active failure frequencies, respectively; f_c represents either f_p or f_a of a breaker; r_c is the repair or replacement time of the component; r_{sw} is the switching time (hours); r_{swb} is the bypass switching time (hours); P_s is the breaker stuck probability; and P_{bypass} is the probability of unsuccessful bypass switching.

a. Passive failure event of one component (switching action not required):

$$f_i = f_p \tag{11.1}$$

$$r_i = r_c \tag{11.2}$$

b. Active failure event of one component (switching action required):

$$f_i = f_a \tag{11.3}$$

$$r_i = r_{sw} \tag{11.4}$$

c. Active failure event of one component plus breaker stuck condition (switching action required):

$$f_i = f_a \times P_s \tag{11.5}$$

$$r_i = r_{sw} \tag{11.6}$$

d. Failure event of one breaker associated with bypass switching:

- For successful bypass switching:

$$f_i = f_c \times (1 - P_{bypass}) \tag{11.7}$$

$$r_i = r_{swb} \tag{11.8}$$

- For unsuccessful bypass switching:

$$f_i = f_c \times P_{bypass} \tag{11.9}$$

$$r_i = r_c \tag{11.10}$$

The failure frequency f_c can be f_p or f_a depending on whether it is a passive or active failure.

e. Overlapping failure events for two components [6]:

$$f_i = \frac{f_1 f_2 (r_1 + r_2)}{8760} \tag{11.11}$$

$$r_i = \frac{r_1 r_2}{r_1 + r_2} \tag{11.12}$$

The failure frequency and recovery duration of each component (f_1 and r_1 or f_2 and r_2) can be one of the four cases indicated in items 2a–2d). Note that

8760 occurs in Equation (11.11) because the unit of r is in hours/recovery and the unit of f is in failures/year.

f. Overlapping failure events for three components. It can be seen that Equations (11.11) and (11.12) are the formulas for calculating equivalent failure frequency and recovery duration of two components in parallel. This concept can be extended to overlapping failure events of three components using the following formulas:

$$f_i = \frac{f_1 f_2 f_3 (r_1 r_2 + r_1 r_3 + r_2 r_3)}{8760 \times 8760} \tag{11.13}$$

$$r_i = \frac{r_1 r_2 r_3}{r_1 r_2 + r_1 r_3 + r_2 r_3} \tag{11.14}$$

It should be appreciated that the situation in which overlapping failure events of three components need to be considered is very rare since the probability of such events is extremely low and negligible in most cases. There is no need to consider overlapping failure events of three components for the example given in this section. However, when the aging failure mode of components (such as transformers) resulting in a very long replacement time needs to be considered, it may be necessary to include third-order failure events in the evaluation.

g. For all the failure events described above:

$$U_i = f_i \cdot r_i \tag{11.15}$$

Note that the unavailability given by Equation (11.15) has the unit of hours/year. $U_i = (f_i \cdot r_i)/8760$ gives the probability value without unit for the unavailability.

It should be noted that outage components in an active failure event, breaker stuck condition, or bypass switching state include not only failed components but also some healthy components that have to be isolated from a fault point. Recognition of outage components and connectivity identification of substation topology after outages are the keys in determining impacted buses and evaluating load curtailments at the buses (see Section 5.4 for details).

3. Each failure event enumerated is a minimum cutset. The minimum cutsets are not mutually exclusive, and disjoint calculations are required. The equation for the total failure probability of a substation configuration using the minimum cutset technique is given by

$$\begin{aligned} U_s &= P(C_1 \cup C_2 \cdots \cup C_n) \\ &= \sum_i P(C_i) - \sum_{i,j} P(C_i \cap C_j) + \sum_{i,j,k} P(C_i \cap C_j \cap C_k) - \cdots \\ &\quad + (-1)^{n-1} P(C_1 \cap C_2 \cdots \cap C_n) \end{aligned} \tag{11.16}$$

where U_s is the failure probability of substation configuration; C_i, C_j, C_k, or C_n represents the *i*th, *j*th, *k*th, or *n*th minimum cutset (each failure event); $P(*)$ represents the probability of a minimum cutset or an intersection of multiple minimum cutsets; and n is the number of failure events enumerated.

In general, it is difficult to perform the disjoint calculations for a substation configuration, particularly when dependent failures, multiple failure modes, and switching actions are considered. Fortunately, the probabilities of intersection of two or more minimum cutsets are generally very low because of small unavailability values of components in a substation configuration. Therefore it is acceptable to approximate the total failure probability by using only the first term in Equation (11.16) from an engineering viewpoint. This means that non-mutual exclusion among all failure events is ignored. Another approximation is to ignore the difference between failure rate and failure frequency, which is generally small for substation components. The two approximations will not result in an effective error for the example in Section 11.2 and many other cases. Therefore, the reliability indices of a substation configuration can be approximately calculated by

$$f_s = \sum_{i=1}^{n} f_i \tag{11.17}$$

$$U_s = \sum_{i=1}^{n} U_i \tag{11.18}$$

$$r_s = \frac{U_s}{f_s} \tag{11.19}$$

where f_s, U_s, and r_s are the failure frequency, duration, and failure probability of substation configuration, respectively; n is the number of failure events.

It is worth noting that the simplified minimum cutset technique should be used with caution in some cases since the conditions of the two approximations may not always be valid.

11.2.2.2 Economic Analysis Approach. The economic analysis can be performed using two approaches.

1. *Relative Comparison.* In the example given, the two substation configurations are compared. If there are more than two configurations, the comparison can be made between any two first and the better one is compared with the third one, and so on. The configuration that has a better reliability (smaller unreliability index) and needs a higher investment cost is used as a reference. By assuming configuration 1 to be the reference, we can calculate the relative differences (%) in the reliability and investment cost between the two configurations by

$$\Delta R = \frac{R_2 - R_1}{R_1} \times 100\% \tag{11.20}$$

$$\Delta I = \frac{I_1 - I_2}{I_1} \times 100\% \tag{11.21}$$

where R_1 and R_2 are the unreliability indices of configurations 1 and 2, respectively; and I_1 and I_2 are the investment costs of the two configurations, respectively. The ΔR and ΔI represent the relative differences (in percent) in the reliability and investment cost between the two configurations. If $\Delta R > \Delta I$, then the percentage in reliability improvement due to configuration 1 versus configuration 2 is greater than the percentage in the investment cost required by configuration 1 versus configuration 2.

2. *Incremental Reliability Cost.* The incremental reliability cost (IRC) is the investment cost for improvement of unit reliability index. The existing system case without any configuration change is used as the base case. In the example given, the base case is the supply system of two circuits in series before the IPP generating station configuration is added. The IRC is equal to the investment cost of a configuration divided by the difference in a reliability index between the base case and the case with the configuration:

$$\text{IRC}_i = \frac{I_i}{R_b - R_i} \tag{11.22}$$

Here, IRC_i is the incremental reliability cost for configuration i, R_b and R_i are the reliability indices for the base case and the case with configuration i, respectively, and I_i is the investment of configuration i. If $\text{IRC}_i > \text{IRC}_j$, this indicates that configuration i requires a higher investment cost for improvement of unit reliability index than does configuration j.

11.2.3 Comparison between the Two Configurations

More details of this application study can be found in Reference 124.

11.2.3.1 Study Conditions and Data. The simplified minimum cutset technique described in Section 11.2.2.1 was used to evaluate the reliability of the two substation configurations shown in Figures 11.1 and 11.2. The study conditions are summarized as follows:

- All possible failure modes were considered in the study, including the passive and active outages, breaker stuck conditions, and unsuccessful switching of bypass switch.
- Both forced and maintenance outages of the breakers were included.
- Forced outages of the disconnect and bypass switches were not considered (i.e., were assumed 100% reliable). The disconnect switches are all in a normally closed state, and the bypass switches are all in a normally opened state. The probability of a switch not remaining in its normal operation state is extremely

low, whereas the frequency of a short-circuit fault on a switch is also very low and can be treated as a part of the active failure frequency for a busbar.

- The maintenance outages of all the switches were considered. The maintenance frequency of all the breakers and switches in the substation was assumed to be the same.
- The live maintenance is implemented for the overhead transmission lines, and thus no maintenance outage is required for the lines. No maintenance activity is assumed for the busbars.
- The forced outage data of transmission lines were based on outage statistics in the previous 10 years obtained from the reliability database of the utility. The forced outage data of circuit breakers and busbars were based mainly on the statistics of Canadian utilities and supplemented by some generic data [6,11,125].
- Manual switching actions were assumed to require 4 h to isolate failed components, and the restoration time for replacing a failed breaker with a spare was assumed to be 3 days. The switching time of the bypass switch in a maintenance activity is reduced to be 2 h considering the fact that the maintenance crew is on site already.

The equipment reliability data are summarized in Tables 11.1 and 11.2.

11.2.3.2 Reliability Results. The two reliability indices of failure frequency and unavailability that use loss of load as the failure criterion were evaluated. Two operation scenarios were considered in the study, in which, islanding operation of the IPP was allowed and was not allowed. *Islanding operation* refers to a case in which, if an upstream circuit trips as a result of a fault, the IPP can be operated alone to avoid a blackout to the customers connected to circuit 2. The generation capacity of the proposed IPP is capable of serving all the loads connected along circuit 2.

The reliability results for the three-breaker ring and two-breaker configurations with and without the islanding operation are summarized in Table 11.3. Note that the unavailability index can be expressed either in the form of probability (no unit) or in hours/year. The form in hours/year is used in this example.

The following observations and analyses can be made:

1. The three-breaker ring configuration results in a slightly higher failure frequency than does the two-breaker configuration in the example. This is basically due to the fact that the three-breaker ring configuration has more components that require maintenance activities. However, the maintenance is not a random outage in nature, and therefore the maintenance outage impact on customers can be minimized as long as the maintenance is scheduled at a less critical time (during a light load) with an advance notice of interruption to customers. Therefore the key index for comparison between the two substation configurations is the unavailability (in hours/year) in this example. It is worth pointing out that the slightly lower failure frequency for the two-breaker configuration versus the three-breaker ring configuration is not a universal conclusion. First,

TABLE 11.1. Equipment Reliability Data (Unplanned Outages)

Equipment	Failure Frequency		Repair Time (hours)	Manual Switching Time (hours)	Spare CB Installation time (hours)	Stuck Probability	Unsuccessful Switching Probability
	Active (failures/year)	Passive (failures/year)					
L1[a]	1.04	—	42.03	4.00	—	—	—
L2[b]	0.26	—	4.85	4.00	—	—	—
Circuit breaker(CB)	0.05	0.004	212.70	4.00	72.0	0.02	—
Busbar	0.02	—	17.90	4.00	—	—	—
Semibusbar[c]	0.01	—	17.90	4.00	—	—	—
Bypass switch	—	—	—	4.00	—	—	0.04

[a]L1 = circuit 1 + circuit 2a.
[b]L2 = Circuit 2b.
[c]*Semibusbar* refers to a physical connection that can be treated as a bus.

TABLE 11.2. Equipment Reliability Data (Maintenance Outages)

Equipment	Maintenance		
	Frequency (failures/year)	Downtime (hours)	Switching(hours)
L1[a]	—	—	—
L2[b]	—	—	—
Circuit breaker	0.125	6.00	—
Busbar	—	—	—
Semibusbar[c]	—	—	—
Bypass switch	0.125	6.00	2.00
Disconnect switch	0.125	6.00	—

[a]L1 = circuit 1 + circuit 2a.
[b]L2 = Circuit 2b.
[c]*Semibusbar* refers to a physical connection that can be treated as a bus.

TABLE 11.3. Reliability Indices of the Two Substation Configurations

Scenario	Three-breaker ring configuration		Two-breaker configuration		Reliability difference[a] (%)	
	Failure frequency (failures/year)	Unavailability (hours/year)	Failure frequency (failures/year)	Unavailability (hours/year)	Failure Frequency (failures/year)	Unavailability (hours/year)
Islanding operation allowed	1.62	3.50	1.57	6.89	−3	97
Islanding operation not allowed	1.62	47.04	1.57	46.44	−3	−1

[a]*Reliability difference* refers to the difference (%) in the index between the two-breaker and three-breaker ring configurations divided by the index for the three-breaker ring configuration.

the results presented above are based on the assumption that the bypass switch used in the two-breaker configuration has a capacity similar to a circuit breaker (capable of breaking a loop/parallel current on a loaded line). This type of switch will most likely have a higher failure probability than will regular switches since it has to be equipped with additional components such as vacuum interrupters. This factor was not considered in the study. Also, the system topology of a radial network having only one main supply source in this example is a crucial origin that degrades the merit of the three-breaker ring configuration. In the case of a looped network with multiple main supply sources, the merit of the three-breaker ring configuration will be enhanced.

2. It is interesting to note that if the islanding operation is not allowed, the three-breaker ring configuration not only provides a little bit worse reliability

compared to the two-breaker configuration in this radial system with one main supply source but also requires one more breaker. This is called a *noncoherence phenomenon* in reliability evaluation. Conceptually, it is similar to the situation of components in series, in which one more component would lead to lower reliability. In actual operation practice, however, the islanding operation is highly preferable because after the upstream supply source is lost, causing a prolonged outage, the IPP generation can still serve the loads on circuit 2 during that time.

3. If the islanding operation is allowed, the three-breaker ring configuration provides much better availability compared to the two-breaker configuration. The unavailability seen by customers for the three-breaker ring configuration is about half that for the two-breaker configuration. This is because the former will offer a much quicker supply restoration after losing the upstream supply. The three-breaker ring configuration can establish the islanding operation right after either the upstream circuit is lost or the relevant busbar fails since the breakers will be automatically tripped to isolate a fault. A restarting process of the IPP generator, which may or may not be needed, requires at most 0.5 h. The two-breaker configuration would require a much longer time to establish the islanding operation mode. If a fault occurs on the upstream circuit, it requires 4 h of average restoration time because a manual isolation process on the line is needed; if a fault occurs on the relevant busbar, it requires 17.9 h of average restoration time because the islanding operation mode cannot be established until the busbar is returned to the normal state.

11.2.3.3 Economic Comparison. The method given in Section 11.2.2.2 was used to perform the economic analysis for the two substation configurations. There are two arrangements for substation equipment: flat and stacked. The investment costs and unavailability values for the base system and the system with each substation configuration in the case of islanding operation allowed are presented in Table 11.4.

The relative differences (in percent) in the unavailability (ΔR) and investment cost (ΔI) between the two substation configurations were calculated using Equations (11.20) and (11.21), and are shown in Table 11.5. The three-breaker ring configuration was used as the reference. It can be seen that the three-breaker ring configuration costs only

TABLE 11.4. Data for Economic Comparison between the Two Substation Configurations

System	Investment Cost ($M)		Unavailability (hours/year)
	Flat Arrangement	Stacked Arrangement	
Base system	—	—	44.97
System with three-breaker ring configuration	36	30	3.50
System with two-breaker configuration	33	29	6.89

TABLE 11.5. Relative Differences between the Two Substation Configurations

Relative Difference in Unavailability	Relative Difference in Investment Cost(%)	
	Flat Arrangement	Stacked Arrangement
97%	8.3	3.3

TABLE 11.6. Incremental Reliability Costs of the Two Substation Configurations

Configuration	Incremental Reliability Cost ($M/unit unavailability improvement)	
	Flat Arrangement	Stacked Arrangement
Three-breaker ring	0.868	0.723
Two-breaker	0.867	0.762

TABLE 11.7. Comparison in Traditional Considerations between the Two Substation Configurations

Criterion	Three-Breaker Ring Configuration	Two-Breaker Configuration
Land	More	Less
Operation flexibility	More	Less
Maintenance requirements	Similar	Similar
Island operation Implementation	Easier to implement	Harder to implement
Safety risk	Lower risk	Higher risk for live line work
Provision of expansion	Easier	More difficult

3.3% (stacked arrangement) or 8.3% (flat arrangement) more than the two-breaker configuration while it results in relative unavailability reduction up to 97%, suggesting that the three-breaker ring configuration is better.

The incremental reliability costs (IRCs) for the two substation configurations were calculated using Equation (11.22), and are shown in Table 11.6. It can be seen that the three-breaker ring configuration requires a lower investment cost for improvement of unit unavailability index than the two-breaker configuration if the stacked arrangement is used, whereas the two configurations will cost almost the same for unit unavailability improvement if the flat arrangement is used.

11.2.3.4 Other Considerations. Other traditional considerations in substation planning include lands needed, operation flexibility, maintenance requirements, islanding operation implementation, safety risk, and potential for future expansion. These factors are compared in Table 11.7.

11.2.4 Summary

This section presents the probabilistic planning method for substation configurations through an actual application.

In the example, the three-breaker ring configuration is superior to the two-breaker configuration in terms of overall reliability and relative economic efficiency. It should be noted that this is not a universal conclusion but is valid only under the specific conditions (the islanding operation and stacked arrangement). It can be seen that if the islanding operation is not allowed, the three-breaker configuration cannot be justified. This suggests that adding more components in a substation does not necessarily guarantee the improvement in reliability. The failure frequency can deteriorate when more breakers are added, whereas the total unavailability is generally reduced as a result of faster isolations of failed components by breakers. Besides, this example also indicates that the reliability of a substation not only depends on its own configuration but also is associated with the topology of the network to which it is connected.

11.3 TRANSFORMER SPARE PLANNING

Substation equipment spare planning has been a challenge for utilities. The practice of most utilities in this area so far is to use a deterministic method, which is based on an engineering judgment. This section provides a reliability-based equipment spare planning method using transformer spares as an example. The concept and method presented can be applied to other equipment.

11.3.1 Description of Problem

Transformers are the most important equipment in substations. Considerable planning issues at the substation level are associated with adequacy of transformers to reliably transfer powers from a transmission network to customers on distribution systems. The transformer spares shared by multiple substations become a popular strategy at utilities, particularly under the competitive environment today. This strategy can provide an acceptable reliability level while saving expensive capital expenditures compared to the traditional $N-1$ principle in each individual substation. The transformer aging status has been a reality in many utility systems. An aged transformer has a much higher failure probability. The repair or replacement process of a transformer often takes a long time. These two factors are also the drivers for the need of transformer spares.

The basic questions in transformer or other equipment spare panning are

- How many spares are needed in total for an equipment group?
- When should each of them be in place?

Generally, there are two reliability-based methods for transformer spare planning. The first one is based on reliability criteria, and the second one is based on probabilistic reliability cost models. In this section, the reliability criterion method is discussed and an example of a transformer group is used to demonstrate the application of the method. The probabilistic-reliability-cost-model based method was discussed in Reference 6.

11.3.2 Method of Probabilistic Spare Planning

11.3.2.1 Basic Procedure. Spares are shared by a group of transformers with the same voltage and similar structures. Each transformer in the group has its unavailability due to failures. When one or more transformers fail, one or more spares must be put in service to ensure normal operation of substations. Therefore, how many spares are needed depends on the requirement for transformer group reliability. With the unavailability of individual transformers, a state enumeration technique can be used to evaluate the failure and success probabilities of a transformer group with and without spares. The spare planning for a transformer group includes the following steps:

1. A transformer group is determined in such a way that each member can be technically replaced by a commonly shared spare transformer and the replacement can be implemented in time based on consideration of geographic locations. In general, a transformer whose single failure during the peak load level will lead to loss of load is selected as a member of the group. However, the transformers that meet the $N-1$ principle in an individual substation can also be included in the group if necessary. In this case, the reliability of the substation will be further enhanced because multiple failed transformers in the substation can be replaced by spares.
2. An appropriate reliability criterion for the transformer group is specified. This will be discussed in Section 11.3.2.3.
3. The unavailability of each transformer in the group is calculated. There are two failure modes for transformers: repairable and aging (end-of-life) failures. In many reliability evaluations for power systems, only the unavailability due to repairable failures is considered. However, the unavailability due to aging failures should be factored into the spare analysis since the aging failure of transformers is one of the main considerations for using spares, particularly for an aged transformer group. The modeling concept for the unavailability due to repairable and aging failures has been addressed in Sections 10.2.2.2 and 10.2.2.3.
4. Individual failure event probabilities and group failure and success probabilities both without and with transformer spares are evaluated. The evaluation method will be discussed in Section 11.3.2.2.
5. Steps 3 and 4 are repeated for all the years in the planning timespan. This is required because the unavailability value of each transformer increases over the years when the aging failure mode is considered.
6. The success probabilities of the transformer group with spares at different years are compared with the specified reliability criterion. The group success probability must be equal to or greater than the acceptable probability. The number of transformer spares and their timing are determined from the comparison. This step will be illustrated using the example given in Section 11.3.3.

11.3.2.2 Reliability Evaluation Technique for a Transformer Group. A state of a transformer group with one transformer failure is called a *first-order failure state*, a state with two transformer failures at the same time is called a *second-order failure state*, and so on. The cumulative probability of all first-order failure states can be calculated by

$$P(1)=\sum_{i=1}^{N}\left[U_i \cdot \prod_{\substack{j=1\\ j\neq i}}^{N}(1-U_j)\right] \tag{11.23}$$

where $P(1)$ represents the cumulative probability of all first order failure states; U_i or U_j is the unavailability of the ith or jth transformer, respectively; and N is the number of transformers in the group. Introduction of the product term of $(1\text{-}U_j)$ ensures that all first-order failure states are mutually exclusive and therefore their state probabilities can be directly summed up.

The cumulative probability of all second-order failure states can be calculated by

$$P(2)=\sum_{i=1}^{N-1}\sum_{k=i+1}^{N}\left[U_i U_k \cdot \prod_{\substack{j=1\\ j\neq i,k}}^{N}(1-U_j)\right] \tag{11.24}$$

where $P(2)$ represents the cumulative probability of all second-order failure states; U_i,U_k, and U_j are the unavailability values of the ith, kth, or jth transformer, respectively.

Similarly, the equations for the cumulative probabilities of all third-, or fourth-, or higher-order failure states can be derived. Obviously, $P(N)$ simply equals the product of unavailability values of all the transformers in the group. The total group failure probability with zero spare P_{0s} is obtained by direct summation

$$P_{0s}=\sum_{i=1}^{N}P(i) \tag{11.25}$$

where $P(i)$ is the cumulative probability of all the ith-order failure states. For a group containing a relatively small number of transformers, all the $P(i)$ in Equation (11.25) can be easily calculated. For a group containing a relative large number of transformers, it is not necessary to enumerate up to the last-order failure state set. Enumerating to which order failure state set depends on the number of transformers in a group. A cutoff threshold of probability can be specified such that the high-order failure state sets with a probability lower than the threshold are neglected.

The group failure probabilities with one, two, and more spares are calculated by

$$P_{is}=P_{0s}-\sum_{j=1}^{i}P(j) \tag{11.26}$$

where P_{is} represents the group failure probability with i spares. For example, $P_{1s} = P_{0s} - P(1)$, which indicates that the probability of first-order failure state set is excluded from the total group failure probability with no spare since these events no longer lead to a group failure as the spare can replace a failed transformer; similarly, $P_{2s} = P_{0s} - P(1) - P(2)$, and so on.

Once the group failure probabilities with zero, one, two, and more spare are obtained, the group success probabilities with zero, one, two, and more spares can be calculated by 1.0 minus the corresponding failure probability.

11.3.2.3 Reliability Criterion. The term *reliability criterion* refers to a target for the transformer group success probability, above which the group reliability is considered acceptable. There are different approaches to specify the acceptable success probability for a transformer group depending on different cases or utility requirements. The acceptable success probability based on the T-SAIDI index is used in the actual example given in the next section. The T-SAIDI (see Section 7.3.3.1 for its definition), which is the average interruption duration per delivery point, has been utilized as a key performance indicator in many utilities, and therefore using the T-SAIDI-based acceptable success probability is consistent with the utilities' overall reliability objective. The utilities set a target value of T-SAIDI on the basis of its historical statistics. The T-SAIDI target can be converted into an acceptable threshold success probability (availability) P_{th} for a group of transformers as follows

$$P_{\text{th}} = 1 - \frac{S \cdot N_D}{8760} \tag{11.27}$$

where S represents the T-SAIDI target of a utility and N_D is the number of delivery points corresponding to the transformer group considered. A delivery point is defined as a bus at the low-voltage side of a stepdown transformation substation. Most substations have one delivery point each, but some may have more than one delivery point depending on the substation structure.

Equation (11.27) is straightforward as the unit of T-SAIDI is expressed in hours per delivery per year. The following example is used for interpretation.

Assume that the T-SAIDI target S at a utility is 2.1 hours per delivery per year and the transformers corresponding to 35 delivery points are considered as a group. The total average interruption duration target for the group is

$$S \times N_D = 2.1 \times 35 = 73.5 \text{ h/year}$$

The acceptable success probability is $1 - 73.5/8760 = 0.9916$ (or 99.16%)

The example above indicates that the availability of 0.9916 is required as a specified reliability criterion for this transformer group in order to align with the company's performance target in T- SAIDI of 2.1 hours per delivery per year.

Conceptually, either of the following conditions should be met in using the T-SAIDI- based reliability criterion because the T-SAIDI is the average load interruption duration:

- The transformers in a substation that meets the $N-1$ principle are not included in the transformer group considered since one transformer failure in the substation will not cause load interruption.
- If the transformers in a substation that meets the $N-1$ principle are included the group, the partial contributions due to single failures of these transformers to the corresponding $P(i)$ are excluded from the calculations.

In an actual application, however, it is acceptable not to strictly meet the two conditions, which will create a relatively conservative (secure) result (spares in service earlier).

11.3.3 Actual Example

More details of this actual example can be found in References 126 and 127.

11.3.3.1 Case Description. The 138/25-kV transformers at BC Hydro, each of which has a capacity between 10 and 30 MVA, are considered as a transformer group that is backed up by 138/25-kV 25-MVA transformer spares. It has been predicted that the largest load supplied by each transformer will not exceed 25 MVA. Three study scenarios are presented in this example. The first one focuses on the fixed turn ratio transformer subgroup, which consists of 34 transformers located in 29 substations (delivery points). The second one focuses on the on-load tap changer (OLTC) transformer subgroup, which consists of 16 transformers located in 12 substations. The third one combines both fixed turn ratio and OLTC transformers altogether, which consists of 50 transformers located in 35 substations. The planning period for the transformer group is the 10 years from 2006 to 2015. Some transformers have been operated for years and reached an old age. Both repairable and aging failure (end-of-life) modes of transformers are considered. The unavailability values of transformers for the repairable failure mode are based on historical failure statistics, whereas the unavailability values for the aging failure mode are calculated from the Weibull distribution model, which is the same as that used for aged equipment in Section 10.2.2.3. The scale and shape parameters in the Weibull distribution model are estimated using the method given in Section 10.2.2.2 and the data of 148 transformers (138 kV) at different ages and with six retired in different years. The mean life and standard deviation for the Weibull distribution can be calculated from the scale and shape parameters [see Equations (A.19) and (A.20) in Appendix A]. The resulting mean life and standard deviation for the 138/25-kV transformers are 57.1 and 14.5 years, respectively.

11.3.3.2 Fixed Turn Ratio Transformer Group. The T-SAIDI target is 2.1 hours per delivery point per year. There are 29 substations (delivery points) in total for this group. The acceptable success probability of the group is calculated using Equation (11.27) as follows:

$$P_{\text{th}} = 1 - \frac{2.1 \times 29}{8760} = 0.993$$

TABLE 11.8. Success Probability of 138/25-kV Fixed Turn Ratio Transformer Group (34 Units) with Different Numbers of Transformer Spares

Year	Number of Transformer Spares			
	0	1	2	3
2006	0.8757	0.9922	0.9997	1.0000[a]
2007	0.8651	0.9908	0.9996	1.0000[a]
2008	0.8537	0.9891	0.9995	1.0000[a]
2009	0.8417	0.9872	0.9993	1.0000[a]
2010	0.8289	0.9849	0.9991	1.0000[a]
2011	0.8154	0.9824	0.9989	0.9999
2012	0.8011	0.9794	0.9986	0.9999
2013	0.7862	0.9761	0.9982	0.9999
2014	0.7706	0.9723	0.9978	0.9999
2015	0.7542	0.9680	0.9972	0.9998

[a]The values of 1.0000 were obtained by rounding to the fourth decimal place.

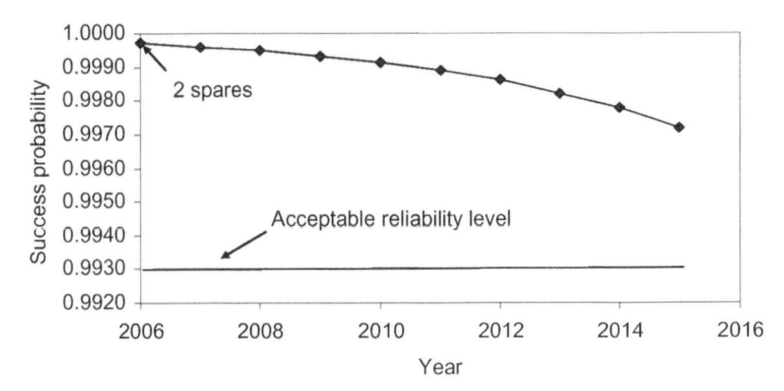

Figure 11.3. Number of transformer spares required to meet the specified reliability criterion for the fixed turn ratio transformer group.

The availability of 0.993 is used as the specified reliability criterion for the 34 fixed turn ratio transformers located in the 29 substations. The majority of substations are single-transformer substations that will share common transformer spares. The transformer group success probability must be at least equal to or greater than the specified reliability criterion of 0.993 all the time during the planning period from 2006 to 2015.

A computer program called SPARE [128] has been designed using the method in Section 11.3.2 and was used for the spare transformer analysis. The results obtained are shown in Table 11.8 and graphically presented in Figure 11.3. Table 11.8 shows the annual success probabilities of the 138/25-kV fixed turn ratio transformer group with and without spare transformers (up to three spares). It is worth noting that the annual success probability decreases over the years since the aging failure probability of transformers increases with years.

It can be seen from Figure 11.3 that two fixed turn ratio transformer spares were needed in 2006. With the two transformer spares, the group will be able to meet the specified reliability criterion (0.993) until the end of the planning period (in 2015).

11.3.3.3 On-Load Tap Changer (OLTC) Transformer Group. The T-SAIDI target is still 2.1 hours per delivery point per year. There are 12 substations (delivery points) in total for this group. The acceptable success probability of the group is calculated using Equation (11.27) as follows:

$$P_{th} = 1 - \frac{2.1 \times 12}{8760} = 0.9971$$

The availability of 0.9971 is used as the specified reliability criterion for the 16 on-load tap changer transformers located in the 12 substations. Similarly, the majority of substations are single-transformer substations that will share common transformer spares. The transformer group success probability must be at least equal to or greater than the specified reliability criterion of 0.9971 all the time during the planning period from 2006 to 2015.

The results obtained using the SPARE program for the 138/25-kV OLTC transformers are shown in Table 11.9 and graphically presented in Figure 11.4. It can be seen from Figure 11.4 that the first OLTC transformer spare was required in 2006 in order for the 138/25-kV OLTC transformer group to meet the specified reliability criterion (0.9971). In 2012, the group with the first transformer spare will no longer meet the specified reliability criterion, and the second OLTC transformer spare will be required.

11.3.3.4 Combined Fixed Turn Ratio and OLTC Transformer Group. An OLTC transformer spare can replace either a fixed turn ratio or an OLTC transformer.

TABLE 11.9. Success Probability of 138/25-kV OLTC Transformer Group (16 Units) with Different Numbers of Transformer Spares

	Number of Transformer Spares		
Year	0	1	2
2006	0.9514	0.9989	1.0000[a]
2007	0.9470	0.9987	1.0000[a]
2008	0.9422	0.9984	1.0000[a]
2009	0.9371	0.9981	1.0000[a]
2010	0.9316	0.9978	1.0000[a]
2011	0.9257	0.9974	0.9999
2012	0.9194	0.9969	0.9999
2013	0.9127	0.9963	0.9999
2014	0.9055	0.9957	0.9999
2015	0.8979	0.9950	0.9998

[a]The values of 1.0000 were obtained by rounding to the fourth decimal place.

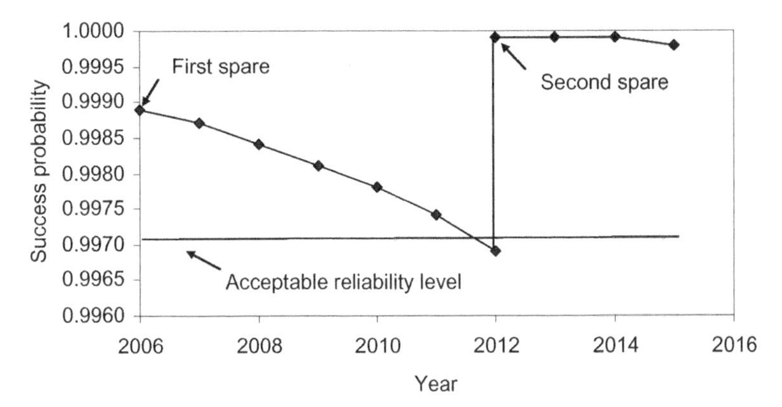

Figure 11.4. Number of transformer spares required to meet the specified reliability criterion for the OLTC transformer group.

TABLE 11.10. Success Probability of 138/25-kV Fixed Turn Ratio and OLTC Transformer Group (50 Units) with Different Numbers of Transformer Spares

	Number of Transformer Spares			
Year	0	1	2	3
2006	0.8331	0.9856	0.9992	1.0000[a]
2007	0.8192	0.9829	0.9989	1.0000[a]
2008	0.8044	0.9799	0.9986	0.9999
2009	0.7887	0.9764	0.9983	0.9999
2010	0.7722	0.9724	0.9978	0.9999
2011	0.7548	0.9678	0.9972	0.9998
2012	0.7366	0.9626	0.9964	0.9997
2013	0.7175	0.9566	0.9955	0.9997
2014	0.6978	0.9499	0.9944	0.9995
2015	0.6772	0.9423	0.9931	0.9994

[a]The values of 1.0000 were obtained by rounding to the fourth decimal place.

The number of OLTC transformer spares needed to back up all the 138/25-kV fixed turn ratio and OLTC transformers can be determined using the same method. The T-SAIDI target is still 2.1 hours per delivery point per year. There are 35 substations (delivery points) in total for the combined group. The acceptable success probability of the group is

$$P_{\text{th}} = 1 - \frac{2.1 \times 35}{8760} = 0.9916$$

The availability of 0.9916 is used as the specified reliability criterion for the 50 transformers (fixed turn ratio and OLTC) located in the 35 delivery points (substations). The results obtained using the SPARE program for this group are shown in Table 11.10 and

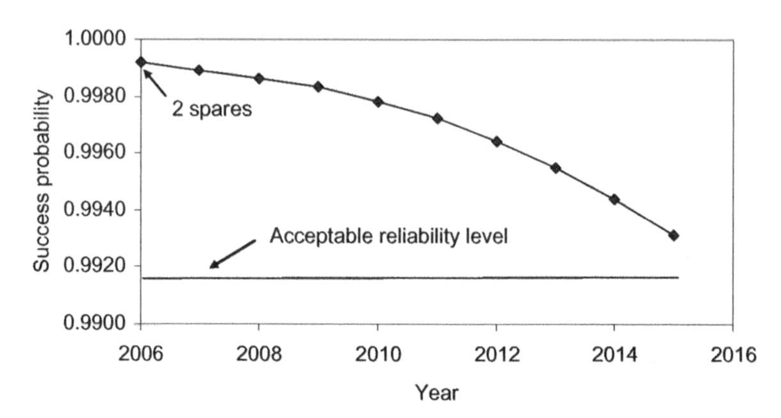

Figure 11.5. Number of transformer spares required to meet the specified reliability criterion for the transformer group composed of both fixed turn ratio and OLTC transformers.

graphically presented in Figure 11.5. It can be seen from Figure 11.5 that two OLTC transformer spares were needed in 2006 to back up both the fixed turn ratio and OLTC transformers, and the transformer group with the two spares will be able to meet the specified reliability criterion until the end of the planning period (in 2015).

11.3.4 Summary

This section presents a transformer spare planning method based on quantified reliability evaluation and its actual application in the utility.

The results indicate that if the spares for the fixed turn ratio and OLTC transformers are considered separately, the utility system would need four 138/25-kV transformer spares (two fixed turn ratio spares and two OLTC spares) in the 10-year period from 2006 to 2015. However, if the OLTC transformer spares are considered to back up both the fixed turn ratio and OLTC transformers, the utility system would need only two OLTC transformer spares to meet the same reliability criterion in the same period. This leads to a considerable saving in the capital investment while still satisfying the specified reliability requirement for the 138/25-kV transformer group.

11.4 CONCLUSIONS

This chapter addressed two main aspects in substation planning using the probabilistic techniques: substation configuration planning and transformer spare planning.

The method for substation configuration planning includes both quantified reliability evaluation and reliability-based economic analysis. A simplified minimum cutset technique is presented for the reliability evaluation of substation configurations. The technique can reduce complexity in the evaluation while maintaining acceptable accuracy when relatively simple substation configurations (such as the ones in the example) are considered. In a general case, the state numeration technique for substation reliability assessment given in Section 5.4 should be applied. The two economic analysis approaches for relative comparison between different configurations have been discussed. These two approaches are appropriate to cases in which the reliability indices do not change over the years. When the load levels and/or substation equipment aging models are incorporated in the reliability assessment of substation configurations, the reliability indices will be increased over the years. In this case, the EENS index should be used and converted into unreliability costs to conduct the more comprehensive economic analysis that requires the use of the present value method. Conceptually, the procedure will be similar to the economic analysis in Section 9.3.

The method for transformer spare planning includes the estimation of unavailability of individual transformers due to both repairable and aging failures, reliability evaluation of a transformer group with and without spares, selection of a reliability criterion for the group, and spare analysis for determining the numbers and timing of transformer spares. In the given example, the reliability criterion based on the T-SAIDI index is used since the T-SAIDI has been used as a key performance indicator in many utilities. It should be emphasized that this is not a unique approach in selecting the acceptable reliability level, and that other principles can be used depending on the reliability objective at a utility.

Substation planning is associated with other issues. The probabilistic planning concepts and methods presented in this chapter can be either directly used or further extended to similar planning problems in substations. For example, a transformer addition or any change of the existing layout in a substation can be viewed as an alternative against the existing configuration, and the proposed probabilistic substation configuration planning method can be applied. The author would like to leave some space for readers to envisage their own ideas.

12

SINGLE-CIRCUIT SUPPLY SYSTEM PLANNING

12.1 INTRODUCTION

A *single-circuit supply system* is a simple system in which one or more delivery points (substations) are supplied by one single-circuit and an interruption of the circuit will cause load curtailment at all the delivery points. For simplicity of discussion in this chapter, it is assumed that one substation contains one delivery point. The delivery points are divided into two categories: single-circuit-supplied ones and multiple-circuit-supplied ones. The first category can be further classified into two subgroups: single-circuit-radial-supplied delivery points, where only one power source is connected to the single-circuit, and single-circuit-network-supplied delivery points, where more than one source is connected to the single-circuit. As shown in Figure 12.1, DS1, DS2, DS3, and TS1 are the single-circuit-radial-supplied delivery points; TS2 is a single-circuit-network-supplied delivery point; and DS4 and DS5 are the multicircuit-supplied delivery points [129]. Note that a "transmission customer" in the figure refers to a single industry customer, whereas a distribution customer refers to a distribution substation that will supply its loads through feeders.

Obviously, delivery points supplied by multiple circuits have much higher reliability than will those supplied by a single-circuit. Single-circuit-network-supplied

Probabilistic Transmission System Planning, by Wenyuan Li

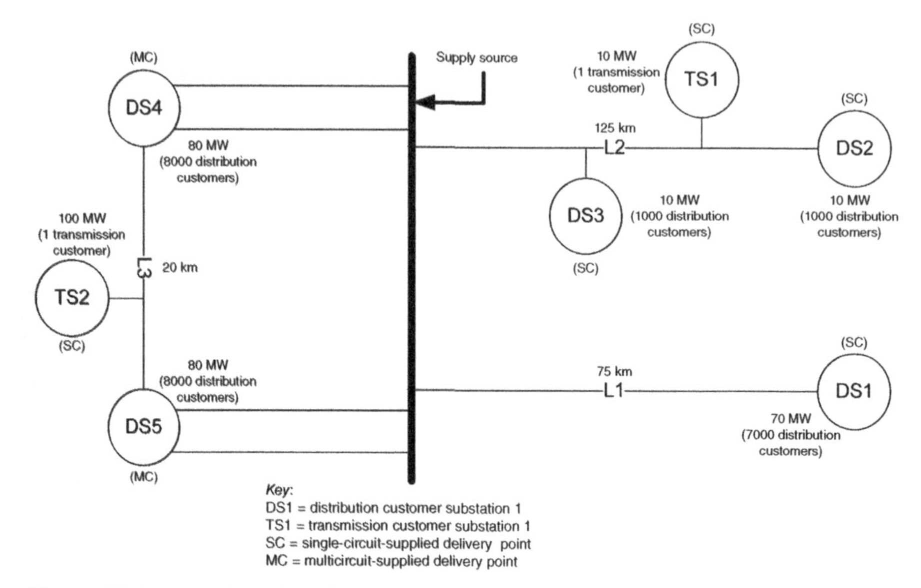

Figure 12.1. Examples of single-circuit-supplied and multicircuit-supplied delivery points.

delivery points have higher availability than do single-circuit-radial-supplied delivery points in terms of outage duration. For example, if circuit L3 shown in Figure 12.1 is interrupted in the upper portion, the load at TS2 will be lost during the outage event. However, it is possible that a crew worker can restore the supply to TS2 by manually isolating the conductor section in the contingency and reenergizing the supply source from the lower portion (via DS5).

There are many single-circuit supply systems in utilities. This may be because some loads located in suburban areas were small and less important historically, and a multicircuit supply could not be justified. However, the loads may gradually grow and become more important over the years with urbanization. The two basic questions facing utilities in single-circuit supply system planning are

- Apparently, the conventional $N-1$ criterion cannot be applied in this situation. How can a reinforcement project for a single-circuit supply system be economically justified?
- It is impossible and unnecessary to reinforce all single-circuit supply systems at the same time because of the constraint in the capital budget. Which single-circuit supply system should be reinforced first?

This chapter presents a probabilistic planning approach to address these issues. The contribution of different system components and subnetwork configurations to overall system reliability performance is analyzed using the statistics of a utility in Section 12.2. The probabilistic planning method for single-circuit supply systems is developed in Section 12.3. An application of the presented method to an actual utility system is demonstrated in Section 12.4.

12.2 RELIABILITY PERFORMANCE OF SINGLE-CIRCUIT SUPPLY SYSTEMS

12.2.1 Delivery Point Reliability Indices

The reliability performance seen by customers widely varies at different delivery points. In general, single-circuit-supplied delivery points have relatively low reliability performance. However, the reliability level at each individual delivery point is quite different because the failure probability of each single-circuit is different. The reliability performance can be measured from various perspectives, such as

- How often does a supply interruption occur (interruption frequency)?
- How long does a supply interruption last (interruption duration)?
- How severe is the supply interruption (magnitude of load lost)?
- How many customers are affected by a supply interruption?

The answers to these questions can be represented using reliability performance matrices. The reliability indices can therefore have various forms, such as a frequency-related index, a duration-related index, or an index that combines all the frequency, duration, and severity factors.

As discussed in Section 7.3.3, historical outage statistics can be used to calculate the reliability indices at delivery points. The following three indices are selected in this chapter to capture different perspectives of supply interruptions to customer delivery points.

1. *Customer Hours Lost (CHL).* This is a measure of cumulative interruption duration associated with a number of customers being interrupted on an annual basis. Its unit is customer-hours per year. The CHL index is a combination of frequency, duration, and number of customers being interrupted. The demerit of this index is that it does not include the amount of loss of loads (MW). Also, even if the interruption frequency and duration of transmission lines remain unchanged, the CHL will increase with the number of customers over the years. Such a situation does not represent real degradation of transmission system reliability. Regardless, the CHL index is very commonly used in the utility industry because of its straightforwardness.
2. *System Average Interruption Duration Index (T-SAIDI).* T-SAIDI is defined in Section 7.3.3.1. Its unit is hours (or minutes) per delivery point per year and represents the average outage duration per delivery point. T-SAIDI is the most popular index in the power industry since many utilities have used it as a corporate key performance indicator to measure overall system reliability. It is essential to appreciate that the T-SAIDI used for transmission systems is different from the SAIDI index used for distribution systems by definition and the prefix T has been used in the book.
3. *Delivery Point Unreliability Index (DPUI).* DPUI is also defined in Section 7.3.3.1. Its unit is system-minutes per year, and it is calculated as the total

energy not supplied because of interruption events at all delivery points normalized by the system peak load. The DPUI is also known as the *severity index*, which indicates the time length in minutes for which it would last if an interruption of total system load occurred at the time of system peak load in order to cause the same amount of cumulative energy not supplied occurring at all delivery points in one year. In other words, the DPUI represents the severity related to the magnitude of energy supply lost as a result of interruptions.

12.2.2 Contributions of Different Components to Reliability Indices

An example is used to show the contributions of different system components and configurations to the reliability indices of overall system based on historical outage statistics [129]. This information is useful for system planners to quantitatively recognize weak components, configurations, and locations in terms of reliability performance.

The three indices (CHL, T-SAIDI, and DPUI) were calculated using the historical data of a utility in the 5 years from 2004 to 2008. The outage data causing customer interruptions were categorized in terms of two types of components (substation and transmission components) and further sub-categorized in terms of three types of delivery points as shown in Figure 12.1 (i.e., single-circuit-radial-supplied delivery points, single-circuit-network-supplied delivery points, and multicircuit-supplied delivery points). The results are presented in a percentage contribution of an index due to each category to the overall system index, as shown in Table 12.1 and Figures 12.2–12.4.

The following observations can be made:

- Transmission-line component interruptions make much higher contributions to customer outages than do substation component interruptions. Transmission-line component interruptions contribute to 75% of the total CHL, 91% of the total T-SAIDI, and 80% of the total DPUI.
- The customer outages due to single-circuit-supplied systems make much higher contributions to the reliability indices than do those due to multicircuit-supplied systems. The impact on the multicircuit-supplied delivery points due to substation

TABLE 12.1. Percentage Contributions of CHL, T-SAIDI, and DPUI Due to Each Category to Overall System Indices

Component Type	Type of Delivery Point Impacted by Outage	CHL (%)	T-SAIDI (%)	DPUI (%)
Transmission-line components	Single-circuit radial	55	74	54
	Single-circuit network	10	13	13
	Multiple-circuit	10	4	13
Substation components	Single-circuit radial	5	3	2
	Single-circuit network	3	2	8
	Multicircuit	17	4	10

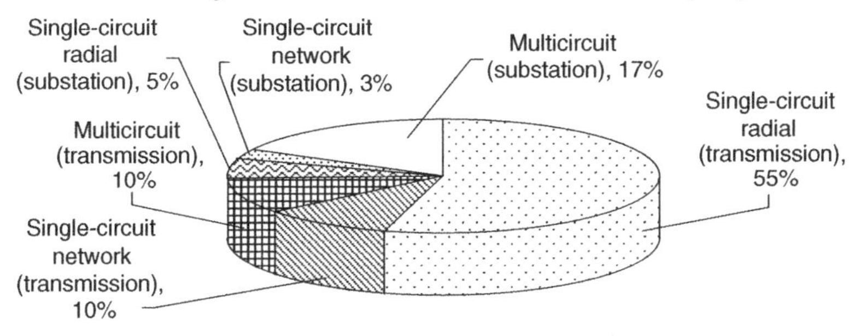

Figure 12.2. Contributions in CHL categorized by types of component outage and impacts on types of delivery point.

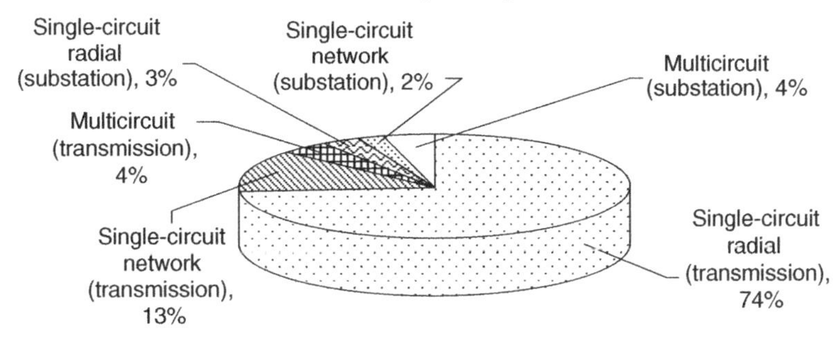

Figure 12.3. Contributions in T-SAIDI categorized by types of component outage and impacts on types of delivery point.

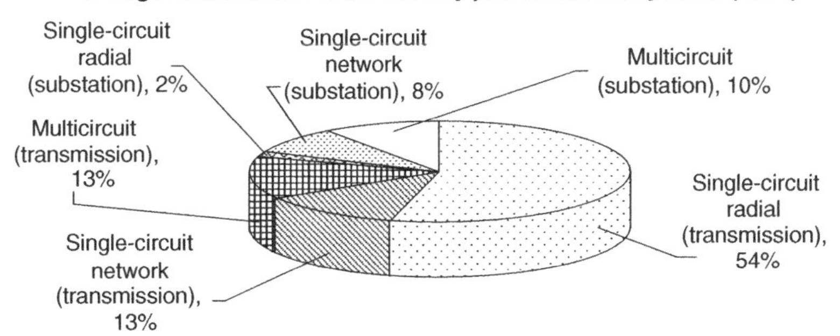

Figure 12.4. Contributions in DPUI categorized by types of component outage and impacts on types of delivery point.

and transmission line component interruptions altogether is only 8% (4% + 4%) of the total T-SAIDI. The remaining 92% of the total T-SAIDI originated from single-circuit-supplied systems. The contributions due to single-circuit-supplied systems to the total CHL and DPUI indices are 73% and 77%, respectively.

- In the contributions of the single-circuit-supplied systems, the customer outages due to transmission-line interruptions dominate over those due to substation component interruptions. The single-circuit transmission-line interruptions (including both radial and network delivery point categories altogether) contribute to 65% of the total CHL index, 87% of the total T-SAIDI index, and 67% of the total DPUI index.
- The single-circuit radial transmission lines make the highest contributions to the reliability indices. In the contribution of 65% for the total CHL index, 55% originates from single-circuit redial transmission lines and 10% from single-circuit network lines. In the contribution of 87% for the total T-SAIDI index, 74% originates from single-circuit radial transmission lines and 13% from single-circuit network lines. In the contribution of 67% for the total DPUI index, 54% originates from single-circuit radial transmission lines and 13% from single-circuit network lines.

Obviously, the customer outages in the transmission system of this utility are caused mainly by interruptions of single-circuit transmission lines, particularly by interruptions of single-circuit radial transmission lines. The contribution percentages to the reliability indices due to different types of components and network configurations in a transmission system vary for different utilities. However, the general pattern should be similar. The analysis results suggest the importance of single-circuit supply system planning.

12.3 PLANNING METHOD OF SINGLE-CIRCUIT SUPPLY SYSTEMS

This topic is discussed in detail in Reference 130. It can be seen from the analysis of reliability performance indices above that the impact of transmission lines greatly dominates over that of substation components. In the following discussions for single-circuit supply systems, we will focus on transmission lines.

12.3.1 Basic and Weighted Reliability Indices

As shown in Section 12.2, the single-circuit supply systems can have different contribution percentages for the three reliability indices (CHL, T-SAIDI, and DPUI) even though they are obtained from the same outages sources. This is due to the fact that each reliability index represents a different reliability perspective. The CHL focuses on the number of customers; the T-SAIDI, on the number of delivery points; and the DPUI, on the outage severity (i.e., magnitude of lost loads). All three indices are therefore used together in order to capture various perspectives associated with power supply outages. The calculation methods of the three basic indices for single-circuit supply systems are presented and demonstrated using an example. Then a composite index based on weightings of the three basic indices is presented for reliability ranking purposes. A circuit that is ranked higher has a greater unreliability impact than do lower-ranked ones.

12.3.1.1 Basic Reliability Indices. The CHL index (in customer-hours per year) for single-circuit supply systems can be calculated by

$$\text{CHL} = \sum_{k=1}^{M} \sum_{i=1}^{f_k} r_{ik} N_k \tag{12.1}$$

where r_{ik} is the interruption duration or restoration time (hours/interruption) in the ith interruption event of the kth circuit, f_k is the number of interruptions of the kth circuit in a given year, N_k is the number of customers supplied by the kth circuit, and M is the number of circuits considered. When $M = 1$, the CHL obtained is the index for one circuit.

For the purpose of ranking single-circuits, the CHL of each circuit should be normalized by the number of interruption events of the circuit. The normalized CHL is the average customer hours lost per event (ACHL) of each circuit. The measure unit of ACHL is customer-hours per interruption per year. The ACHL for all single-circuit supply systems considered can be calculated by

$$\text{ACHL} = \sum_{k=1}^{M} \frac{\sum_{i=1}^{f_k} r_{ik} N_k}{f_k} \tag{12.2}$$

where r_{ik}, f_k, N_k and M are the same as defined in Equation (12.1). If there is no interruption for the kth circuit in a given year (i.e., $f_k = 0$), this circuit is excluded from the calculation for that year.

The T-SAIDI (in hours per delivery point per year) for single-circuit supply systems can be calculated by

$$\text{T-SAIDI} = \frac{\sum_{k=1}^{M} \sum_{i=1}^{f_k} r_{ik} N_{Dk}}{N_{DT}} \tag{12.3}$$

where r_{ik}, f_k, and M are the same as defined in Equation (12.1); N_{Dk} is the number of delivery points on the kth circuit; and N_{DT} is the number of delivery points in the whole system. Note that N_{DT} is not the sum of the numbers of delivery points on the circuits considered. Equation (12.3) has the same implication as Equation (7.44), except that the counting considerations for single-circuit supply systems are utilized in Equation (12.3).

The DPUI (in system-minutes per year) for single-circuit supply systems can be calculated by

$$\text{DPUI} = \frac{60 \cdot \sum_{k=1}^{M} \sum_{i=1}^{f_k} r_{ik} L_{ik}}{P_{\text{sys}}} \tag{12.4}$$

where r_{ik} f_k, and M are the same as defined in Equation (12.1); L_{ik} is the actual total load (MW) on the kth circuit that is curtailed in the ith interruption event; and P_{sys} is the annual peak load (MW) of the whole system. The 60 is introduced to convert the unit in hours into the unit in minutes. Similarly, Equation (12.4) has the same implication as Equation (7.48) except for different counting considerations.

In Equations (12.1)–(12.4), when $M = 1$, the index is the one for an individual circuit. An index for all single-circuit supply systems considered can be used to reflect the overall performance of a circuit group. The percentage contribution of the index of each circuit to the index of the whole circuit group can be easily calculated.

The indices are calculated on a yearly basis in Equations (12.1)–(12.4). When multiple years are considered, an average index per year can be calculated. Obviously, the historical data of each individual interruption event can be directly used in the equations above. If the average interruption frequency and average interruption duration in one year are used, the indices can be expressed as follows:

$$\text{CHL} = \sum_{k=1}^{M} \bar{f}_k \cdot \bar{r}_k \cdot N_k \tag{12.5}$$

$$\text{ACHL} = \sum_{k=1}^{M} \bar{r}_k \cdot N_k \tag{12.6}$$

$$\text{T-SAIDI} = \frac{\sum_{k=1}^{M} \bar{f}_k \cdot \bar{r}_k \cdot N_{Dk}}{N_{DT}} \tag{12.7}$$

$$\text{DPUI} = \frac{60 \cdot \sum_{k=1}^{M} \bar{f}_k \cdot \bar{r}_k \cdot L_k}{P_{sys}} \tag{12.8}$$

Here, $\bar{f}_k$ and $\bar{r}_k$ are the average interruption frequency (interruptions/year) and average interruption duration (hours/interruption) for the kth circuit, respectively; L_k is the total average load (MW) supplied by the kth circuit; and N_k, N_{Dk}, N_{DT}, M, and P_{sys} are the same as defined earlier. The $\bar{f}_k$ and $\bar{r}_k$ can be the average values based on the data of either each circuit in multiple interruption events during several years or a group of circuits at the same voltage level and/or under similar operation and environment conditions. In cases where there is insufficient statistical data for some circuits, it is preferable to use the approach of utilizing the average $\bar{f}_k$ and $\bar{r}_k$ of a group of circuits. The use of the average interruption frequency and duration can reduce the effect due to uncertainty of statistics data of individual circuits. Also, the average interruption frequency and duration can be used to estimate the indices for future years.

The system shown in Figure 12.1 is used to illustrate how to calculate the ACHL, T-SAIDI, and DPUI [129]. It is assumed that the load factor for all the seven delivery points in Figure 12.1 is 0.55. The total system peak is 360 MW. There are three interruption events occurring in a given year. The first interruption event causes line L1 to be out of service for 10 h. The second interruption event causes line L2 to be out of service for 10 h. The third interruption event occurs on the upper portion of line L3 and lasts for 10 h, but the load supply is restored from the lower portion of L3 within 2 h by separating the upper and lower portions of the line.

ACHL Index (in customer-hours per interruption per year)

Overall system ACHL(sys) = (7000 × 10)/1 + [(1000 + 1000 + 1) × 10]/1 + (1 × 2)/1 = 90,012

L1 contribution ACHL(L1) = (7000 × 10)/1 = 70,000
[78% contribution to system ACHL(sys)]

L2 contribution ACHL(L2) = [(1000 + 1000 + 1) × 10]/1 = 20,010
[22% contribution to system ACHL(sys)]

L3 contribution ACHL(L3) = (1 × 2)/1 = 2
[rounded to 0.0% contribution to system ACHL(sys)]

T-SAIDI index (in hours per delivery point per year)

Overall system T-SAIDI(sys) = (10 × 1 + 10 + 3 + 2 × 1)/7 = 6.00

L1 contribution T-SAIDI(L1) = (10 × 1)/7 = 1.43
[24% contribution to system T-SAIDI(sys)]

L2 contribution T-SAIDI(L2) = (10 × 3)/7 = 4.28
[71% contribution to system T-SAIDI(sys)]

L3 contribution T-SAIDI(L3) = (2 × 1)/7 = 0.29
[5% contribution to system T-SAIDI(sys)]

Note that the denominator in calculating the T-SAIDI index is the total number of delivery points in the whole system, including those supplied by multiple circuits.

DPUI Index (in system-minutes per year)

Overall system DPUI(sys) = 60 × 0.55 × (10 × 70 + 10 × 30 + 2 × 100)/360 = 110.0

L1 contribution DPUI(L1) = 60 × 0.55 × (10 × 70)/360 = 64.2
[58% contribution to system DPUI(sys)]

L2 contribution DPUI(L2) = 60 × 0.55 × (10 × 30)/360 = 27.5
[25% contribution to system DPUI(sys)]

L3 contribution DPUI(L3) = 60 × 0.55 × (2 × 100)/360 = 18.3
[17% contribution to system DPUI(sys)]

The reliability performance of the three circuits can be ranked using the percentage contribution of the index of each circuit to the corresponding system index. For example, if the T-SAIDI index is adopted as the criterion in single-circuit supply system planning, constructing a second circuit in parallel with line L2 will provide more improvements in the system T-SAIDI than will enhancing line L1 or L3. However, utilizing the T-SAIDI alone does not address the severity of load curtailments due to interruption events. The failure of line L1, which supplies a large delivery point (DS1), makes a smaller contribution to the system T-SAIDI index but has the greatest impact on the number of customers affected (the highest ACHL) and the severity of interruption events (the highest DPUI). In order to reflect all the perspectives in reliability performance, the weighted reliability index criterion using a combination of the three reliability indices can be used.

12.3.1.2 *Weighted Reliability Index.* The weighting factors are introduced to create a composite index using the three basic reliability indices. It should be emphasized that the percentage contributions of the three indices should be used in weighting instead of using the absolute values of the indices because they have different units and cannot be directly summed up. The weighted reliability index (WRI) for each single-circuit can be calculated by

$$\text{WRI} = \%(\text{ACHL}) \cdot W_1 + \%(\text{T-SAIDI}) \cdot W_2 + \%(\text{DPUI}) \cdot W_3 \tag{12.9}$$

where %(ACHL), %(T-SAIDI), or %(DPUI) represents the percentage contribution of each circuit's index to the whole system index, respectively, for each of the three basic indices; W_1, W_2, and W_3 are the weighting factors; and $W_1 + W_2 + W_3 = 1.0$.

Since the interruption frequency and duration are the common parts in all three reliability indices, the effects of the weighting factors are focused on the number of affected customers (ACHL), the number of delivery points (T-SAIDI), and the magnitude of lost loads (DPUI). Selection of the weighting factors is a management decision of the utility depending on its reliability objective. Since the weighted reliability index is based on the percentage contribution of each single-circuit, it can be used to rank the performance of single-circuit supply systems and to determine an initial short list of circuits for reinforcement. A threshold for the WRI can be specified as a cutoff criterion. The single-circuit supply systems whose WRI is greater than the threshold are the candidates in the initial short list considered for reinforcement.

The weighting factors for the %(ACHL), %(T-SAIDI), and %(DPUI) used in the example shown in Figure 12.1 are 0.3, 0.4, and 0.3, respectively. The percentage contributions of each circuit's index to the system index and the weighted reliability indices (%) in the given example are shown in Table 12.2. It can be seen that if the T-SAIDI is used as a reliability criterion, L2 is ranked at a higher priority for enhancement than are L1 and L3, whereas if the weighted reliability index containing all the three reliability measures is used, L1 is ranked at a higher priority for enhancement than are L2 and L3.

TABLE 12.2. Percentage Contributions in Basic Indices and WRI for Each Single-Circuit in Example

Circuit	Percentage Contribution to System Index (%)			Weighted Reliability Index (%)
	%ACHL	%T-SAIDI	%DPUI	
L1	78	24	58	50.4
L2	22	71	25	42.5
L3	0	5	17	7.1

12.3.2 Unit Incremental Reliability Value Approach

Although the weighted reliability index can be directly used for ranking the reliability performance of circuits, it does not include any economic consideration for system reinforcement. This section presents the ranking criteria that combine a reliability index with the investment cost for reinforcement.

12.3.2.1 Annual Capital Investment. In principle, a single-circuit supply system can be reinforced by adding a second circuit. For a single-circuit supplying only one delivery point, access to a second power source can also be a reinforcement alternative. However, such an alternative does not apply to the case in which multiple substations are supplied by a single-circuit. For generality, addition of a second circuit for reinforcement is used as an example in the following discussions. On one hand, the second circuit will greatly improve the system reliability. On the other hand, it requires a capital investment.

Using the capital return factor discussed in Section 6.3.4.3, the equivalent annual capital investment (ACI) of a new circuit can be obtained from its total capital investment (TCI) as

$$\mathrm{ACI} = \mathrm{TCI} \cdot \frac{r(1+r)^n}{(1+r)^n + 1} \tag{12.10}$$

where ACI and TCI represent the equivalent annual capital investment (k$/year) and the total capital investment (k$) at the beginning, respectively; r is the discount rate; and n is the useful life of a new circuit.

In ranking the priority of reinforcement projects for single-circuit supply systems, an approximate cost estimate is generally acceptable. For example, an average investment cost per km can be applied to the reinforcement projects of circuits at the same voltage level. If necessary, however, more accurate costs for each reinforcement project can be individually estimated through detailed investigation.

12.3.2.2 Unit Incremental Reliability Value. The capital investment required to reinforce the system will lead to system reliability improvement. The unit incremental reliability value (UIRV) concept [131,132] is utilized to evaluate the incremental investment cost for unit reliability improvement. The UIRV is defined as follows:

$$\mathrm{UIRV}_k = \frac{\mathrm{ACI}_k}{\Delta R_k} \tag{12.11}$$

Here, ACI_k is the annual capital investment cost required by a reinforcement for the kth circuit, and ΔR_k represents the reliability improvement due to the reinforcement in a yearly average reliability metric over the planning timespan. Conceptually, the reliability metric can be any index. It is noted that ΔR_k is expressed in a percentage contribution of a circuit's index to the corresponding system index. In other words, the improvement in %(ACHL), or %(T-SAIDI), or %(DPUI), or WRI index as defined in Equation (12.9) is used as ΔR_k.

If a single-circuit supply system is reinforced by adding a second circuit, the ACI_k is simply the equivalent annual capital investment of the second circuit addition. The ΔR_k is the difference in the selected percentage index between the existing single-circuit and the configuration with the second circuit addition. The equivalent interruption frequency and duration for a simultaneous outage of two circuits can be calculated using the parallel reliability formulas [see Equations (11.11) and (11.12) in Chapter 11], and then Equations (12.5)–(12.8) are still valid for calculating the basic reliability indices for the configuration with the second circuit addition. In most cases, however, the probability of a simultaneous outage of two circuits is very low. Therefore, the reliability index after the second circuit addition is much smaller than that of the existing single-circuit (often smaller than 1%) and can be neglected in the sense of relative comparison. In this case, the ΔR_k can be approximately considered to be the selected percentage index of the existing single-circuit. This approximation basically will not create an effective error in ranking the priority of single-circuits.

Although the UIRV can be directly used for ranking, it is suggested to use the following relative UIRV index (RUIRV), which has no unit and therefore can be combined with another ranking index, the relative cost/benefit ratio (RCBR) (see Section 12.3.3):

$$\text{RUIRV}_k = \frac{\text{UIRV}_k}{\sum_{k=1}^{M_c} \text{UIRV}_k} \tag{12.12}$$

Here, UIRV_k is the unit incremental reliability value for reinforcement of the kth circuit, and M_c is the number of circuits considered as the candidates in the short list for reinforcement.

12.3.3 Benefit/Cost Ratio Approach

The deterministic $N-1$ criterion is not applicable for judging whether a single-circuit supply system should be reinforced. The benefit/cost analysis based on reliability worth assessment can be used for this purpose.

12.3.3.1 Expected Damage Cost. The EENS (expected energy not supplied) index (in MWh/year) for each circuit can be calculated by

$$\text{EENS}_k = \overline{f}_k \cdot \overline{r}_k \cdot L_k \tag{12.13}$$

where $\overline{f}_k$ and $\overline{r}_k$ are the average interruption frequency (interruptions per year) and duration (hours per interruption) for the kth circuit, respectively; L_k is the total annual average load (MW) supplied by the kth circuit. When the EENS for different years in the future is estimated, L_k for each year should be different and is based on the load forecast and load factor of customers supplied by the circuit.

The EDC (expected damage cost) index (in k$/year) for each circuit can be calculated by

$$\mathrm{EDC}_k = \mathrm{EENS}_k \cdot \mathrm{UIC}_k \tag{12.14}$$

where UIC_k represents the composite unit interruption cost (in \$/kWh) for the kth circuit.

The composite UIC can be calculated using the unit interruption costs of different customer sectors and the composition of customers supplied by the circuit. This is similar to the approach shown in the example given in Section 9.3.4.1. The other methods for estimating the UIC given in Section 5.3.1 can also be used.

12.3.3.2 Benefit/Cost Ratio. In the benefit/cost analysis, the cost is the capital investment (CI) of the second circuit for reinforcing a single-circuit supply system, whereas the benefit is the reduction in the expected damage cost (EDC) created by the second circuit addition. The benefit/cost ratio is calculated by

$$\mathrm{BCR}_k = \frac{\Delta \mathrm{EDC}_k}{\mathrm{CI}_k} \tag{12.15}$$

where CI_k is the capital investment of the second circuit addition for the kth circuit over the planning timespan; $\Delta\mathrm{EDC}_k$ is the difference in the EDC index between the existing single-circuit (the kth circuit) and the configuration with the second circuit addition in the same planning length. Equations (12.13) and (12.14) are still valid for calculation of the EDC of the configuration with the second circuit if $\bar{f}_k$ and $\bar{r}_k$ are replaced by the equivalent interruption frequency and duration for a simultaneous outage of two circuits, which can be calculated using the parallel reliability formulas.

The BCR can be used to judge whether the reinforcement for a single-circuit supply system is economically justifiable. Conceptually, if the BCR is greater than 1.0, it is financially acceptable. Otherwise, it cannot be justified. In actual applications, it is usual practice to select a number larger than 1.0 (such as 1.5 or 2.0) as a threshold in order to cover the uncertainty in input data and the errors that may be caused by approximations in calculations.

It should be emphasized that the EENS and EDC indices are different for each year in the future since L_k usually increases over the years. A planning timespan (such as 10–20 years) needs to be considered. By using the present value method, Equation (12.15) can be expressed in a detailed form

$$\mathrm{BCR}_k = \frac{\sum_{j=1}^{m}\{\Delta\mathrm{EDC}_{kj}/[(1+r)^{j-1}]\}}{\sum_{j=1}^{m}\{\mathrm{ACI}_{kj}/[(1+r)^{j-1}]\}} \tag{12.16}$$

where $\Delta\mathrm{EDC}_{kj}$ and ACI_{kj} are the annual reduction of EDC and equivalent annual investment in year j for reinforcement of the kth circuit, respectively; r is the discount rate; and m is the number of years considered in the planning timespan. Note that the reference year in Equation (12.16) is year 1 and the reference year in the corresponding equations of Chapter 6 is the year 0.

The cost/benefit ratio (CBR) is the reciprocal of the BCR:

$$\mathrm{CBR}_k = \frac{\mathrm{CI}_k}{\Delta\mathrm{EDC}_k} \tag{12.17}$$

Obviously, CBR is in the same form of UIRV. The CBR represents the incremental investment cost for unit improvement in EDC. Therefore the CBR can be also used to rank the priority of single-circuits for reinforcement. Like the UIRV, the relative CBR for a circuit should be used and is defined as

$$\mathrm{RCBR}_k = \frac{\mathrm{CBR}_k}{\sum_{k=1}^{M_c}\mathrm{CBR}_k} \tag{12.18}$$

where CBR_k is the cost/benefit ratio for reinforcement of the kth circuit and M_c is the number of circuits considered as the candidates in the short list for reinforcement.

In actual applications, the two relative indices of RUIRV and RCBR can be directly summed up to obtain a combined relative contribution index (CRCI):

$$\mathrm{CRCI}_k = \mathrm{RUIRV}_k + \mathrm{RCBR}_k \tag{12.19}$$

In Equation (12.19), the RUIRV and RCBR have been equally weighted, although different weighting factors can be introduced to calculate the CRCI if necessary. Apparently, the value of CRCI ranges between 0.0 and 2.0 since the value of RUIRV or RCBR ranges between 0.0 and 1.0. In fact, the CRCI index is often much smaller than 1.0. A circuit with a smaller CRCI requires a lower investment cost for unit reliability improvement.

12.3.4 Procedure of Single-Circuit Supply System Planning

As mentioned in Section 12.1, two basic questions in single-circuit supply system planning are: (1) how to justify the reinforcement for a single-circuit supply system and (2) which single-circuit supply system should be reinforced first.

The single-circuit supply system planning procedure includes the following steps:

1. The ACHL, T-SAIDI, and DPUI indices of single-circuit-supplied delivery points for all single-circuit supply systems are calculated using historical outage data.
2. The weighted reliability indices (WRIs) are calculated using the percentage contributions of ACHL, T-SAIDI, and DPUI indices in the form of an individual single-circuit's index to the corresponding system index.
3. The cash flows of $\Delta\mathrm{EDC}_{kj}$ and ACI_{kj} for individual single supply systems in the planning timespan are calculated to obtain their benefit/cost ratios (BCRs).
4. A short list is created from the information in steps 2 and 3. The single-circuit supply systems whose WRIs are smaller than the specified threshold should be removed from the candidate list for reinforcement. Each candidate in the short

list should be financially justifiable. If the BCR of a single-circuit supply system is smaller than the selected threshold of BCR, it should also be removed from the short list.

5. The RUIRV and RCBR indices for the individual single-circuit supply systems are calculated.
6. The CRCI indices for the individual single-circuit supply systems are calculated.
7. The priority of the single-circuit supply systems in the short list for reinforcement is ranked using their CRCI indices.
8. One or more single-circuit supply systems ranked at the top are selected for reinforcement depending on the constraint of investment budget.

The procedure is summarized in Figure 12.5.

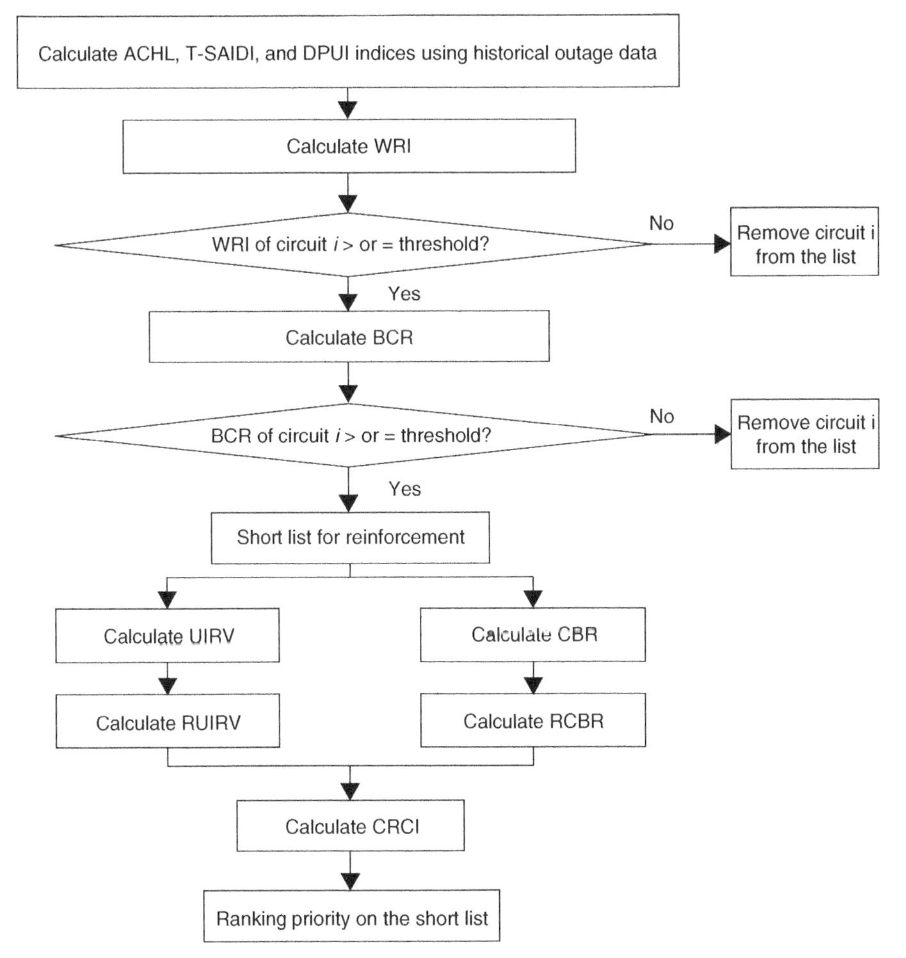

Figure 12.5. Procedure of single-circuit supply system planning.

12.4 APPLICATION TO ACTUAL UTILITY SYSTEM

An actual application to BC Hydro system is presented in detail in References 129 and 130.

12.4.1 Short List Based on Weighted Reliability Index

As pointed out in Section 12.1, the single-circuit-supplied delivery points are divided into two categories: single-circuit-radial-supplied delivery points and single-circuit-network-supplied delivery points, as demonstrated in Figure 12.1. Accordingly, all single-circuit supply systems in the BC Hydro system are divided into two groups designated as a single-circuit radial supply group and a single-circuit network supply group.

Tables 12.3 and 12.4 present the weighted reliability indices and the percentage contribution indices %(ACHL), %(T-SAIDI), and %(DPUI) for the single-circuit radial

TABLE 12.3. Circuit Ranking in Single-Circuit Radial Supply Group Based on Weighted Reliability Index

		Percent Contribution to BC Hydro Overall System Index			
Circuit	Voltage (kV)	%(ACHL)	%(T-SAIDI)	%(DPUI)	Weighted Reliability Index (%)
LR1-138	138	16.20	34.17	16.21	*23.39*
LR2-138	138	15.52	0.21	1.12	*5.08*
LR3-138	138	2.38	2.55	10.23	*4.80*
LR4-138	138	8.02	1.02	3.33	*3.81*
LR5-138	138	4.25	2.72	3.69	*3.47*
LR6-60	60	2.40	3.04	3.26	*2.92*
LR7-60	60	0.37	3.65	1.42	*2.00*
LR8-60	60	0.00	4.73	0.08	*1.92*
LR9-138	138	0.40	3.61	0.90	*1.83*
LR10-60	60	1.45	1.79	2.12	*1.79*
LR11-60	60	1.05	2.45	1.09	*1.62*
LR12-60	60	1.81	0.75	2.49	*1.59*
LR13-230	230	0.00	0.49	4.52	*1.55*
LR14-60	60	0.43	1.68	1.79	1.34
LR15-138	138	1.35	1.34	1.16	1.29
LR16-60	60	2.07	1.21	0.40	1.22
LR17-60	60	0.51	2.04	0.53	1.13
LR18-60	60	1.18	0.91	1.37	1.13
LR19-60	60	0.00	1.51	1.48	1.05
LR20-230	230	0.28	0.70	2.28	1.05
LR21-60	60	0.54	1.46	0.77	0.98
LR22-60	60	1.96	0.70	0.42	0.99
LR23-60	60	1.50	0.11	0.27	0.57
LR24-60	60	1.34	0.29	0.13	0.56
LR25-138	138	0.01	1.08	0.05	0.45
LR26-230	230	0.86	0.26	0.30	0.45

TABLE 12.3. *Cont'd*

Circuit	Voltage (kV)	Percent Contribution to BC Hydro Overall System Index			Weighted Reliability Index (%)
		%(ACHL)	%(T-SAIDI)	%(DPUI)	
LR27-230	230	0.52	0.43	0.39	0.44
LR28-138	138	0.52	0.57	0.16	0.43
LR29-138	138	0.33	0.39	0.56	0.42
LR30-60	60	0.31	0.64	0.19	0.41
LR31-60	60	0.45	0.35	0.46	0.41
LR32-230	230	0.00	0.07	1.20	0.39
LR33-138	138	0.00	0.77	0.20	0.37
LR34-138	138	0.00	0.18	0.94	0.35
LR35-138	138	0.20	0.53	0.21	0.33
LR36-138	138	0.07	0.32	0.49	0.30
LR37-138	138	0.00	0.56	0.17	0.27
LR38-60	60	0.04	0.61	0.05	0.27
LR39-60	60	0.00	0.66	0.01	0.27
LR40-60	60	0.17	0.22	0.44	0.27
LR41-138	138	0.53	0.02	0.12	0.20
LR42-60	60	0.00	0.13	0.25	0.13
LR43-138	138	0.24	0.06	0.04	0.10
LR44-138	138	0.00	0.04	0.18	0.07
LR45-230	230	0.17	0.01	0.02	0.06
LR46-60	60	0.00	0.09	0.06	0.05
LR47-138	138	0.00	0.10	0.04	0.05
LR48-230	230	0.00	0.04	0.00	0.01
LR49-60	60	0.00	0.02	0.00	0.01
LR50-138	138	0.00	0.00	0.01	0.00
LR51-138	138	0.00	0.00	0.00	0.00
LR52-60	60	0.00	0.00	0.00	0.00
LR53-230	230	0.00	0.00	0.00	0.00
LR54-60	60	0.00	0.00	0.00	0.00
LR55-138	138	0.00	0.00	0.00	0.00
LR56-138	138	0.00	0.00	0.00	0.00
LR57-60	60	0.00	0.00	0.00	0.00
LR58 60	60	0.00	0.00	0.00	0.00

and network supply groups, respectively. All circuits in the two tables are listed in descending order of weighted reliability index. The indices were calculated from historical data in the 5 years from 2004 to 2008.

The weighted reliability index does not automatically determine a cutoff threshold above which a reinforcement project can be considered. Determination of a threshold depends on multiple factors, including the reliability performance target of the utility and the constraint on the investment budget for reinforcing single-circuit supply systems. The threshold value of 1.5% is used to create an initial short list in the example.

TABLE 12.4. Circuit Ranking in Single-Circuit Network Supply Group Based on Weighted Reliability Index

Circuit	Voltage (kV)	Percent Contribution to BC Hydro Overall System Index			Weighted Reliability Index (%)
		%(ACHL)	%(T-SAIDI)	%(DPUI)	
LN1-138	138	2.47	0.67	1.76	*1.54*
LN2-60	60	0.49	1.15	0.58	0.78
LN3-60	60	1.46	0.23	0.43	0.66
LN4-138	138	0.24	0.53	0.68	0.49
LN5-60	60	0.41	0.41	0.32	0.39
LN6-138	138	0.33	0.17	0.58	0.34
LN7-60	60	0.39	0.24	0.28	0.30
LN8-138	138	0.13	0.28	0.41	0.27
LN9-230	230	0.00	0.20	0.59	0.26
LN10-60	60	0.00	0.24	0.43	0.23
LN11-60	60	0.00	0.25	0.26	0.18
LN12-138	138	0.00	0.29	0.18	0.17
LN13-60	60	0.00	0.29	0.17	0.16
LN14-60	60	0.06	0.21	0.18	0.16
LN15-60	60	0.20	0.13	0.11	0.15
LN16-60	60	0.33	0.06	0.09	0.15
LN17-60	60	0.03	0.18	0.16	0.13
LN18-60	60	0.09	0.18	0.06	0.12
LN19-60	60	0.21	0.03	0.07	0.10
LN20-138	138	0.03	0.12	0.06	0.07
LN21-138	138	0.22	0.00	0.01	0.07
LN22-60	60	0.00	0.15	0.01	0.06
LN23-60	60	0.00	0.08	0.11	0.06
LN24-60	60	0.02	0.13	0.01	0.06
LN25-60	60	0.02	0.04	0.01	0.03
LN26-138	138	0.00	0.03	0.00	0.01
LN27-60	60	0.01	0.01	0.01	0.01
LN28-230	230	0.00	0.00	0.02	0.01
LN29-60	60	0.00	0.00	0.00	0.00

This implies that if any circuit contributes 1.5% or above to the entire BC Hydro transmission system reliability in terms of the weighted reliability index, it should be considered for further investigation using the benefit/cost analysis.

The rationale for proposing 1.5% as the threshold in this example is based on the comparison between the weighted reliability indices in Tables 12.3 and 12.4. It is obviously shown that the contribution of the single-circuit radial supply group to the weighted reliability index of the entire BC Hydro transmission system is much greater than the contribution of the single-circuit network supply group. The greatest percentage contribution in the weighted reliability index shown in Table 12.4 is only 1.54% on the circuit LN1-138, whereas the greatest percentage contribution shown in Table

12.3 is 23.39% on the circuit LR1-138. Clearly, there is a considerable gap in the impact on the overall system reliability between the two groups. In order to bridge this gap, the threshold value of 1.5% is proposed to bring the unreliability contribution of the circuits at the top in Table 12.3 to be close to the maximum weighted reliability index shown in Table 12.4, which is 1.54% on the circuit LN1-138. Conceptually, this implies that the future reliability of the existing single-circuit radial supply group will be close to the reliability of the single-circuit network supply group after reinforcing the single-circuit radial supply systems at the top in Table 12.3. Once this target is achieved, a new threshold value can be reestablished using a more aggressive target to rank the remaining single-circuit supply systems that include many circuits in the single-circuit network supply group. Based on the threshold of 1.5%, 13 circuits from the single-circuit radial supply group (italicized entries in WRI column in Table 12.3) and 1 circuit from the single-circuit network supply group (italicized entry in WRI column in Table 12.4) are selected as the candidates in the initial short list for further investigation.

12.4.2 Financial Justification of Reinforcement

The 14 single-circuit supply systems ranked at the top using the weighted reliability index composed the initial short list. Each candidate in the list must be financially justifiable for reinforcement. The average outage data of both the single-circuit radial and network supply groups are used in the benefit/cost analysis. Using the average outage data rather than using the outage data of individual circuits can reduce the uncertainty of data in predicting the future reliability performance of the circuits, particularly for those circuits with very few records of outage events. The average outage data of the single-circuits are based on the statistics in the 5 years from 2004 to 2008 and are shown in Table 12.5. The line lengths and the annual investment costs due to reinforcement by adding a second circuit for the identified 14 single-circuits are presented in Table 12.6. The annual investment costs were calculated using Equation (12.10). The load factors and load forecasts of customers on the 14 circuits are given in Table 12.7.

The benefit/cost analysis was conducted in the planning timeframe of the 10 years from 2009 to 2018. The EENS indices for each year were evaluated using Equation (12.13) and shown in Table 12.8. The composite UICs of each circuit, which were estimated using the unit interruption costs of customer sectors and composition of customers on each circuit, are also given in Table 12.8. The EDC indices were calculated using Equation (12.14). The reduction (ΔEDC) in the EDC index due to adding a second circuit represents the benefit created by reinforcing each single-circuit supply

TABLE 12.5. Average Outage Data of Single-Circuits in Radial and Network Supply Groups

Voltage Class (kV)	Interruption Frequency (interruptions/100 km/year)	Repair Time (h/interruption)	Unavailability (per 100 km)
60	4.60	20.29	0.01065
138	1.31	16.91	0.00252
230	1.04	16.57	0.00197

TABLE 12.6. Length and Annual Investment Cost for Reinforcement of the 14 Circuits

Circuit	Voltage (kV)	Length (km)	Annual Investment Cost (k$/year)
LR1-138	138	196.63	2411
LR2-138	138	79.56	1918
LR3-138	138	39.74	958
LR4-138	138	41.36	997
LR5-138	138	142.29	3430
LR6-60	60	132.47	2353
LR7-60	60	105.59	1876
LR8-60	60	65.00	1155
LR9-138	138	174.34	4203
LR10-60	60	84.00	1492
LR11-60	60	102.35	1818
LR12-60	60	30.40	540
LR13-230	230	59.99	1903
LN1-138	138	105.25	2537

system. The present values (in 2009 dollars) of the benefits and investment costs due to reinforcement of the 14 circuits in the planning period of the 10 years are presented in Table 12.9.

The benefit/cost ratio (BCR) is obtained by dividing the present value of benefit by the present value of investment cost. The BCR values for reinforcement of the 14 circuits are also given in Table 12.9. The threshold of BCR is 1.5 in the example. The reinforcement project for each single-circuit supply system is economically justifiable if its BCR is equal to or greater than 1.5. It can be seen from Table 12.9 that the two circuits LR8-60 and LR9-138 cannot be economically justified since their BCR values are even smaller than 1.0. These two circuits should be excluded from the short list.

12.4.3 Ranking Priority of Single-Circuit Systems

The number of candidate circuits in the short list has been reduced to 12 from 14, based on the benefit/cost analysis. The reinforcements of the 12 circuits not only provide larger contributions to reliability improvement of the overall system than others not in the list but also are all economically justifiable. However, we cannot implement the reinforcement projects of all 12 circuits at the same time. It is necessary to rank their priority to obtain an idea about which one should be reinforced first.

The T-SAIDI index has been used by the utility as a key performance indicator for years and therefore was selected for calculating the UIRV indices of the 12 single-circuits in the updated short list. The relative %(T-SAIDI) index rather than the absolute T-SAIDI value was used. The UIRV index was obtained using the equivalent annual capital investment cost divided by the yearly average reduction in the %(T-SAIDI) index due to addition of a second circuit and was divided by 10. The division by 10 was taken because the reductions in %(T-SAIDI) for some circuits were estimated to be <1%. This leads to the UIRV values to be expressed in the unit of k$/year for 0.1%

TABLE 12.7. Load Forecasts (MW) and Load Factors of Customers on the 14 Circuits

Circuit	Load Factor	Peak Load Supplied by Each Circuit in 10 years (MW)									
		2009	2010	2011	2012	2013	2014	2015	2016	2017	2018
LR1-138	0.517	72.6	73.2	73.6	73.8	74.1	74.3	74.6	74.8	74.9	75.1
LR2-138	0.592	79.7	81.9	83.6	84.4	85.2	86.6	87.6	88.0	88.4	88.8
LR3-138	0.682	120.0	121.2	123.5	124.6	123.1	124.9	127.3	128.3	128.4	128.4
LR4-138	0.532	55.4	55.7	56.0	56.4	56.5	56.7	56.9	57.2	57.3	57.4
LR5-138	0.572	57.7	60.8	62.1	62.4	62.4	62.8	63.2	63.5	63.7	63.8
LR6-60	0.504	76.5	77.1	77.5	77.8	78.0	78.3	78.5	78.7	78.9	79.1
LR7-60	0.538	11.1	11.2	11.2	11.2	11.3	11.3	11.3	11.4	11.4	11.4
LR8-60	0.508	0.3	0.3	0.3	0.3	0.3	0.3	0.3	0.3	0.3	0.3
LR9-138	0.555	5.3	5.3	5.3	5.4	5.4	5.4	5.4	5.4	5.4	5.5
LR10-60	0.531	23.5	24.5	24.5	24.6	24.6	24.7	24.9	25.0	25.1	25.3
LR11-60	0.486	13.3	13.3	13.3	13.3	13.4	13.4	13.4	13.4	13.5	13.5
LR12-60	0.578	17.7	17.8	17.9	18.0	18.1	18.2	18.2	18.3	18.4	18.5
LR13-230	0.600	119.2	120.5	122.9	124.0	124.4	124.4	127.0	128.0	128.0	128.0
LN1-138	0.508	29.2	29.3	29.4	29.5	29.6	29.7	29.8	29.9	30.0	30.0

TABLE 12.8. EENS and UIC Values of the 14 Circuits

Circuit	UIC ($/kWh)	Expected Energy Not Supplied (MWh/year) for Each Circuit in the 10 Years									
		2009	2010	2011	2012	2013	2014	2015	2016	2017	2018
LR1-138	11.09	1631	1644	1653	1659	1665	1670	1677	1681	1684	1688
LR2-138	7.33	830	853	871	879	887	902	912	916	920	924
LR3-138	15.73	718	726	740	746	737	748	763	769	769	769
LR4-138	10.13	269	271	272	274	275	276	277	278	279	279
LR5-138	11.71	1037	1095	1118	1123	1123	1130	1137	1142	1145	1149
LR6-60	10.27	4759	4799	4821	4837	4854	4870	4883	4896	4907	4919
LR7-60	12.48	590	592	593	595	597	598	601	603	604	606
LR8-60	15.76	9	9	9	9	9	9	9	9	9	9
LR9-138	12.67	113	114	114	114	115	115	115	116	116	117
LR10-60	11.30	979	1018	1020	1022	1025	1030	1035	1041	1046	1051
LR11-60	7.92	615	616	617	618	619	621	622	623	625	626
LR12-60	6.72	291	292	294	295	297	298	299	300	301	302
LR13-230	15.76	740	749	763	770	773	773	789	795	795	795
LN1-138	10.40	422	424	425	427	429	430	431	432	433	435

TABLE 12.9. Present Values of Benefits and Capital Investment Costs and BCRs for Reinforcement of the 14 Circuits

Circuit	Present Value of Benefit (k$)	Present Value of Investment Cost (k$)	Benefit/Cost Ratio (BCR)
LR1-138	143,851	18,314	7.9
LR2-138	50,577	14,571	3.5
LR3-138	91,490	7,278	12.6
LR4-138	21,677	7,575	2.9
LR5-138	101,942	26,059	3.9
LR6-60	388,397	22,293	17.4
LR7-60	58,137	17,770	3.3
LR8-60	1,136	10,939	0.1
LR9-138	11,330	31,929	0.4
LR10-60	90,267	14,136	6.4
LR11-60	38,275	17,225	2.2
LR12-60	15,525	5,116	3.0
LR13-230	94,821	16,628	5.7
LN1-138	34,755	23,581	1.5

TABLE 12.10. UIRV and CBR Indices for Reinforcement of the 12 Circuits

Circuit	Annual Investment Cost (k$/year)	Reduction in (T-SAIDI) Index (%)	UIRV (k$/year for 0.1% T-SAIDI Improvement)	CBR
LR1-138	2411	5.65	42.7	0.13
LR2-138	1918	0.38	504.7	0.29
LR3-138	958	0.48	199.6	0.08
LR4-138	997	0.20	498.5	0.35
LR5-138	3430	2.73	125.6	0.26
LR6-60	2353	9.38	25.1	0.06
LR7-60	1876	2.14	87.7	0.31
LR10-60	1492	0.85	175.5	0.16
LR11-60	1818	4.14	43.9	0.45
LR12-60	540	0.31	174.2	0.33
LR13-230	1903	0.11	1730.0	0.18
LN1-138	2537	0.50	507.4	0.68

T-SAIDI improvement. The annual investment costs, reductions in the %(T-SAIDI) index, and UIRV indices for the 12 circuits are summarized in Table 12.10. The CBR index (the reciprocal of BCR) can be used as another ranking criterion. The CBR indices for reinforcement of the 12 circuits, which were calculated by dividing the present values of investment cost by the present values of benefit listed in Table 12.9, are also given in Table 12.10.

It can be observed from the results in Table 12.10 that reinforcing the LR6-60 circuit would reduce the T-SAIDI contributed by the single-circuit supply system by ~9.4%. The UIRV for reinforcement of the circuit LR6-60 is $25k/year for 0.1%

TABLE 12.11. RUIRV, RCBR, and CRCI Indices for Reinforcement of the 12 Circuits

Circuit	RUIRV (Based on T-SAIDI)	RCBR (Based on EDC)	CRCI Ranking Index	Rank
LR6-60	0.006	0.018	0.024	1
LR1-138	0.010	0.040	0.050	2
LR3-138	0.049	0.024	0.073	3
LR10-60	0.043	0.049	0.091	4
LR5-138	0.031	0.079	0.110	5
LR7-60	0.021	0.095	0.116	6
LR12-60	0.042	0.101	0.143	7
LR11-60	0.011	0.137	0.148	8
LR2-138	0.123	0.088	0.211	9
LR4-138	0.121	0.107	0.228	10
LN1-138	0.123	0.207	0.331	11
LR13-230	0.420	0.055	0.475	12

T-SAIDI reduction, which is the lowest unit investment cost for unit reliability improvement among all the listed circuits. Also, its cost/benefit ratio is the lowest. The circuit LR13-230 results in the highest UIRV ($1730k/year for a 0.1% T-SAIDI reduction), and the circuit LN1-138 has the largest CBR (0.68). The observations enable us to make a judgment that the circuit LR6-60 should be reinforced first whereas the circuits LR13-230 and LN1-138 should be ranked last.

However, there is inconsistency if the UIRV and CBR are separately used to rank the priority of all the circuits. This is because the UIRV and CBR indices represent different reliability perspectives. The UIRV indices reflect the incremental cost for unit improvement in T-SAIDI, and the CBR indices provide the information on the incremental cost for unit improvement in the EDC index. In order to overcome this disadvantage, the relative contribution indices of RUIRV and RCBR were calculated using Equations (12.12) and (12.18) and combined into the CRCI indices for ranking. The RUIRV, RCBR, and CRCI indices for reinforcement of the 12 circuits in the increasing order of CRCI are given in Table 12.11. The circuit with a smaller CRCI has a higher priority for reinforcement.

It can be seen that the circuit LR6-60 has the smallest CRCI (0.024) since both its RUIRV and RCBR values are the smallest, and therefore it ranks at the top of the list over the other 11 circuits. This implies that the LR6-60 reinforcement project can maximize the improvements of both T-SAIDI and EDC while minimizing the investment cost for enhancing reliability. The circuit LR13-230 has the largest value of CRCI and is consequently ranked at the bottom of the list. It is worth to noting that LR13-230 is a single radial circuit serving one transmission customer, which is the largest single customer in this study. The RCBR (0.055) of LR13-230 based on the EDC is smaller than that of other 7 circuits in the list. This means that the improvement of EDC after the reinforcement is relatively significant. However, the RUIRV (0.420) of LR13-230 based on the T-SAIDI is the largest because the T-SAIDI reduction due to reinforcing this circuit is very small: only 0.11% T-SAIDI reduction, as shown in Table 12.10. For this reason, the combined CRCI ranking index of the circuit LR13-230 becomes the largest (0.475), meaning that this reinforcement can be beneficial only from the EDC perspective (as a large industrial customer) but is not beneficial from the T-SAIDI improvement viewpoint. It can be noted that the ranking order obtained using the CRCI index for the circuits in the middle rows in the list is different from that based on the RUIRV or RCBR. This indicates that using a single reliability or economic index for ranking is not sufficient in decisionmaking for single-circuit supply system planning.

12.5 CONCLUSIONS

Conceptually, it is apparent that single-circuit supply systems are the weakest network configurations in a transmission system in terms of power supply reliability. The reliability performance indices based on historical outage statistics can quantitatively provide the information on contributions of different components and subnetwork configurations to the overall transmission system unreliability. This information is useful for enhancement direction in transmission planning. An actual example is used for demonstration.

A decision on reinforcement of single-circuit supply systems has been a challenging issue in transmission planning for years because the deterministic $N - 1$ criterion is not applicable in this case. This chapter proposed a probabilistic reinforcement planning method for single-circuit supply systems, which includes the following three main steps:

- An initial short list is determined. The three reliability performance indices, which are called the ACHL, T-SAIDI, and DPUI, are calculated using historical outage statistics. The three indices not only provide the information about the existing reliability status of each individual circuit in different performance perspectives but also are used to develop the weighted reliability index (WRI) for establishing the initial short list for reinforcement.
- The short list is modified. A benefit/cost analysis, which is based on the reduction in the expected damage cost and the investment cost, is conducted for reinforcement of each circuit in the list. Any circuit whose reinforcement cannot be economically justified in the analysis is removed from the initial short list.
- A set of circuits recommended for reinforcement are identified. The UIRV and CBR indices, which represent the incremental cost for unit reliability improvement in the different metrics, are estimated first. The RUIRV and RCBR indices in a relative contribution form of UIRV and CBR indices are calculated to develop the CRCI index for ranking the priority of the circuits.

All single-circuit supply systems at BC Hydro are used as an example to demonstrate the application of the probabilistic planning method presented in this chapter. A total of 12 out of 87 single-circuit supply systems are recognized for reinforcement. The 12 circuits are most critical in light of the weighted reliability index, and all of them are financially justifiable. The ranking order for the 12 circuits using the CRCI index indicates the priority for reinforcement.

APPENDIX A

ELEMENTS OF PROBABILITY THEORY AND STATISTICS

A.1 PROBABILITY OPERATION RULES

A.1.1 Intersection

Given two events A and B, we calculate the probability of simultaneous occurrence of A and B as

$$P(A \cap B) = P(A)P(B \mid A) \tag{A.1}$$

where $P(B \mid A)$ is the conditional probability of B occurring given that A has occurred. If A and B are independent, Equation (A.1) becomes

$$P(A \cap B) = P(A)P(B) \tag{A.2}$$

This can be generalized to the case of N events. If N events are independent of each other, the following probability equation holds:

$$P(A_1 \cap A_2 \cap \cdots \cap A_N) = P(A_1)P(A_2)\cdots P(A_N) \tag{A.3}$$

Probabilistic Transmission System Planning, by Wenyuan Li

A.1.2 Union

Given two events A and B, we calculate the probability of occurrence of either A, B, or both as

$$P(A \cup B) = P(A) + P(B) - P(A \cap B) \tag{A.4}$$

If A and B are mutually exclusive, then $P(A \cap B) = 0$. This can be generalized to the case of N events. If N events are mutually exclusive, the following probability equation holds:

$$P(A_1 \cup A_2 \cup \cdots \cup A_N) = P(A_1) + P(A_2) + \cdots + P(A_N) \tag{A.5}$$

A.1.3 Conditional Probability

If the events $\{B_1, B_2, \ldots, B_N\}$ represent a full and mutually exclusive set, that is, $P(B_1) + P(B_2) + \cdots + P(B_N) = 1.0$ and $P(B_i \cap B_j) = 0.0$ ($i \neq j$; $i, j = 1, 2, \ldots, N$), then for any event A, we have

$$P(A) = \sum_{i=1}^{N} P(B_i) P(A \mid B_i) \tag{A.6}$$

This equation is often used in the case of two conditional events. If B_1 and B_2 are mutually exclusive and $P(B_1) + P(B_2) = 1.0$, then

$$P(A) = P(B_1) P(A \mid B_1) + P(B_2) P(A \mid B_2) \tag{A.7}$$

A.2 FOUR IMPORTANT PROBABILITY DISTRIBUTIONS

A.2.1 Binomial Distribution

This is a discrete distribution. Given that the probability of success in a trial is p, the probability of m successes in n trials is

$$P_m = C_n^m P^m (1-p)^{n-m} \tag{A.8}$$

where C_n^m represents the number of combinations of m items from n items, which is calculated by

$$C_n^m = \frac{n!}{m!(n-m)!} \tag{A.9}$$

The mean and variance of the binomial distribution are np and $np(1-p)$, respectively.

A.2.2 Exponential Distribution

The density function of exponential distribution is

$$f(x)=\lambda \exp(-\lambda x) \qquad (x \geq 0) \tag{A.10}$$

The cumulative distribution function is

$$\begin{aligned} F(x) &= 1-\exp(-\lambda x) \\ &= \lambda x - \frac{(\lambda x)^2}{2!} + \frac{(\lambda x)^3}{3!} - \cdots \end{aligned} \tag{A.11}$$

When $\lambda x \ll 1$, Equation (A.11) is approximated by

$$F(x)=\lambda x \tag{A.12}$$

The mean and variance of the exponential distribution are $1/\lambda$ and $1/\lambda^2$, respectively.

A.2.3 Normal Distribution

The density function of normal distribution is

$$f(x)=\frac{1}{\sigma\sqrt{2\pi}}\exp\left[-\frac{(x-\mu)^2}{2\sigma^2}\right] \qquad (-\infty \leq x \leq \infty) \tag{A.13}$$

where μ and σ^2 are the mean and variance of the normal distribution.

With the following substitution

$$z=\frac{x-\mu}{\sigma} \tag{A.14}$$

Equation (A.13) becomes

$$f(z)=\frac{1}{\sqrt{2\pi}}\exp\left[-\frac{z^2}{2}\right] \qquad (-\infty \leq z \leq \infty) \tag{A.15}$$

Equation (A.15) is the density function of standard normal distribution.

There is no explicitly analytical expression for the cumulative distribution function of normal distribution. The area $Q(z)$ under the standard normal density function curve shown in Figure A.1 can be found from the following polynomial approximation for $z \geq 0$:

$$Q(z)=f(z)\cdot[b_1 t + b_2 t^2 + b_3 t^3 + b_4 t^4 + b_5 t^5] \tag{A.16}$$

where $t = 1/(1+rz)$ and

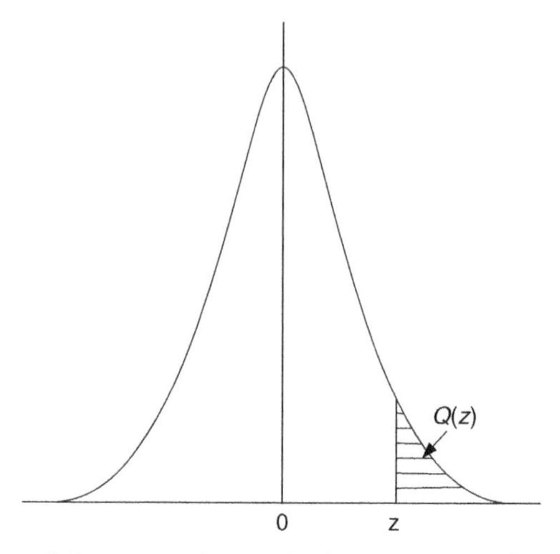

Figure A.1. Area under standard normal density function.

$r = 0.2316419$

$b_1 = 0.31938153$ $\quad b_2 = -0.356563782$ $\quad b_3 = 1.781477937$

$b_4 = -1.821255978$ $\quad b_5 = 1.330274429$

The maximum error of Equation (A.16) is smaller than 7.5×10^{-8}.

A.2.4 Weibull Distribution

The density function of the Weibull distribution is

$$f(x) = \frac{\beta x^{\beta-1}}{\alpha^{\beta}} \exp\left[-\left(\frac{x}{\alpha}\right)^{\beta}\right] \qquad (\infty > x \geq 0, \quad \beta > 0, \quad \alpha > 0) \tag{A.17}$$

The cumulative distribution function of the Weibull distribution is

$$F(x) = 1 - \exp\left[-\left(\frac{x}{\alpha}\right)^{\beta}\right] \qquad (\infty > x \geq 0, \quad \beta > 0, \quad \alpha > 0) \tag{A.18}$$

The mean and variance of the Weibull distribution can be calculated from the scale (α) and shape (β) parameters as follows:

$$\mu = \alpha \Gamma\left(1 + \frac{1}{\beta}\right) \tag{A.19}$$

$$\sigma^2 = \alpha^2 \left[\Gamma\left(1 + \frac{2}{\beta}\right) - \Gamma^2\left(1 + \frac{1}{\beta}\right)\right] \tag{A.20}$$

where $\Gamma(\bullet)$ is the gamma function, which is defined by

$$\Gamma(x) = \int_0^\infty t^{x-1} e^{-t} dt \tag{A.21}$$

A.3 MEASURES OF PROBABILITY DISTRIBUTION

Distributions of random variables can be described using one or more parameters called the *numerical characteristics*. The most useful numerical characteristics are the mathematical expectation (mean), variance or standard deviation, covariance, and correlation coefficients.

A.3.1 Mathematical Expectation

If a random variable X has the probability density function $f(x)$ and the random variable Y is a function of X, that is, $y = y(x)$, then the mathematical expectation or mean value of Y is defined as

$$E(Y) = \int_{-\infty}^{\infty} y(x) f(x) dx \tag{A.22}$$

As a special case of the general definition, the mean of the random variable X is

$$E(X) = \int_{-\infty}^{\infty} x f(x) dx \tag{A.23}$$

For a discrete random variable, Equations (A.22) and (A.23) become Equations (A.24) and (A.25), respectively:

$$E(Y) = \sum_{i=1}^{n} y(x_i) p_i \tag{A.24}$$

$$E(X) = \sum_{i=1}^{n} x_i p_i \tag{A.25}$$

Here, x_i is the ith value of X, p_i is the probability of x_i, and n is the number of discrete values of X.

A.3.2 Variance and Standard Deviation

The variance of a random variable X with the probability density function $f(x)$ is defined as

$$V(X) = E([X - E(X)]^2) = \int_{-\infty}^{\infty} [x - E(X)]^2 f(x) dx \tag{A.26}$$

If X is a discrete random variable, Equation (A.26) becomes

$$V(X) = \sum_{i=1}^{n} [x_i - E(X)]^2 p_i \tag{A.27}$$

The variance is an indicator for the dispersion degree of possible values of X from its mean. The square root of the variance is known as the *standard deviation* and is often expressed by the notation $\sigma(X)$.

A.3.3 Covariance and Correlation Coefficient

Given an N-dimensional random vector $(X_1, X_2, \ldots, X_N)$, the covariance between any two elements X_i and X_j is defined as

$$\begin{aligned} c_{ij} &= E\{[X_i - E(X_i)][X_j - E(X_j)]\} \\ &= E(X_i X_j) - E(X_i)E(X_j) \end{aligned} \tag{A.28}$$

The covariance is often expressed using the notation $\text{cov}(X_i, X_j)$. The covariance between an element and itself is its variance:

$$\text{cov}(X_i, X_i) = V(X_i) \tag{A.29}$$

The correlation coefficient between X_i and X_j is defined as

$$\rho_{ij} = \frac{\text{cov}(X_i, X_j)}{\sqrt{V(X_i)}\sqrt{V(X_j)}} \tag{A.30}$$

The absolute value of ρ_{ij} is smaller than or equal to 1.0. If $\rho_{ij} = 0$, then X_i and X_j are not correlated; if $\rho_{ij} > 0$, then X_i and X_j are positively correlated; and if $\rho_{ij} < 0$, then X_i and X_j are negatively correlated.

A.4 PARAMETER ESTIMATION

A.4.1 Maximum Likelihood Estimation

The objective of point estimation is to estimate single values of parameters of probability distribution. The maximum likelihood estimator is the most popular point estimation method and can be described as follows.

A likelihood function L is constructed by

$$L(\theta_1, \ldots, \theta_k) = \prod_{i=1}^{n} f(x_i, \theta_1, \ldots, \theta_k) \tag{A.31}$$

where x_i is the ith sample value of population variable X, n is the number of sample values, θ_j ($j = 1,\ldots,k$) is the jth parameter in the probability distribution of the variable X, and f represents its density function.

The parameters θ_j ($j = 1,\ldots,k$) can be estimated by solving Equation (A.32) or (A.33) (i.e., maximizing the likelihood function L or $\ln L$):

$$\frac{\partial L}{\partial \theta_j} = 0 \qquad (j = 1,\ldots,k) \tag{A.32}$$

$$\frac{\partial \ln L}{\partial \theta_j} = 0 \qquad (j = 1,\ldots,k) \tag{A.33}$$

The $\ln L$ and L will reach maximum value at the same values of the parameters θ_j. The advantage of using the natural logarithm function $\ln L$ is that the product form of density functions in L is transformed into the sum form.

A.4.2 Mean, Variance, and Covariance of Samples

Let $(x_1,\ldots,x_n)$ be the n samples of a population variable X. The sample mean is calculated by

$$\overline{X} = \frac{1}{n}\sum_{i=1}^{n} x_i \tag{A.34}$$

where $\overline{X}$ is an unbiased estimate of the population mean.

The sample variance is calculated by

$$s^2 = \frac{1}{n-1}\sum_{i=1}^{n}(x_i - \overline{X})^2 \tag{A.35}$$

where s^2 is an unbiased estimate of the population variance.

Let $(x_1,\ldots,x_n)$ and $(y_1,\ldots,y_n)$ be the samples of population variables X and Y. The sample covariance between X and Y is calculated by

$$s_{xy} = \frac{1}{n-1}\sum_{i=1}^{n}(x_i - \overline{X})(y_i - \overline{Y}) \tag{A.36}$$

where s_{xy} is an unbiased estimate of the population covariance.

Once the estimates of sample variances and covariance of X and Y are obtained, the estimate of correlation coefficient between X and Y can be calculated using Equation (A.30).

A.4.3 Interval Estimation

Let θ_j represent an unknown parameter of probability distribution of a population variable X. An interval $[\theta^*_{j1}, \theta^*_{j2}]$ can be estimated using the samples of X, and the following equation holds:

$$p(\theta_{j1}^* \leq \theta_j \leq \theta_{j2}^*) = 1 - \alpha \tag{A.37}$$

The interval $[\theta_{j1}^*, \theta_{j2}^*]$ is a confidence interval of θ_j. The quantity $1 - \alpha$ is called the *confidence degree* and α is called the *significance level.*

Equation (A.37) signifies that a random confidence interval contains the unknown parameter with probability $1 - \alpha$. It should be appreciated that the confidence degree is not the probability that the random parameter falls in a fixed interval. In other words, the confidence interval varies depending on sample size and how the estimation function is constructed.

A.5 MONTE CARLO SIMULATION

A.5.1 Basic Concept

The basic idea of Monte Carlo simulation is to create a series of experimental samples using a random number sequence generator. According to the central limit theorem or the law of large numbers, the sample mean can be used as an unbiased estimate of mathematical expectation of a random variable following any distribution when the number of samples is large enough. The sample mean is also a random variable, and its variance is an indicator of estimation accuracy. The variance of the sample mean is $1/n$ of the population variance, where n is the number of samples:

$$V(\bar{X}) = \frac{1}{n} V(X) \tag{A.38}$$

Therefore, the standard deviation of sample mean is calculated by

$$\sigma = \sqrt{V(\bar{X})} = \frac{\sqrt{V(X)}}{\sqrt{n}} \tag{A.39}$$

Equation (A.39) indicates that there are two measures for reducing the standard deviation of an estimate (i.e., sample mean) in Monte Carlo simulation: increasing the number of samples or decreasing the sample variance. Many variance reduction techniques have been developed to improve the effectiveness of Monte Carlo simulation. It is important to appreciate that the variance cannot be reduced to zero in any actual simulation, and therefore it is always necessary to consider a reasonable and sufficiently large number of samples.

Monte Carlo simulation creates a fluctuating convergence process, and there is no guarantee that a few more samples will definitely lead to a smaller error. However, it is true that the error bound or confidence range decreases as the number of samples increases. The accuracy level of Monte Carlo simulation can be measured using the coefficient of variance, which is defined as the standard deviation of the estimate divided by the estimate:

$$\eta = \frac{\sqrt{V(\bar{X})}}{\bar{X}} \tag{A.40}$$

The coefficient of variance η is often used as a convergence criterion in Monte Carlo simulation.

A.5.2 Random-Number Generator

Generating a random number is a key step in Monte Carlo simulation. Theoretically, a random number generated by a mathematical method is not really random and is called a pseudorandom number. A pseudorandom number sequence should be statistically tested to ensure its randomness, which includes uniformity, independence, and a long repeat period.

There are different algorithms for generating a random number sequence in the interval [0,1]. The most popular algorithm is the mixed congruent generator, which is given by the following recursive relationship:

$$x_{i+1} = (ax_i + c)(\text{mod}\, m) \tag{A.41}$$

where a is the multiplier, c is the increment, and m is the modulus; a, c, and m have to be non-negative integers. The module operation (mod m) means that

$$x_{i+1} = (ax_i + c) - mk_i \tag{A.42}$$

where k_i is the largest positive integer from $(ax_i + c)/m$.

Given an initial value x_0 that is called a "seed," Equation (A.41) generates a random number sequence that lies uniformly in the interval $[0,m]$. A random number sequence uniformly distributed in the interval [0,1] can be obtained by

$$R_i = \frac{x_i}{m} \tag{A.43}$$

Obviously, the random number sequence generated using Equation (A.41) will repeat itself in at most m steps and is periodic. If the repeat period equals m, it is called a *full period*. Different choices of the parameters a, c, and m produce a large impact on statistical features of random numbers. According to numerous statistical tests, the following two sets of parameters provide satisfactory statistical features of generated random numbers:

$$m = 2^{31} \qquad a = 314159269 \qquad c = 453806245$$
$$m = 2^{35} \qquad a = 5^{15} \qquad c = 1$$

A.5.3 Inverse Transform Method

A *random variate* refers to a random number sequence following a given distribution. The mixed congruent generator generates a random number sequence following a uniform distribution in the interval [0,1]. The inverse transform method is commonly used to generate random variates following other distributions. The method includes the following two steps:

1. Generate a uniformly distributed random number sequence R in the interval [0,1].
2. Calculate the random variate that has the cumulative probability distribution function $F(x)$ by $X = F^{-1}(R)$

A.5.4 Three Important Random Variates

A.5.4.1 Exponential Distribution Random Variate. The cumulative probability distribution function of exponential distribution is

$$F(x) = 1 - e^{-\lambda x} \tag{A.44}$$

A uniform distribution random number R is generated so that

$$R = F(x) = 1 - e^{-\lambda x} \tag{A.45}$$

Using the inverse transform method, we have

$$X = F^{-1}(R) = -\frac{1}{\lambda}\ln(1 - R) \tag{A.46}$$

Since $(1 - R)$ distributes uniformly in the same way as R in the interval [0,1], Equation (A.46) equivalently becomes

$$X = -\frac{1}{\lambda}\ln(R) \tag{A.47}$$

where R is a uniform distribution random number sequence and X follows the exponential distribution.

A.5.4.2 Normal Distribution Random Variate. There exists no analytical expression for the inverse function of the normal cumulative distribution function. The following approximate expression can be used. Given an area $Q(z)$ under the normal density distribution curve as shown in Figure A.1, the corresponding z can be calculated by

$$z = s - \frac{\sum_{i=0}^{2} c_i s^i}{1 + \sum_{i=1}^{3} d_i s^i} \tag{A.48}$$

where

$$s = \sqrt{-2\ln Q} \tag{A.49}$$

and

$$c_0 = 2.515517 \qquad c_1 = 0.802853 \qquad c_2 = 0.010328$$
$$d_1 = 1.432788 \qquad d_2 = 0.189269 \qquad d_3 = 0.001308$$

The maximum error of Equation (A.48) is smaller than 0.45×10^{-4}.

The algorithm for generating the normal distribution random variate includes the following two steps:

1. Generate a uniform distribution random number sequence R in the interval [0,1].
2. Calculate the normal distribution random variate X by

$$X = \begin{cases} z & \text{if} \quad 0.5 < R \le 1.0 \\ 0 & \text{if} \quad R = 0.5 \\ -z & \text{if} \quad 0 \le R < 0.5 \end{cases} \tag{A.50}$$

where z is obtained from Equation (A.48) and Q in Equation (A.49) is given by

$$Q = \begin{cases} 1 - R & \text{if} \quad 0.5 < R \le 1.0 \\ R & \text{if} \quad 0 \le R \le 0.5 \end{cases} \tag{A.51}$$

A.5.4.3 Weibull Distribution Random Variate. By using the inverse transform method, let a uniform distribution random number R equal the Weibull cumulative probability distribution function given in Equation (A.18):

$$R = F(x) = 1 - \exp\left[-\left(\frac{x}{\alpha}\right)^{\beta}\right] \tag{A.52}$$

Equivalently, we obtain

$$X = \alpha[-\ln(1 - R)]^{1/\beta} \tag{A.53}$$

Since $(1 - R)$ distributes uniformly in the same way as R in the interval [0,1], Equation (A.53) becomes

$$X = \alpha(-\ln R)^{1/\beta} \tag{A.54}$$

where R is a uniform distribution random number sequence and X follows the Weibull distribution.

APPENDIX B

ELEMENTS OF FUZZY MATHEMATICS

B.1 FUZZY SETS

B.1.1 Definition of Fuzzy Set

Let U be a traditional set and its member be denoted by x. A fuzzy set A on U is defined as a set of ordered pairs and is expressed by

$$A = \{(x, \mu_A(x)) \mid x \in U\} \tag{B.1}$$

where $\mu_A(x)$, whose value varies from 0 to 1, is called the *membership function* of A. If U is a discrete set with n members, A can be represented as

$$A = \frac{\mu_A(x_1)}{x_1} + \cdots + \frac{\mu_A(x_n)}{x_n} = \sum_{i=1}^{n} \frac{\mu_A(x_i)}{x_i} \tag{B.2}$$

where + or Σ indicates the union of the members in A and $\mu_A(x_i)$ is the membership grade of x_i. If U is a continuous set, A can be represented as

$$A = \int_U \frac{\mu_A(x)}{x} \tag{B.3}$$

where $\int$ indicates the union of the members in A. Note that the horizontal bar in Equations (B.2) and (B.3) is not a quotient but a delimiter.

An α cutset of A, denoted by A_α, is defined as

$$A_\alpha = \{x \in U \mid \mu_A(x) \geq \alpha, \alpha \in [0,1]\} \tag{B.4}$$

For any fuzzy set A, the membership function can be expressed using its α cutset by

$$\mu_A(x) = \sup_{\alpha \in [0,1]} \min(\alpha, \mu_{A_\alpha}(x) | x \in U) \tag{B.5}$$

In particular, A is probabilistic fuzzy set if $\mu_A(x)$ is a random variable defined on a probabilistic space.

B.1.2 Operations of Fuzzy Sets

Assume that A and B are two fuzzy sets. A new fuzzy set C can be obtained using the following operations:

Intersection $C = A \cap B$:

$$\mu_C(x) = \min(\mu_A(x), \mu_B(x)) \tag{B.6}$$

Union $C = A \cup B$:

$$\mu_C(x) = \max(\mu_A(x), \mu_B(x)) \tag{B.7}$$

Complement $C = \bar{A}$:

$$\mu_C(x) = 1 - \mu_A(x) \tag{B.8}$$

Algebraic product $C = A \bullet B$:

$$\mu_C(x) = \mu_A(x) \cdot \mu_B(x) \tag{B.9}$$

The majority of relation laws of crisp sets hold for fuzzy sets. These include

Commutativity	$A \cup B = B \cup A$	$A \cap B = B \cap A$
Associativity	$(A \cup B) \cup C = A \cup (B \cup C)$	$(A \cap B) \cap C = A \cap (B \cap C)$
Distributivity	$A \cup (B \cap C) = (A \cup B) \cap (A \cup C)$	$A \cap (B \cup C) = (A \cap B) \cup (A \cap C)$
Absorption	$A \cup (A \cap B) = A$	$A \cap (A \cup B) = A$
DeMorgan's law	$\overline{A \cap B} = \bar{A} \cup \bar{B}$	$\overline{A \cup B} = \bar{A} \cap \bar{B}$

However, it should be noted that the following two laws are not the same as those for crisp sets (where W represents the universal set and ϕ represents the empty set):

$$A \cup \bar{A} \neq W \qquad A \cap \bar{A} \neq \phi$$

B.2 FUZZY NUMBERS

B.2.1 Definition of Fuzzy Number

A fuzzy number is a special type of fuzzy set. A fuzzy set A, which is defined on the real number space R, is convex if the following inequality holds (where $x_1, x_2 \in U$, $\lambda \in [0,1]$):

$$\mu_A(\lambda x_1 + (1-\lambda)x_2) \geq \min(\mu_A(x_1), \mu_A(x_2)) \tag{B.10}$$

A *fuzzy number* is defined as a convex, normalized fuzzy set with a piecewise continuous membership function. Under this definition, it is clear that the α cutset A_α of a fuzzy number A is an interval: $A_\alpha = [a_l(\alpha), a_u(\alpha)]$ with $a_l(\alpha) \leq a_u(\alpha)$.

It is obvious that $a_l(\alpha)$ and $a_u(\alpha)$ are both a monotonically decreasing function of α. Therefore, operations of fuzzy numbers can be performed using calculations of intervals.

B.2.2 Arithmetic Operation Rules of Fuzzy Numbers

For two given fuzzy numbers $A_\alpha = [a_l(\alpha), a_u(\alpha)]$ and $B_\alpha = [b_l(\alpha), b_u(\alpha)]$, the following operation rules apply.

B.2.2.1 *Addition.*

$$(A+B)_\alpha = [a_l(\alpha) + b_l(\alpha), a_u(\alpha) + b_u(\alpha)] \tag{B.11}$$

B.2.2.2 *Subtraction.*

$$(A-B)_\alpha = [a_l(\alpha) - b_u(\alpha), a_u(\alpha) - b_l(\alpha)] \tag{B.12}$$

B.2.2.3 *Multiplication.*

$$\begin{aligned}(AB)_\alpha = [&\min(a_l(\alpha)\cdot b_l(\alpha), a_u(\alpha)\cdot b_l(\alpha), a_l(\alpha)\cdot b_u(\alpha), a_u(\alpha)\cdot b_u(\alpha),\\ &\max(a_l(\alpha)\cdot b_l(\alpha), a_u(\alpha)\cdot b_l(\alpha), a_l(\alpha)\cdot b_u(\alpha), a_u(\alpha)\cdot b_u(\alpha)]\end{aligned} \tag{B.13}$$

If A and B are defined on positive monotonic real number space, (B.13) becomes

$$(AB)_\alpha = [a_l(\alpha)\cdot b_l(\alpha), a_u(\alpha)\cdot b_u(\alpha)] \tag{B.14}$$

In particular, if H is a positive constant number, we have

$$(HA)_\alpha = [Ha_l(\alpha), Ha_u(\alpha)] \tag{B.15}$$

B.2.2.4 Division.

$$\left(\frac{A}{B}\right)_\alpha = [a_l(\alpha), a_u(\alpha)] \cdot \left[\frac{1}{b_u(\alpha)}, \frac{1}{b_l(\alpha)}\right] \tag{B.16}$$

where $b_1(\alpha) \neq 0$ and $b_u(\alpha) \neq 0$. Otherwise, one or both ends of the interval are extended to ∞.

B.2.2.5 Maximum and Minimum Operations.

$$(A \vee B)_\alpha = [a_l(\alpha) \vee b_l(\alpha), a_u(\alpha) \vee b_u(\alpha)] \tag{B.17}$$

$$(A \wedge B)_\alpha = [a_l(\alpha) \wedge b_l(\alpha), a_u(\alpha) \wedge b_u(\alpha)] \tag{B.18}$$

where $\vee$ or $\wedge$ denotes finding maximum or minimum.

Note that when the arithmetic operation rules are used, the fuzzy numbers in the operation must be independent of each other.

B.2.3 Functional Operation of Fuzzy Numbers

If $C = f(A, B)$ is a monotonic function of A and B on the real number space R, then its α cutset is calculated by

$$\begin{aligned} C_\alpha = [&\min\{f((a_l(\alpha), b_l(\alpha)), f(a_u(\alpha), b_l(\alpha)), f(a_l(\alpha), b_u(\alpha)), f(a_u(\alpha), b_u(\alpha))\}, \\ &\max\{f((a_l(\alpha), b_l(\alpha)), f(a_u(\alpha), b_l(\alpha)), f(a_l(\alpha), b_u(\alpha)), f(a_u(\alpha), b_u(\alpha))\}] \end{aligned} \tag{B.19}$$

It is apparent that this is a calculation to find the minimum and maximum in all possible combinations. A similar rule can be extended to a function containing more fuzzy variables. In this situation, the number of combinations can be very large. If the fuzzy variables in a function are not independent, the arithmetic operation rules given in Section B.2.2 are invalid and the general rule given in Equation (B.19) must be applied. For example, if a fuzzy number occurs in more than one term in a functional expression, the repeated use of the arithmetic operation rules is generally incorrect because the independence condition is violated.

Theoretically, a general function of fuzzy numbers is still a fuzzy set, but there is no guarantee that it is a fuzzy number. However, the arithmetic operations of independent fuzzy numbers on the positive real number space result in a fuzzy number.

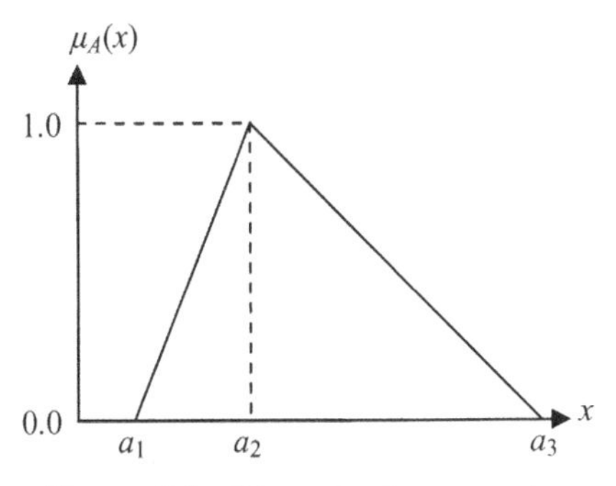

Figure B.1. Triangular fuzzy number.

B.3 TWO TYPICAL FUZZY NUMBERS IN ENGINEERING APPLICATIONS

B.3.1 Triangular Fuzzy Number

The triangular fuzzy number, which is often denoted by $A = (a_1,a_2,a_3)$, is defined by the following membership function:

$$\mu_A(x) = \begin{cases} (x-a_1)/(a_2-a_1) & \text{if} \quad a_1 \le x \le a_2 \\ (a_3-x)/(a_3-a_2) & \text{if} \quad a_2 \le x \le a_3 \\ 0 & \text{if} \quad x \le a_1 \text{ or } x \ge a_3 \end{cases} \tag{B.20}$$

This membership function is shown in Figure B.1.

The α cutset A_α of the triangular fuzzy number is calculated by

$$A_\alpha = [a_1 + \alpha(a_2 - a_1), a_3 - \alpha(a_3 - a_2)] \tag{B.21}$$

For two triangular fuzzy numbers $A = (a_1,a_2,a_3)$ and $B = (b_1,b_2,b_3)$, it can be shown that their sum and difference are still triangular fuzzy numbers:

$$A + B = (a_1 + b_1, a_2 + b_2, a_3 + b_3) \tag{B.22}$$

$$A - B = (a_1 - b_3, a_2 - b_2, a_3 - b_1) \tag{B.23}$$

However, their product AB is no longer a triangular fuzzy number.

B.3.2 Trapezoidal Fuzzy Number

The trapezoidal fuzzy number, which is often denoted by $A = (a_1,a_2,a_3,a_4)$, is defined by the following membership function:

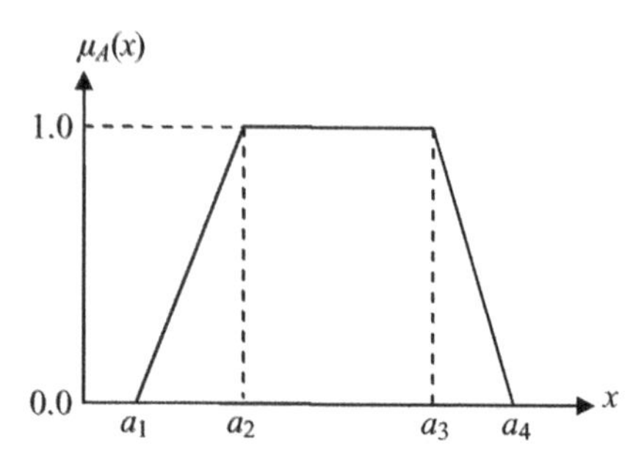

Figure B.2. Trapezoidal fuzzy number.

$$\mu_A(x) = \begin{cases} \dfrac{x - a_1}{a_2 - a_1} & \text{if} \quad a_1 \le x \le a_2 \\ 1 & \text{if} \quad a_2 \le x \le a_3 \\ \dfrac{a_4 - x}{a_4 - a_3} & \text{if} \quad a_3 \le x \le a_4 \\ 0 & \text{if} \quad x \le a_1 \text{ or } x \ge a_4 \end{cases} \tag{B.24}$$

This membership function is shown in Figure B.2.

The α cutset A_α of the trapezoidal fuzzy number is calculated by

$$A_\alpha = [a_1 + \alpha(a_2 - a_1), a_4 - \alpha(a_4 - a_3)] \tag{B.25}$$

For two trapezoidal fuzzy numbers $A = (a_1,a_2,a_3,a_4)$ and $B = (b_1,b_2,b_3,b_4)$, it can be shown that their sum and difference are still trapezoidal fuzzy numbers:

$$A + B = (a_1 + b_1, a_2 + b_2, a_3 + b_3, a_4 + b_4) \tag{B.26}$$

$$A - B = (a_1 - b_4, a_2 - b_3, a_3 - b_2, a_4 - b_1) \tag{B.27}$$

Similarly, the product AB is no longer a trapezoidal fuzzy number.

B.4 FUZZY RELATIONS

B.4.1 Basic Concepts

If the relation between two sets X and Y is vague and cannot be described by "yes" or "no," it is a fuzzy relation. Such a fuzzy relation is a fuzzy set, which is denoted by $\tilde{R}(X,Y)$. The membership function of $\tilde{R}(X,Y)$ is expressed by $\mu_{\tilde{R}}(x_i, y_j)$, where x_i and y_j are the elements of X and Y, respectively. The fuzzy relation between the two sets X and Y can be expressed using the following matrix:

$$\begin{bmatrix} \mu_{\tilde{R}}(x_1, y_1) & \mu_{\tilde{R}}(x_1, y_2) & \cdots & \mu_{\tilde{R}}(x_1, y_n) \\ \mu_{\tilde{R}}(x_2, y_1) & \mu_{\tilde{R}}(x_2, y_2) & & \mu_{\tilde{R}}(x_2, y_n) \\ \vdots & \vdots & \vdots & \vdots \\ \mu_{\tilde{R}}(x_m, y_1) & \mu_{\tilde{R}}(x_m, y_2) & & \mu_{\tilde{R}}(x_m, y_n) \end{bmatrix}$$

Specifically, $\tilde{R}(X, X)$ is the fuzzy relation between the elements in a set X and can be expressed using an $n \times n$ square matrix $\tilde{R}$ where n is the number of elements.

B.4.1.1 Reflexivity. A fuzzy relation $\tilde{R}(X, X)$ is reflexive if and only if

$$\mu_{\tilde{R}}(x_i, x_i) = 1 \qquad \forall x_i \in X \tag{B.28}$$

This corresponds to a fuzzy relation matrix $\tilde{R}$ in which each diagonal element equals to 1.

B.4.1.2 Symmetry. A fuzzy relation $\tilde{R}(X, X)$ is symmetric if and only if

$$\mu_{\tilde{R}}(x_i, x_j) = \mu_{\tilde{R}}(x_j, x_i) \qquad \forall x_i, x_j \in X \tag{B.29}$$

This corresponds to a symmetric fuzzy relation matrix $\tilde{R}$.

B.4.1.3. Resemblance. If a fuzzy relation $\tilde{R}(X, X)$ is both reflexive and symmetric, it is called a *resemblance* relation.

B.4.1.4 Transitivity. A fuzzy relation $\tilde{R}(X, X)$ is transitive if and only if

$$\mu_{\tilde{R}}(x_i, x_k) \geq \sup_{x_j} \min(\mu_{\tilde{R}}(x_i, x_j), \mu_{\tilde{R}}(x_j, x_k)) \qquad \forall x_i, x_j, x_k \in X \tag{B.30}$$

This corresponds to $\tilde{R} \circ \tilde{R} \subset \tilde{R}$, where the symbol of $\tilde{R} \circ \tilde{R}$ denotes the self-multiplication of the relation matrix $\tilde{R}$ [see Equation (B.34) below].

B.4.1.5 Equivalence. If a resemblance fuzzy relation $\tilde{R}(X, X)$ satisfies transitivity, it is an equivalence relation.

B.4.2 Operations of Fuzzy Matrices

A fuzzy matrix can be denoted by $A = [a_{ij}]$. For the two fuzzy matrices with the same dimension $A = [a_{ij}]$ and $B = [b_{ij}]$, the following operation rules apply:

Intersection $C = [c_{ij}] = A \cap B$:

$$c_{ij} = \min[a_{ij}, b_{ij}] = a_{ij} \wedge b_{ij} \tag{B.31}$$

Union $C = [c_{ij}] = A \bigcup B$:

$$c_{ij} = \max[a_{ij}, b_{ij}] = a_{ij} \vee b_{ij} \tag{B.32}$$

Complement $C = [c_{ij}] = \bar{A}$:

$$c_{ij} = [1 - a_{ij}] \tag{B.33}$$

Product $C = A \circ B$:

$$C_{ij} = \max_k \min[a_{ik}, b_{kj}] = \vee_k [a_{ik} \wedge b_{kj}] \tag{B.34}$$

Equation (B.34) signifies that each element in the fuzzy matrix C is calculated using a rule similar to that for multiplication of crisp matrices, except that the product of two elements is replaced by taking the minimum one and the addition of two products is replaced by taking the maximum one. Note that in general, $A \circ B \neq B \circ A$ even if A and B are both square matrices of the same size.

A square fuzzy relation matrix $\tilde{R}$ represents the first-order relation. $\tilde{R}_2 = \tilde{R} \circ \tilde{R}$ represents the second-order fuzzy relation. Similarly, the following equation represents the nth-order fuzzy relation:

$$\tilde{R}_n = \underbrace{\tilde{R} \circ \tilde{R} \circ \tilde{R} \cdots \tilde{R}}_{n \text{ self-multiplications}}$$

If a fuzzy set contains n elements and $\tilde{R}$ is a resemblance relation matrix on the set, then the $(n - 1)$th-order fuzzy relation matrix $\tilde{R}_{n-1}$ not only maintains reflexivity and symmetry but also has transitivity. In other words, $\tilde{R}_{n-1}$ obtained from self-multiplications of a resemblance relation matrix is an equivalence relation matrix and satisfies

$$\tilde{R}_{n-1} = \tilde{R}_n = \tilde{R}_{n+1} = \cdots = \tilde{R}_{n+m} \tag{B.35}$$

where m is any positive integer. The equivalence relation matrix can be used for fuzzy clustering.

APPENDIX C

ELEMENTS OF RELIABILITY EVALUATION

C.1 BASIC CONCEPTS

C.1.1 Reliability Functions

Reliability can be represented using a probability distribution as

$$R(t) = \int_t^\infty f(t)dt \tag{C.1}$$

where $f(t)$ is the failure probability density function.

The inverse operation of (C.1) gives

$$f(t) = \frac{-dR(t)}{dt} \tag{C.2}$$

Unreliability can be expressed as

$$Q(t) = 1 - R(t) = \int_0^t f(t)dt \tag{C.3}$$

Probabilistic Transmission System Planning, by Wenyuan Li

The mean time to failure can be calculated by

$$\text{MTTF} = \int_0^\infty tf(t)dt = \int_0^\infty R(t)dt \tag{C.4}$$

The failure rate is defined as

$$\lambda(t) = \frac{f(t)}{R(t)} \tag{C.5}$$

By substituting Equation (C.2) into Equation (C.5), we can also represent reliability using the failure rate function as

$$R(t) = \exp\left[-\int_0^t \lambda(t)dt\right] \tag{C.6}$$

C.1.2 Model of Repairable Component

Components are classified into nonrepairable and repairable categories. A power system component is repairable in the normal life period and can die at the end-of-life stage. For a repairable component, reliability and availability are two different concepts. *Reliability* is the probability that a component has not failed as of a given time point, whereas *availability* is the probability that a component is found available at a given time point although it may have experienced failures and repairs before this point. If the exponential distribution is used to model the failure and repair processes, a repairable component has constant failure and repair rates. Figure C.1 shows the state transition model of a repairable component.

The failure rate λ, repair rate μ, and failure frequency f are calculated by

$$\lambda = \frac{1}{d} \tag{C.7}$$

$$\mu = \frac{1}{r} \tag{C.8}$$

$$f = \frac{1}{d+r} \tag{C.9}$$

where d and r are the mean time to failure and mean time to repair, respectively.

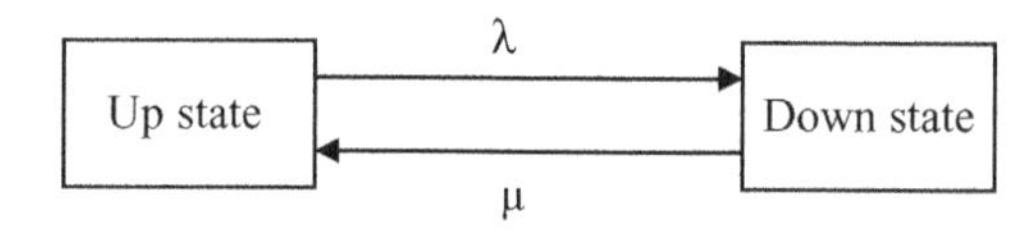

Figure C.1. Two-state model of a repairable component.

The average unavailability is calculated by

$$U = \frac{\lambda}{\lambda + \mu} = f \cdot r \tag{C.10}$$

The failure rate and failure frequency have the following relationship:

$$f = \frac{\lambda}{1 + \lambda r} \tag{C.11}$$

In most engineering applications, f and λ are numerically close since λ and r are both small values.

Note that in the equations above, the unit of λ, μ, or f is in occurrences/year and the unit of d or r is in years.

C.2 CRISP RELIABILITY EVALUATION

C.2.1 Series and Parallel Networks

Components are said to be *in series* if only one needs to fail for the network failure, or they must be all up for the network success. Components are said to be *in parallel* if they must all fail for the network failure, or only one needs to be up for the network success.

In this section, U, λ, μ, f, and r are the same as defined in Section C.1.2 except that the subscript 1 or 2 represents component 1 or 2. A denotes availability, which equals $1 - U$. The subscripts "se" and "pa" represent series and parallel networks, respectively.

C.2.1.1 Series Network. Consider the case of two repairable components in series as shown in Figure C.2.

From the definition of the series network, the following relationships hold:

$$U_{se} = U_1 + U_2 - U_1 U_2 \tag{C.12}$$

$$\lambda_{se} = \lambda_1 + \lambda_2 \tag{C.13}$$

$$A_{se} = A_1 A_2 \tag{C.14}$$

The equivalent repair time and failure frequency for the series network can be derived and expressed as follows:

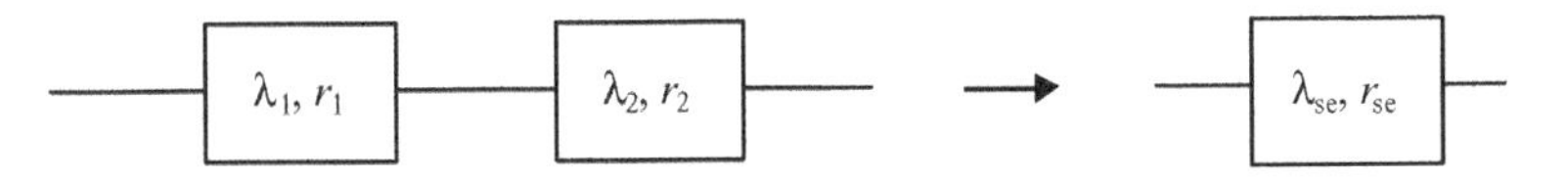

Figure C.2. Series network and its equivalence.

$$r_{se} = \frac{\lambda_1 r_1 + \lambda_2 r_2 + \lambda_1 r_1 \lambda_2 r_2}{\lambda_1 + \lambda_2} \quad \text{(C.15)}$$

$$f_{se} = f_1(1 - f_2 r_2) + f_2(1 - f_1 r_1) \quad \text{(C.16)}$$

In most engineering applications, because failure rates (λ) and repair times (r) are small values, Equation (C.15) can be approximated by

$$r_{se} \approx \frac{\lambda_1 r_1 + \lambda_2 r_2}{\lambda_1 + \lambda_2} \quad \text{(C.17)}$$

C.2.1.2 Parallel Network. Figure C.3 shows the case of two repairable components in parallel.

From the definition of the parallel network, the following relationships hold:

$$U_{pa} = U_1 U_2 \quad \text{(C.18)}$$

$$\mu_{pa} = \mu_1 + \mu_2 \quad \text{(C.19)}$$

$$A_{pa} = A_1 + A_2 - A_1 A_2 \quad \text{(C.20)}$$

The equivalent repair time, failure rate, and failure frequency for the parallel network can be derived and expressed as:

$$r_{pa} = \frac{r_1 r_2}{r_1 + r_2} \quad \text{(C.21)}$$

$$\lambda_{pa} = \frac{\lambda_1 \lambda_2 (r_1 + r_2)}{1 + \lambda_1 r_1 + \lambda_2 r_2} \quad \text{(C.22)}$$

$$f_{pa} = f_1 f_2 (r_1 + r_2) \quad \text{(C.23)}$$

In most engineering applications, because $\lambda r \ll 1$, Equation (C.22) can be approximated by

$$\lambda_{pa} = \lambda_1 \lambda_2 (r_1 + r_2) \quad \text{(C.24)}$$

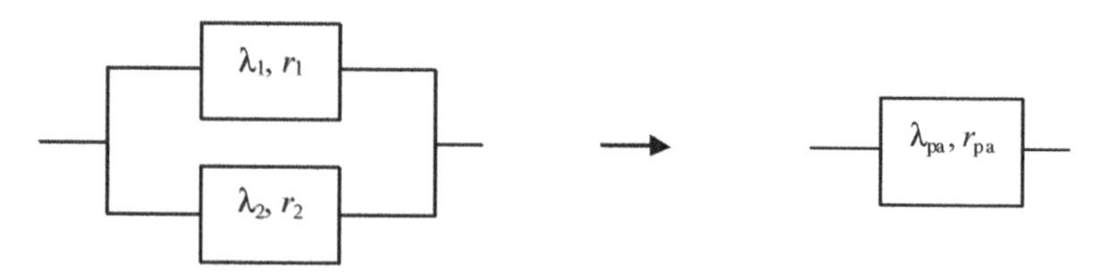

Figure C.3. Parallel network and its equivalence.

C.2.2 Minimum Cutsets

A *cutset* is defined as a set of components that, when failed, results in failure of the network. A *minimum cutset* is defined as a set of components that, when failed, results in failure of the network but when any component in the set has not failed, does not result in network failure.

Obviously, the failure probability or unavailability of a network can be calculated by

$$\begin{aligned} U &= P(C_1 \cup C_2 \cdots \cup C_n) \\ &= \sum_i P(C_i) - \sum_{i,j} P(C_i \cap C_j) + \sum_{i,j,k} P(C_i \cap C_j \cap C_k) - \cdots \\ &\quad + (-1)^{n-1} P(C_1 \cap C_2 \cdots \cap C_n) \end{aligned} \tag{C.25}$$

where U is the failure probability or unavailability of the network, C_i represents the ith minimum cutset, and n is the number of minimum cut sets.

In most engineering applications, two approximations are often adopted: (1) it is unnecessary to enumerate all minimum cutsets since the failure probabilities of components are generally small and thus the probability of higher-order cutsets can be very low—in other words, high-order minimum cutsets can be ignored from enumeration; and (2) the probabilities of intersection of two or more minimum cutsets are usually extremely low in many cases and therefore the effects of non-mutual exclusion among minimum cutsets are negligible. The second approximation signifies that it is often acceptable to consider only the first term in Equation (C.25).

C.2.3 Markov Equations

The Markov equation method can be used to solve both time-dependent and limiting state probabilities. The former is associated with a set of differential equations and the latter, with a set of algebraic equations. The Markov equation method for limiting state probabilities is illustrated using a two-component network as an example. The procedure includes the following steps:

1. A state space diagram is constructed according to the transitions between component states, as shown in Figure C.4, in which λ and μ are failure and repair rates of components, respectively.
2. The transition matrix is built based on the state space diagram as follows:

$$\boldsymbol{T} = \begin{array}{c} \\ 1 \\ 2 \\ 3 \\ 4 \end{array} \begin{array}{c} \begin{array}{cccc} 1 & 2 & 3 & 4 \end{array} \\ \begin{bmatrix} 1-(\lambda_1+\lambda_2) & \lambda_1 & \lambda_2 & 0 \\ \mu_1 & 1-(\mu_1+\lambda_2) & 0 & \lambda_2 \\ \mu_2 & 0 & 1-(\mu_2+\lambda_1) & \lambda_1 \\ 0 & \mu_2 & \mu_1 & 1-(\mu_1+\mu_2) \end{bmatrix} \end{array} \tag{C.26}$$

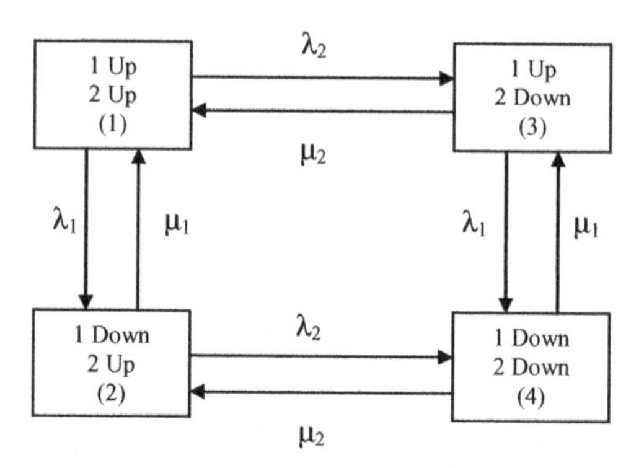

Figure C.4. State space diagram of two repairable components.

3. The Markov principle, which states that the limiting state probabilities would not change in the further transition process, is applied:

$$\boldsymbol{PT} = \boldsymbol{P} \tag{C.27}$$

Here, $\boldsymbol{P}$ is the limiting network state probability row vector whose component P_i is the probability of the ith state, and $\boldsymbol{T}$ is the transition matrix. Equation (C.27) can be rewritten as follows:

$$\begin{bmatrix} -(\lambda_1+\lambda_2) & \mu_1 & \mu_2 & 0 \\ \lambda_1 & -(\mu_1+\lambda_2) & 0 & \mu_2 \\ \lambda_2 & 0 & -(\mu_2+\lambda_1) & \mu_1 \\ 0 & \lambda_2 & \lambda_1 & -(\mu_1+\mu_2) \end{bmatrix} \begin{bmatrix} P_1 \\ P_2 \\ P_3 \\ P_4 \end{bmatrix} = \begin{bmatrix} 0 \\ 0 \\ 0 \\ 0 \end{bmatrix} \tag{C.28}$$

4. The full probability condition, which states that the sum of probabilities of all network states should be 1, is used to replace any equation in Equation (C.28). For instance, the first equation in Equation (C.28) is replaced to yield

$$\begin{bmatrix} 1 & 1 & 1 & 1 \\ \lambda_1 & -(\mu_1+\lambda_2) & 0 & \mu_2 \\ \lambda_2 & 0 & -(\mu_2+\lambda_1) & \mu_1 \\ 0 & \lambda_2 & \lambda_1 & -(\mu_1+\mu_2) \end{bmatrix} \begin{bmatrix} P_1 \\ P_2 \\ P_3 \\ P_4 \end{bmatrix} = \begin{bmatrix} 1 \\ 0 \\ 0 \\ 0 \end{bmatrix} \tag{C.29}$$

5. The Markov matrix equation obtained in step 4 is solved using a linear algebraic algorithm. For the given example, the solution is as follows:

$$P_1 = \frac{\mu_1\mu_2}{(\mu_1+\lambda_1)(\mu_2+\lambda_2)} \tag{C.30}$$

$$P_2 = \frac{\lambda_1 \mu_2}{(\mu_1 + \lambda_1)(\mu_2 + \lambda_2)} \tag{C.31}$$

$$P_3 = \frac{\mu_1 \lambda_2}{(\mu_1 + \lambda_1)(\mu_2 + \lambda_2)} \tag{C.32}$$

$$P_4 = \frac{\lambda_1 \lambda_2}{(\mu_1 + \lambda_1)(\mu_2 + \lambda_2)} \tag{C.33}$$

Once the probability P_i of the ith state is obtained by the Markov method, the frequency f_i of encountering the state is calculated by

$$f_i = P_i \sum_{k=1}^{M_i} \lambda_k \tag{C.34}$$

where λ_k is the departing transition (failure or repair) rate; M_i is the number of the transition rates departing from the ith state.

The probability P, frequency f, and duration D of a state or state set satisfy the following generic relationship:

$$P = f \cdot D \tag{C.35}$$

The mean duration of residing a state or state set can be calculated from the state probability and frequency.

C.3 FUZZY RELIABILITY EVALUATION

C.3.1 Series and Parallel Networks Using Fuzzy Numbers

In the nofuzzy reliability evaluation, the unreliability U_{pa} of a parallel network with n components can have a general expression as follows [see Equation (C.18)]:

$$U_{pa} = \prod_{i=1}^{n} U_i \tag{C.36}$$

Here, U_i is the unreliability of the ith component, which represents the failure probability of each component when a nonrepairable network is evaluated, or the unavailability of each component when a repairable network is evaluated.

It is assumed that the unreliability of each component is modeled by a triangular fuzzy number: $U_i = (a_{i1}, a_{i2}, a_{i3})$. Its α cut is calculated by

$$(U_i)_\alpha = [a_{i1} + \alpha(a_{i2} - a_{i1}), a_{i3} - \alpha(a_{i3} - a_{i2})] \tag{C.37}$$

Using the rule given in Equation (B.14), the α cut of the unreliability for a parallel network can be directly calculated by

$$\begin{aligned}(U_{\text{pa}})_\alpha &= \prod_{i=1}^{n}(U_i)_\alpha \\ &= \left[\prod_{i=1}^{n}(a_{i1}+\alpha(a_{i2}-a_{i1})), \prod_{i=1}^{n}(a_{i3}-\alpha(a_{i3}-a_{i2}))\right]\end{aligned} \tag{C.38}$$

In the nonfuzzy reliability evaluation, the unreliability of a series network with n components can have a general expression as follows [see Equation (C.14)]:

$$U_{\text{se}} = 1-\prod_{i=1}^{n}A_i = 1-\prod_{i=1}^{n}(1-U_i) \tag{C.39}$$

Using the rules given in Equations (B.12) and (B.14), the α cut of the unreliability for a series network is calculated by

$$\begin{aligned}(U_{\text{se}})_\alpha &= [1,1]-\prod_{i=1}^{n}\{[1,1]-(U_i)_\alpha\} \\ &= [1,1]-\prod_{i=1}^{n}[1-a_{i3}+\alpha(a_{i3}-a_{i2}), 1-a_{i1}-\alpha(a_{i2}-a_{i1})] \\ &= \left[1-\prod_{i=1}^{n}\{1-a_{i1}-\alpha(a_{i2}-a_{i1})\}, 1-\prod_{i=1}^{n}\{1-a_{i3}+\alpha(a_{i3}-a_{i2})\}\right]\end{aligned} \tag{C.40}$$

It should be pointed out that in transmission system reliability evaluation, the input data are often given by outage frequencies and repair times of components. In this case, Equation (C.10) is used to calculate the unavailability of a component from the outage frequency and repair time. If the outage frequency and repair time are expressed using triangular fuzzy numbers, the arithmetic operation rules in Section B.2.2 can be applied to Equation (C.10). Note that the membership function of unavailability is no longer triangular, even if both the outage frequency and repair time have triangular membership functions. This does not affect calculations as the arithmetic operation rules can be individually applied to each discrete membership function grade (i.e., α cut).

C.3.2 Minimum Cutset Approach Using Fuzzy Numbers

It is difficult to apply fuzzy calculations to the terms in Equation (C.25), which are associated with nonmutual exclusion among minimum cutsets. As pointed out in Section C.2.2, however, the effects of nonmutual exclusion are negligible in most engineering applications, including transmission reliability evaluation. Therefore, an approximate method can be considered.

A second-order or higher-order minimum cutset contains more than one component. Each minimum cutset is composed of components in parallel since all the components in the set must fail for the set failure. All minimum cutsets are in series as only one of them needs to fail for the network failure. As such, a combination of minimum

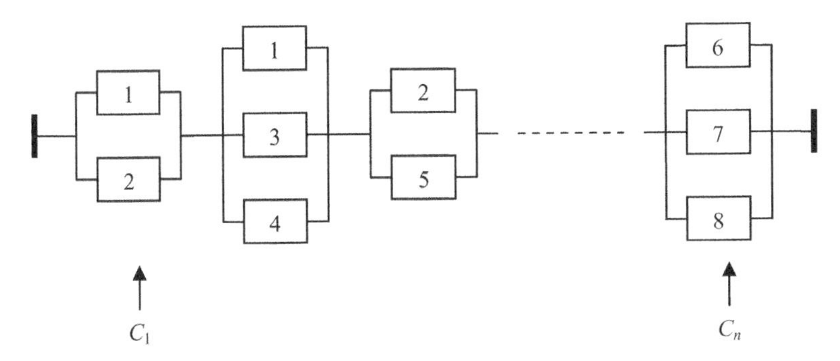

Figure C.5. Approximate parallel and series network for minimum cut sets.

cutsets in series and components of each cutset in parallel can be used to model the network. This is shown in Figure C.5, in which each block represents a component and each group of components in parallel represents a minimum cutset. It can be seen that the second-order and higher-order minimum cutsets may share the same component(s), resulting in their nonmutual exclusion. If the effect of nonmutual exclusion is ignored, the series network reliability formulas can be approximately applied to the minimum cutsets.

If only the failure probability or unavailability of the network needs to be calculated, the fuzzy formulas for parallel and series networks in Section C.3.1 can be directly used. If multiple reliability indices of the network, including frequency, duration, and probability, are needed, the operation rules for fuzzy numbers can be applied to the formulas in Section C.2.1. The approximate minimum cutset method for evaluating multiple reliability indices using fuzzy numbers includes the following two steps:

1. The membership functions of multiple indices for each minimum cutset are calculated using the reliability formulas for a parallel network in Section C.2.1.2 and the operation rules for fuzzy numbers, in Section B.2.2. The basic formulas are Equations (C.18), (C.21), and (C.23) or (C.24), where each variable is a fuzzy number. It is noted that the arithmetic operation rules for fuzzy numbers cannot be directly applied to Equation (C.21) because the variables r_1 and r_2 in this equation appear twice. Otherwise, a much larger uncertainty (wider fuzzy range) in results would be created. Equation (C.21) should be changed to the following form before the interval calculations of the two variables at each α cut are conducted:

$$r = \frac{1}{1/r_1 + 1/r_2} \quad \text{(C.41)}$$

2. The membership functions of multiple network reliability indices are calculated using the reliability formulas for a series network in Section C.2.1.1 and the operation rules for fuzzy numbers, in Section B.2.2. The basic formulas are Equations (C.12) or (C.14), (C.13), and (C.17), where each variable is a fuzzy

number. When Equation (C.12) is used, the term U_1U_2 is ignored so that the addition rule for the fuzzy members can be directly applied. Ignorance of the term U_1U_2 would not create an effective error since it is generally very small. The arithmetic operation rules for fuzzy numbers cannot be directly applied to Equation (C.17) because the variables λ_1 and λ_2 in the equation occur twice and an appropriate form to avoid the double occurrence cannot be found. In this case, the general rule given in Equation (B.19) should be used. An intermediate variable approach to reduce the computing burden in applying Equation (B.19) has been presented in Chapter 8.

In each step, the formulas for two elements (either components or equivalent components of minimum cutsets) are used repeatedly. Any two elements are considered first to obtain the indices for an equivalent element, and then the equivalent element and the third one are considered, and so on.

C.3.3 Fuzzy Markov Models

These models are discussed in detail in Reference 133 (Chapter 9).

C.3.3.1 Approach Based on Analytical Expressions. This approach includes the following two steps:

1. The analytical expressions of probabilities, frequencies, or duration indices for each network state are obtained using the crisp Markov equation method in Section C.2.3.
2. The membership functions for the reliability indices of each network state are calculated from the membership functions of input data (transition rates).

For instance, the membership functions for the probabilities of the four states in the two-component network shown in Figure C.4 can be calculated using the membership functions of transition rates (λ and μ) and Equations (C.30)–(C.33). Note that these equations must be rearranged in such a form that every fuzzy rate variable occurs only once in the expressions before applying the operation rules for fuzzy numbers. For example, for the probability of state 1, Equation (C.30) should be rearranged into

$$P_1 = \frac{1}{1+\lambda_1/\mu_1} \cdot \frac{1}{1+\lambda_2/\mu_2} \tag{C.42}$$

Similar rearrangements for Equations (C.31) to (C.33) are needed.

When the membership function of the probability of a state set is calculated, we must obtain the analytical expression of probability of the state set first. It cannot be calculated by using the membership functions of state probabilities and the fuzzy addition rule because this implies that some fuzzy transition rates are indirectly used more than once. For example, the α cut for the state set composed of states 1 and 2 shown in Figure C.4 is not equal to the sum of the α cuts for these two states:

$$(P_{1\cup 2})_\alpha \neq (P_1)_\alpha + (P_2)_\alpha \tag{C.43}$$

This approach can be applied only to very simple networks since it is difficult to obtain an analytical expression of state indices, even for a network that contains more than three repairable components. Also, it will be impossible for a relatively large network to arrange the expression in an appropriate form to avoid double occurrences of fuzzy transition rates.

C.3.3.2 Approach Based on Numerical Computations.

This approach can be applied to any network modeled by Markov equations. It is assumed that a network has n states with m transition rates λ_i ($i = 1,\ldots,m$). Let $(P_k)_\alpha = [\underline{P_k(\alpha)}, \overline{P_k(\alpha)}]$ denote the α cut of the fuzzy probability of state k, and let $(\lambda_i)_\alpha = [\underline{\lambda_i(\alpha)}, \overline{\lambda_i(\alpha)}]$ denote the α cut of the fuzzy transition rate λ_i. The maximum and minimum bounds (confidence interval) of the fuzzy probability of state k at the membership function grade α can be obtained by respectively solving the following two optimization problems with the same constraints but a different objective function:

$$\underline{P_k(\alpha)} = \min P_k \quad \text{and} \quad \overline{P_k(\alpha)} = \max P_k \tag{C.44}$$

subject to

$$(\mathbf{T}^T - \mathbf{I})\mathbf{P}^T = \mathbf{0} \tag{C.45}$$

$$P_1 + \cdots + P_n = 1 \tag{C.46}$$

$$\underline{\lambda_i(\alpha)} \leq \lambda_i \leq \overline{\lambda_i(\alpha)} \qquad (i = 1,\ldots,m) \tag{C.47}$$

In these equations, $\mathbf{P}$ is the limiting state probability row vector and $\mathbf{T}$ is the transition matrix, as defined in Section C.2.3; P_i is the ith component of $\mathbf{P}$; $\mathbf{I}$ is the unit matrix; and the superscript T represents transposition of a matrix or vector.

Obviously, Equation (C.45) represents the Markov equations, and Equation (C.46) is the full probability condition. If the membership function of probability of a state set is required, the two objective functions in Equation (C.44) are respectively replaced by

$$\underline{P_G(\alpha)} = \min \sum_{j \in G} P_j \quad \text{and} \quad \overline{P_G(\alpha)} = \max \sum_{j \in G} P_j \tag{C.48}$$

where G is the set of states considered; $\underline{P_G(\alpha)}$ and $\overline{P_G(\alpha)}$ are the minimum and maximum bounds at the α cut of the fuzzy probability for the state set G.

REFERENCES

1. North American Electric Reliability Corporation (NERC), *Reliability Standards*, available at http://www.nerc.com/.
2. British Columbia Transmission Corporation (BCTC), *Mandatory Reliability Standards Manual*, Report SPA2008-71, July 31, 2008.
3. Western Electricity Coordinating Council (WECC), *Reliability Standards and Due Process*, available at http://www.wecc.biz/standards.
4. NERC report, *Available Transfer Capability: Definitions and Determination*, June 1996.
5. J. Sun and W. Li, "Remedial action schemes (RAS) in power systems," paper presented at International Symposium on Prospect of Power Systems, Hong Kong, Nov. 6–9, 2005.
6. W. Li, Risk *Assessment of Power Systems—Models, Methods, and Applications*, IEEE Press–Wiley, 2005.
7. W. Li and P. Choudhury, "Probabilistic transmission planning," *IEEE Power & Energy Mag*. **5**(5), 46–53 (Sept./Oct. 2007).
8. W. Li and F. P. P. Turner, "Development of probabilistic transmission planning methodology at BC hydro," *Proceedings of Probabilistic Methods Applied to Power Systems* (PMAPS) *1997 International Conference*, 1997, pp. 25–31.
9. W. Li, Probabilistic Reliability Planning Guidelines, BCTC Technical Report, BCTC-SPPA-R011, June 2006.
10. R. Billinton and W. Li, *Reliability Assessment of Electric Power Systems Using Monte Carlo Methods*, Plenum Press, New York and London, 1994.
11. R. Billinton and R. N. Allan, *Reliability Evaluation of Power Systems*, 2nd ed., Plenum Press, New York and London, 1996.
12. Task Force for Probability and Statistics at the Computing Center of Chinese Academy of Sciences, *Computations in Probability and Statistics*, Science Press, Beijing, 1979.
13. Y. Chen and H. Zhang, *Prediction Techniques and Applications*, Mechanical Industry Press, Beijing, 1985.
14. G. E. P. Box, G. M. Jenkins, and G. C. Reinsel, *Time Series Analysis—Forecasting and Control*, 3rd ed., Prentice-Hall, Englewood Cliffs, NJ, 1994.
15. J. A. Freeman and D. M. Skapura, *Neural Networks—Algorithms, Applications, and Programming Techniques*, Addison-Wesley, 1991.

16. Ronaldo R. B. de Aquino, Otoni Nóbrega Neto, Milde M. S. Lira, Aida A. Ferreira, Manoel A. Carvalho Jr., Geane B. Silva, and Josinaldo B. de Oliveira, "Development of an artificial neural network by genetic algorithm to mid-term load forecasting," *Proceedings of International Joint Conference on Neural Networks*, Orlando, FL, Aug. 12–17, 2007, pp. 1726–1731.
17. J. W. Taylor and R. Buizza, "Neural network load forecasting with weather ensemble predictions," *IEEE Transactions on Power Systems* **17**(3), 626–632 (Aug. 2002).
18. F. Zhang and X. Zhou, "Gray-regression variable weight combination model for load forecasting," *Proceedings of 2008 International Conference on Risk Management & Engineering Management*, Beijing, Nov. 4–6, 2008, pp. 311–316.
19. K. Song, S. Ha, J. Park, D. Kweon, and K. Kim, "Hybrid load forecasting method with analysis of temperature sensitivities," *IEEE Trans. Power Syst.* **21**(2), 869–876 (May 2006).
20. M. S. Aldenderfer and R. K. Blashfield, *Cluster Analysis*, Sage Publications, Newbury Park, CA, 1984.
21. H. Spath, *Cluster Analysis Algorithms for Data Reduction and Classification of Objects*, Halsted, New York, 1980.
22. W. Li, J. Zhou, X. Xiong, and J. Lu, "A statistic-fuzzy technique for clustering load curves," *IEEE Trans. Power Syst.* **22**(2), 890–891 (May 2007).
23. Task Force on Mathematics Handbook Editing, *Mathematics Handbook*, People's Education Press, Beijing, 1979.
24. F. Zhou and Q. Cheng, "A survey on the powers of fuzzy matrices and FBAMS," *Intnatl. J. Comput. Cognition* **2**(2), 1–25 (June 2004).
25. W. Li and R. Billinton, "Effects of bus load uncertainty and correlation in composite system adequacy evaluation," *IEEE Trans. Power Syst.* **6**(4), 1522–1529 (1991).
26. Z. Wang, *Elements of Probability Theory and Its Application*, Science Press, Beijing, 1979.
27. Electric Power Research Institute (EPRI) report, *Load Modeling for Power Flow and Transient Stability Studies*, Report EL-5003, Project 849-7, 1987.
28. IEEE Task Force on Load Representation for Dynamic Performance, "Load representation for dynamic performance analysis," *IEEE Trans. Power Syst.* **8**(2), 472–482 (May 1993).
29. IEEE Task Force on Load Representation for Dynamic Performance, "Standard load models for power flow and dynamic performance simulation," *IEEE Trans. Power Syst.* **10**(3), 1302–1312 (May 1995).
30. EPRI report, *Extended Transient-Midterm Stability Program*, EPRI Project 1208-9, Dec. 1992.
31. P. Zhang and S. T. Lee, "Probabilistic load flow computation using the method of combined cumulants and Gram-Charlier expansion," *IEEE Trans. Power Syst.* **19**(1), 676–682 (Feb. 2004).
32. J. M. Morales and J. Perez-Ruiz, "Point estimate schemes to solve the probabilistic power flow," *IEEE Trans. Power Syst.* **22**(4), 1594–1601 (Nov. 2007).
33. H. P. Hong, "An efficient point estimate method for probabilistic analysis," *Reliab. Eng. Syst. Safety* **59**(3), 261–267 (March 1998).
34. J. E. Wilkins, "A note on skewness and kurtosis," *The Annals of Mathematical Statistics*, Vol. **15**, 1944, pp 333–335.
35. W. Li, *Secure and Economic Operation of Power Systems—Models and Methods*, Chongqing University Publishing House, 1989.

36. A. V. Fiacco and G. P. McCormick, *Nonlinear Programming: Sequential Unconstrained Minimization Techniques*, Wiley, New York, 1968.
37. IEEE PES publication, *Optimal Power Flow: Solution Techniques, Requirements, and Challenges*, IEEE Tutorial 96TP111-0, 1996.
38. S. Mehrotra, "On the implementation of a primal-dual interior point method," *SIAM J. Optimization* **2**(4), 575–601 (1992).
39. IEEE PES Mini Lecture Task Force, *Interior Point Applications to Power Systems*, PICA99, Santa Clara, CA, 1999.
40. J. H. Holland, *Adaptation in Natural and Artificial Systems*, University of Michigan Press, 1975.
41. T. Bäck, *Evolutionary Algorithms in Theory and Practice—Evolution Strategies, Evolutionary Programming, Genetic Algorithms*, Oxford University Press, New York, 1996.
42. J. Kennedy and R. Eberhart, "Particle swarm optimization," *Proceedings of IEEE International Conference on Neural Networks*, Piscataway, NJ, 1995, pp. 1942–1948.
43. S. Kirkpatrick, C. D. Gelatt, and M. P. Vecchi, "Optimization by simulated annealing," *Science* **220**(4598), 671–680 (May 13, 1983).
44. H. Pohlheim, *Genetic and Evolutionary Algorithm Toolbox for Use with MATLAB Documentation*, available at http://www.geatbx.com/docu/algindex.html.
45. Y. Shi and R. Eberhart, "A modified particle swarm optimizer," *Proceedings of the IEEE International Conference on Evolutionary Computation*, Piscataway, NJ, 1998, pp. 69–73.
46. M. Clerc and J. Kennedy, "The particle swarm—explosion, stability, and convergence in a multidimensional complex space," *IEEE Trans. Evolut. Comput.* **6**(1), 58–73 (2002).
47. R. Mendes, J. Kennedy, and J. Neves, "The fully informed particle swarm: Simpler, maybe better," *IEEE Trans. Evolut. Comput.* **8**(3), 204–210 (2004).
48. K. R. C. Mamandur and G. J. Berg, "Efficient simulation of line and transformer outages in power systems," *IEEE Trans. PAS* **101**(10), 3733–3741 (1982).
49. A. P. S. Meliopoulos, C. S. Cheng, and F. Xia, "Performance evaluation of static security analysis methods," *IEEE Trans. Power Syst.* **9**(3), 1441–1449 (Aug. 1994).
50. A. J. Wood and B. F. Wollenberg, *Power Generation, Operation and Control*, 2nd ed., Wiley, New York, 1996.
51. V. Ajjarapu and C. Christy, "The continuation power flow: A tool for steady state voltage stability analysis," *IEEE Trans. Power Syst.* **7**(1), 416–423 (Feb. 1992).
52. H. Mori and T. Kojima, "Hybrid continuation power flow with linear-nonlinear predictor," *Proceedings of 2004 International Conference on Power System Technology* (PowerCon), Nov. 21–24, 2004, pp. 969–974.
53. S. H. Li and H. D. Chiang, "Nonlinear predictors and hybrid corrector for fast continuation power flow," *IET Proceed. Generation Transmiss. Distrib.* **2**(3), 341–354 (May 2008).
54. B. Gao, G. K. Morison, and P. Kundur, "Voltage stability evaluation using modal analysis," *IEEE Trans. Power Syst.* **7**(4), 1529–1542 (Nov. 1992).
55. P. Kundur, *Power System Stability and Control*, McGraw-Hill, 1994.
56. J. Chai, N. Zhu, A. Bose, and D. J. Tylavsky, "Parallel Newton type methods for power system stability analysis using local and shared memory multiprocessors," *IEEE Trans. Power Syst.* **6**(4), 1539–1544 (Nov. 1991).

57. W. Li, *Methods for Determining Unit Interruption Cost*, BC Hydro, CCT-R-009, Jan. 19, 2000.

58. R. Billinton, G. Wacker, and G. Tollefson, *Assessment of Reliability Worth in Electric Power Systems in Canada*, report for NSERC Project STR0045005, June 1993.

59. EPRI report, *Outage Cost Estimation Guidebook*, Report TR-106082, Dec. 1995.

60. M. J. Sullivan, M. Mercurio, J. Schellenberg, and M. A. Freeman, *Estimated Value of Service Reliability for Electric Utility Customers in the United States*, Report LBNL-2132E, prepared for Office of Electricity Delivery and Energy Reliability, US Department of Energy, June 2009.

61. R. Billinton, H. Chen, and J. Zhou, "Individual generating station reliability assessment," *IEEE Trans. Power Syst.* **14**(4), 1238–1244 (Nov. 1999).

62. M. Vega and H. G. Sarmiento, "Algorithm to evaluate substations reliability with cut and path sets," *IEEE Trans. Industry Appl.* **44**(6), 1851–1858 (Nov./Dec. 2008).

63. J. Lu, W. Li, and W. Yan, "State enumeration technique combined with a labeling bus set approach for reliability evaluation of substation configuration in power systems," *Electric Power Syst. Res.* **77**(5–6), 401–406 (April 2007).

64. EPRI report, *Framework for Stochastic Reliability of Bulk Power System*, Report TR-110048, Palo Alto, CA 1998.

65. CIGRE Task Force 38-03-10, "Composite power system reliability analysis," report presented at CIGRE Symposium on Electric Power System Reliability, Sept. 16–18, 1991.

66. IEEE tutorial course textbook, *Electric Delivery System Reliability Evaluation*, 05TP175, March 2005.

67. R. Billinton, M. Fotuhi-Firuzabad, and L. Bertling, "Bibliography on the application of probability methods in power system reliability evaluation: 1996–1999," *IEEE Trans. Power Syst.* **16**(4), 595–602 (Nov. 2001).

68. C. Singh and J. Mitra, "Composite system reliability evaluation using state space pruning," *IEEE Trans. Power Syst.* **12**(1), 471–479 (Feb. 1997).

69. W. Li and R. Billinton, "Common cause outage models in power system reliability evaluation," *IEEE Trans. Power Syst.* **18**(2), 966–968 (May 2003).

70. R. Billinton and W. Li, "A hybrid approach for reliability evaluation of composite generation and transmission systems using Monte Carlo simulation and enumeration technique," *IEE Proceed. C* **138**(3), 233–241 (May 1991).

71. R. Billinton and W. Li, "A novel method for incorporating weather effects in composite system adequacy evaluation," *IEEE Trans. Power Syst.* **6**(3), 1154–1160 (1992).

72. R Billinton and W. Li, "Consideration of multi-state generating unit models in composite system adequacy assessment using Monte Carlo simulation," *Can. J. Electric. Comput. Eng.* **17**(1), 24–28 1992.

73. R. Billinton and W. Li, "Direct incorporation of load variations in Monte Carlo simulation of composite system adequacy," proceedings of Inter-RAMQ Conference for the Electric Power Industry, 27–34, Philadelphia, PA, Aug. 25–28, 1992.

74. R. Billinton and W. Li, "Composite system reliability assessment using a Monte Carlo approach," paper presented at 3rd International Conference on Probabilistic Methods Applied to Power Systems (PMAPS), London, July 3–5, 1991.

75. W. Li and R. Billinton, "A minimum cost assessment method for composite generation and transmission system expansion planning," *IEEE Trans. Power Syst.* **8**(2), 628–635, 1993.

76. W. Li, "Monte Carlo reliability evaluation for large scale composite generation and transmission systems," *J. Chongqing Univ.* **12**(3), 92–98 (May 1989).

77. R. Billinton and W. Li, "A system state transition sampling method for composite system reliability evaluation," *IEEE Trans. Power Syst.* **8**(3), 761–770 (1993).
78. J. Yu, W. Li, and W. Yan, "Risk assessment of static voltage stability," *Proceed. CSEE* **29**(28), 40–46 (Oct. 2009).
79. E. Vaahedi, W. Li, T. Chia, and H. Dommel, "Large scale probabilistic transient stability assessment using B.C. Hydro's on-line tools," *IEEE Trans. Power Syst.* **15**(2), 661–667 (May 2000).
80. W. Li and J. Lu, "Monte Carlo method for probabilistic transient stability assessment," *Proceed. CSEE* **25**(10), 18–23 (May 2005).
81. C. S. Park, *Contemporary Engineering Economics*, 3rd ed., Prentice-Hall, Englewood Cliffs, NJ, 2002.
82. J. L. Riggs, D. D. Bedworth, and S. U. Randhawa, *Engineering Economics*, 4th ed., McGraw-Hill, New York, 1996.
83. E. L. Grant, W. G. Ireson, and R. S. Leavenworth, *Principles of Engineering Economy*, 8th ed., Wiley, New York, 1990.
84. J. Fu, *Technology Economics*, Qinghua University Press, Beijing, 1986.
85. W. Li, E. Vaahedi, and P. Choudhury, "Power system equipment aging," *IEEE Power & Energy* **4**(3), 52–58 (May/June 2006).
86. J. Wu, Y. Qin, and D. Zhang, *Power Systems*, Power Industry Press, Beijing, 1980.
87. W. Ji, *Design Manual of Power Systems*, Power Industry Press, Beijing, 1998.
88. ABB Electric System Technology Institute, *Electrical Transmission and Distribution Reference Book*, Raleigh, NC, 1997.
89. IEEE Std 738-2006, *IEEE Standard for Calculating the Current-Temperature of Bare Overhead Conductors*, 2007.
90. EPRI report, EPRI Underground Transmission Systems Reference Book, 2006 ed., Report 1014840, Palo Alto, CA, 2007.
91. International Electrotechnical Commission (IEC) standard book, *Electric Cables—Calculation of the Current Rating—Current Rating Equations (100% Load Factor) and Calculation of Losses—Current Sharing between Parallel Single-Core Cables and Calculation of Circulating Current Losses*, IEC publication 60287, 2002.
92. W. Li, Architecture Design and Calculation Method of Load Coincidence Factor Application, BCTC report, BCTC-SPPA-R012, April 1, 2007.
93. W. Li, H. C. Jonas, S. Yan, B. Corns, P. Choudhury, and E. Vaahedi, "Reliability decision management systems: Experiences at BCTC", proceedings of 20th Canadian Conference on Electrical and Computer Engineering (CCECE 2007), paper No. 023, Vancouver, April 22–26, 2007.
94. CEA report, *2005 Forced Outage Performance of Transmission Equipment—Equipment Reliability Information System*, 2007.
95. W. Li, J. Zhou, and X. Hu, "Comparison of transmission equipment outage performance in Canada, USA and China," *Proceedings of IEEE Canada Electric Power and Energy Conference 2008*, paper No. 1360, Vancouver, Oct. 6–7, 2008.
96. CEA report, *2007 Bulk Electricity System Delivery Point Interruptions & Significant Power Interruptions* (composite participant version), Dec. 2008.
97. W. Li, J. Zhou, K. Xie, and X. Xiong, "Power system risk assessment using a hybrid method of fuzzy set and Monte Carlo simulation," *IEEE Trans. Power Syst.* **23**(2), 336–343 (May 2008).
98. J. E. Freund, *Mathematical Statistics*, Prentice-Hall, Englewood Cliffs, NJ, 1962.

99. N. R. Mann, R. E. Schafer, and N. D. Singpurwalla, *Methods for Statistical Analysis of Reliability and Life Data*, Wiley, New York, 1974.

100. W. Li, J. Zhou, and X. Xiong, "Fuzzy models of overhead power line weather-related outages," *IEEE Trans. Power Syst.* **23**(3), 1529–1531 (Aug. 2008).

101. W. Li, X. Xiong, and J. Zhou, "Incorporating fuzzy weather-related outages in transmission system reliability assessment," *IET Proceed. Generation, Transmiss. Distribut.* **3**(1), 26–37 (Jan. 2009).

102. W. Li, J. Zhou, J. Lu, and W. Yan, "Incorporating a combined fuzzy and probabilistic load model in power system reliability assessment," *IEEE Trans. Power Syst.* **22**(3), 1386–1388 (Aug. 2007).

103. R. Billinton, S. Kumar, et al., "A reliability test system for educational purpose: Basic data," *IEEE Trans. Power Sys.*, **4**(3), 1238–1244 (1989).

104. W. Li, P. Choudhury, and J. Gurney, "Probabilistic reliability planning: Method and a project case at BCTC," *Proceedings of 2008 PMAPS*, paper No. 005, Puerto Rico, May 25–29, 2008.

105. W. Li, *Expected Energy Not Served (EENS) Study for Vancouver Island Transmission Reinforcement Project, Part I: Reliability Improvements Due to VITR*, Report BCTC-SPPA-R009A, Dec. 8, 2005 available at http://www.bctc.com/transmission_system/engineering_studies_data/studies/probabilistic_studies/selected_tech_reports.htm.

106. W. Li, *Expected Energy Not Served (EENS) Study for Vancouver Island Transmission Reinforcement Project, Part II: Comparison between VITR and Sea Breeze HVDC Light Options*, Report BCTC-SPPA-R009B, Dec. 23, 2005, available at http://www.bctc.com/transmission_system/engineering_studies_data/studies/probabilistic_studies/selected_tech_reports.htm.

107. W. Li, *Expected Energy Not Served (EENS) Study for Vancouver Island Transmission Reinforcement Project, Part IV: Effects of Existing HVDC on VI Power Supply Reliability*, Report BCTC-SPPA-R009D, Jan. 9, 2006, available at http://www.bctc.com/transmission_system/engineering_studies_data/studies/probabilistic_studies/selected_tech_reports.htm.

108. British Columbia Utilities Commission's decision document: *In the Matter of BCTC—an Application for a Certificate of Public Convenience and Necessity for the Vancouver Island Transmission Reinforcement Project*, July 7, 2006, available at http://www.bcuc.com/DecisionIndex.aspx.

109. W. Li, *MCGSR Program: User's Manual*, BC Hydro, Canada, Dec. 2001.

110. W. Li, *Probability Distribution of HVDC Capacity and Impacts of Two Key Components*, Report BCTC-SPA-R003, May 5, 2004.

111. W. Li, Y. Mansour, J. K. Korczynski, and B. J. Mills, "Application of transmission reliability assessment in probabilistic planning of BC Hydro Vancouver South Metro System," *IEEE Trans. Power Syst.* **10**(2), 964–970 (1995).

112. W. Li, Y. Mansour, B. J. Mills, and J. K. Korczynski, "Composite system reliability evaluation and probabilistic planning: BC Hydro's practice," paper presented at 1995 Canadian Electric Association Conference, Vancouver, March 1995.

113. W. Li, *Reliability Assessment and Probabilistic Planning of North Metro System Alternatives*, BC Hydro, STS-940-R5-008, Jan. 26, 1996.

114. W. Wangdee, *Benefit/Cost Analysis Based on Reliability and Transmission Loss Considerations for Central Vancouver Island Transmission Project*, BCTC Report 2007-REL-03.2, Sept. 27, 2007, available at http://www.bctc.com/transmission_system/engineering_studies_data/studies/probabilistic_studies/selected_tech_reports.htm.

115. W. Li, *MECORE Program: User's Manual*, BC Hydro, Canada, Dec. 2001.

116. W. Li, *PLOSS Program: User's Manual*, BC Hydro, Canada, Dec. 2001.

117. W. Li, J. Zhou., J. Lu, and W. Yan, "A probabilistic analysis approach to making decision on retirement of aged equipment in transmission systems," *IEEE Trans. Power Deliv.* **22**(3), 1891–1896 (July 2007).

118. W. Li, "Evaluating mean life of power system equipment with limited end-of-life failure data," *IEEE Trans. Power Syst.* **19**(1), 236–242 (Feb. 2004).

119. R. Billinton and R. N. Allan, *Reliability Evaluation of Engineering Systems—Concepts and Techniques*, Plenum Press, New York, 1992.

120. W. Li, "Incorporating aging failures in power system reliability evaluation," *IEEE Trans. Power Syst.* 918–923 (Aug. 2002).

121. W. Li, P. Choudhury, D. Gillespie, and J. Jue, "A risk evaluation based approach to replacement strategy of aged HVDC components and its application at BCTC," *IEEE Trans. Power Deliv.* **22**(3), 1834–1840 (July 2007).

122. W. Li, "Probability distribution of HVDC capacity considering repairable and aging failures," *IEEE Trans. Power Deliv.* **21**(1), 523–525 (Jan. 2006).

123. J. Zhou, W. Li, J. Lu, and W. Yan, "Incorporating aging failure mode and multiple capacity state model of HVDC system in power system reliability assessment," *Electric Power Syst. Res.* **77**(8), 910–916 (June 2007).

124. W. Wangdee, W. Li, P. Choudhury, and W. Shum, "Reliability study of two station configurations for connecting IPPs to a radial supply system," *Proceedings of IEEE Canada Electric Power and Energy Conference 2008*, paper No. 1362, Vancouver, Oct. 6–7, 2008.

125. C. R. Heising, A. L. J. Janssen, W. Lanz, E. Colombo, and E. N. Dialynas, "Summary of CIGRE 13.06 working group world wide reliability data and maintenance cost data on high voltage circuit breakers above 63 kV," *Proceedings of IEEE Industry Applications Society Annual Meeting, 1994*, Vol. **3**, pp. 2226–2234.

126. W. Wangdee, W. Li, W. Shum, and P. Choudhury, "Applying probabilistic methods in determining the number of spare transformers and their timing requirements," proceedings of 20th Canadian Conference on Electrical and Computer Engineering (CCECE 2007), paper No. 097, Vancouver, April 22–26, 2007.

127. W. Li, *Risk Based Asset Management—Applications at Transmission Companies*, Chapter 7 of IEEE tutorial course textbook 07TP183, June 2007.

128. W. Li, *SPARE Program: User's Manual*, BC Hydro, Canada, April 2001.

129. W. Wangdee and W. Li, *Reliability Improvement Planning Strategy for Single Circuit Supply Systems in BCTC*, BCTC Report SPPA2009-104, May 25, 2009.

130. W. Li, W. Wangdee, and P. Choudhury, "A probabilistic planning approach to single circuit supply systems," *Proceedings of 11th International Conference on PMAPS*, paper No. P0124, Singapore, June 14–17, 2010.

131. W. Li, J. Lu, and W. Yan, "Reliability component in transmission service pricing," Paper conf72a29, *Proceedings of the 7th International Power Engineering Conference (IPEC)*, Singapore, Nov. 29–Dec. 2, 2005.

132. W. Li, J. Zhou, J. Lu, and W. Yan, "Incorporating reliability component in transmission service rate design," *Proceed. CSEE* **26**(13), 43–49 (2006).

133. M. E. El-Hawary, and V. Miranda, et al., *Electric Power Applications of Fuzzy Systems*, IEEE Press, New York, 1998.

INDEX

Probabilistic Transmission System Planning, by Wenyuan Li

1. *Principles of Electric Machines with Power Electronic Applications, Second Edition*
M. E. El-Hawary

2. *Pulse Width Modulation for Power Converters: Principles and Practice*
D. Grahame Holmes and Thomas Lipo

3. *Analysis of Electric Machinery and Drive Systems, Second Edition*
Paul C. Krause, Oleg Wasynczuk, and Scott D. Sudhoff

4. *Risk Assessment of Power Systems: Models, Methods, and Applications*
Wenyuan Li

5. *Optimization Principles: Practical Applications to the Operations of Markets of the Electric Power Industry*
Narayan S. Rau

6. *Electric Economics: Regulation and Deregulation*
Geoffrey Rothwell and Tomas Gomez

7. *Electric Power Systems: Analysis and Control*
Fabio Saccomanno

8. *Electrical Insulation for Rotating Machines: Design, Evaluation, Aging, Testing, and Repair*
Greg Stone, Edward A. Boulter, Ian Culbert, and Hussein Dhirani

9. *Signal Processing of Power Quality Disturbances*
Math H. J. Bollen and Irene Y. H. Gu

10. *Instantaneous Power Theory and Applications to Power Conditioning*
Hirofumi Akagi, Edson H. Watanabe, and Mauricio Aredes

11. *Maintaining Mission Critical Systems in a 24/7 Environment*
Peter M. Curtis

12. *Elements of Tidal-Electric Engineering*
Robert H. Clark

13. *Handbook of Large Turbo-Generator Operation Maintenance, Second Edition*
Geoff Klempner and Isidor Kerszenbaum

14. *Introduction to Electrical Power Systems*
Mohamed E. El-Hawary

15. *Modeling and Control of Fuel Cells: Disturbed Generation Applications*
M. Hashem Nehrir and Caisheng Wang

16. *Power Distribution System Reliability: Practical Methods and Applications*
Ali A. Chowdhury and Don O. Koval

17. *Introduction to FACTS Controllers: Theory, Modeling, and Applications*
Kalyan K. Sen and Mey Ling Sen

18. *Economic Market Design and Planning for Electric Power Systems*
James Momoh and Lamine Mili

19. *Operation and Control of Electric Energy Processing Systems*
James Momoh and Lamine Mili

20. *Restructured Electric Power Systems: Analysis of Electricity Markets with Equilibrium Models*
Xiao-Ping Zhang

21. *An Introduction to Wavelet Modulated Inverters*
S. A. Saleh and M. Azizur Rahman

22. *Probabilistic Transmission System Planning*
Wenyuan Li

23. *Control of Electric Machine Drive Systems*
Seung-Ki Sul

24. *High Voltage and Electrical Insulation Engineering*
Ravindra Arora and Wolfgang Mosch

25. *Practical Lighting Design with LEDs*
Ron Lenk and Carol Lenk

Forthcoming Titles

Electricity Power Generation: The Changing Dimensions
Digambar M. Tagare

Power Conversion and Control of Wind Energy Systems
Bin Wu, Yongqiang Lang, Navid Zargan, and Samir Kouro

Electric Distribution Systems
Abdelhay A. Sallam and O. P. Malik

Doubly Fed Induction Machine: Modeling and Control for Wind Energy Generation Applications
Gonzalo Abad, Jesus Lopez, Miguel Rodriguez, Luis Marroyo, and Grzegorz Iwanski

Maintaining Mission Critical Systems, Second Edition
Peter M. Curtis

Printed and bound by CPI Group (UK) Ltd, Croydon, CR0 4YY
16/04/2025
14658601-0005